PLANNING WUHAN 100 YEARS

武汉百年——规划图记

第二版 VERSION II

武汉市规划研究院 WPDI

陈韦 武洁 成钢 梅子凡 编著

中国建筑工业出版社

图书在版编目（CIP）数据

武汉百年规划图记/陈韦等编著. —2版. —北京：中国建筑工业出版社，2019.11
ISBN 978-7-112-24395-2

Ⅰ.①武… Ⅱ.①陈… Ⅲ.①城市规划-城市史-武汉 Ⅳ.①TU984.263.1

中国版本图书馆CIP数据核字（2019）第233340号

责任编辑：刘　丹　陆新之　吴宇江
书籍设计：陆　澜
责任校对：赵　菲

武汉百年规划图记（第二版）

武汉市规划研究院

陈韦　武洁　成钢　梅子凡　编著

*

中国建筑工业出版社出版、发行（北京海淀三里河路9号）
各地新华书店、建筑书店经销
北京锋尚制版有限公司制版
北京雅昌艺术印刷有限公司印刷

*

开本：965×1270毫米　1/16　印张：20¼　插页：1　字数：981千字
2019年11月第二版　2019年11月第二次印刷
定价：600.00元
ISBN 978-7-112-24395-2
（34876）

版权所有　翻印必究
如有印装质量问题，可寄本社退换
（邮政编码100037）

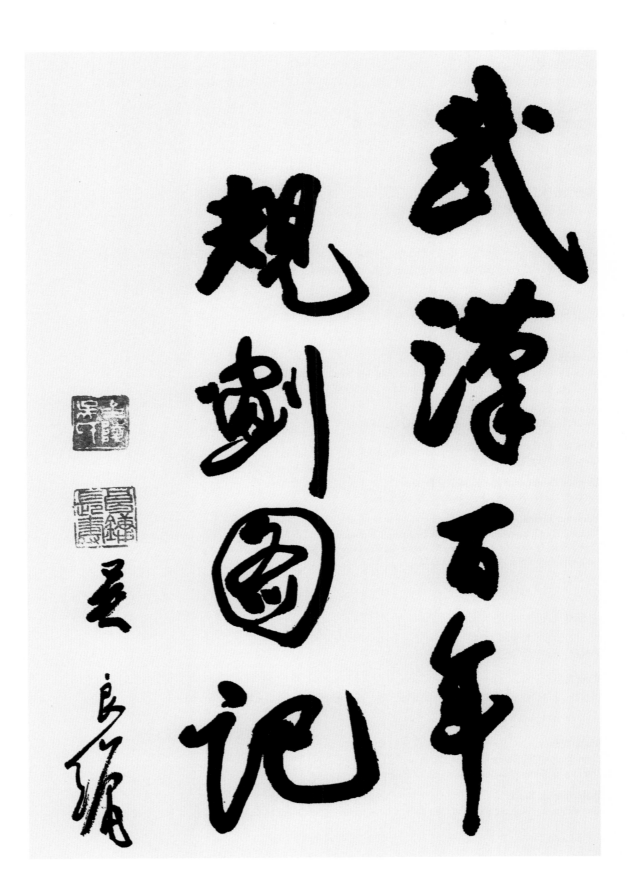

两院院士、清华大学教授吴良镛先生为本书题词

Mr. Wu Liangyong, an academician of the two academies and a professor at Tsinghua University, is pleased to write an inscription for the book

编写委员会

学术委员会
盛洪涛　陈北平　严　春　刘奇志　马文涵　周　强　杨维祥　田　燕　吴立群　陈明宝　徐承军　何孝齐

学术顾问
吴之凌　殷　毅　胡忆东　何　梅　汪　勰　叶　青

编委会主任
陈　韦　彭伟宏

编委会副主任
刘　平　胡　飞　武　洁　肖志中　丘永东

编　委
徐　昊　童建平　吕维娟　何灵聪　蔡三秀　郎秉花　汪　云　杨　昔　李海军　常四铁　黄晓芳　商　渝
汪波宁　董　菲　王　磊　宋中英　雷学锋　戴　时　郭希盛　穆　霖

WRITING COMMITTEE

Academic Committee
Sheng Hongtao, Chen Beiping, Yan Chun, Liu Qizhi, Ma Wenhan, Zhou Qiang, Yang Weixiang, Tian Yan, Wu Liqun, Chen Mingbao, Xu Chengjun, He Xiaoqi

Academic Consultant
Wu Zhiling, Yin Yi, Hu Yidong, He Mei, Wang Xie, Ye Qing

Director of the Editorial Board
Chen Wei, Peng Weihong

Deputy Director of the Editorial Board
Liu Ping, Hu Fei, Wu Jie, Qiu Yongdong, Xiao Zhizhong

Editorial Board
Xu Hao, Tong Jianping, Lv Weijuan, He Lingcong, Cai Sanxiu, Lang Binghua, Wang Yun, Yang Xi, Li Haijun, Chang Sitie, Huang Xiaofang, Shang Yu, Wang Boning, Dong Fei, Wang Lei, Song Zhongying, Lei Xuefeng, Dai Shi, Guo Xisheng, Mu Lin

主要编写人员
陈　韦　武　洁　成　钢　梅子凡
朱林艳　王　莹　宋中英　李海军

参编人员
常四铁　吕维娟　商　渝　朱志兵　袁建峰　游　畅　李晓慧

Main Writer
Chen Wei, Wu Jie, Cheng Gang, Mei Zifan
Zhu Linyan, Wang Ying, Song Zhongying, Li Haijun

Participating Staff
Chang Sitie, Lv Weijuan, Shang Yu, Zhu Zhibing, Yuan Jianfeng, You Chang, Li Xiaohui

再版说明
REVISION NOTES

本书为2009年版的修订版。利用修订机会，对全书进行了几点修改、补充和订正：根据相关文献资料重新梳理了武汉规划发展历史，对历史内容进行了深度挖掘；补充了近十年的重大规划内容；补充了英文对照版本；订正了2009年版中存在的一些疏漏。

本书立足于城乡规划发展轨迹，讲述1889年以来武汉城市建设和规划发展历程，全书正文按照武汉城市空间格局发展形态分为四章，分别讲述三镇独立建设时期、三镇融合规划时期、主城内聚发展时期、区域一体化时期的规划四部分内容，反映了武汉三镇由独立建设发展逐渐融合统一，进而飞跃转型的主要过程。

本书严谨治史，抛砖引玉，以期作为规划历史学术理论专著，为城乡规划工作者提供一定的研究参考价值。

The book is a revision of the 2009 edition. In this revision, the following modifications, additions and corrections have been made: according to relevant literature, the history of Wuhan's planning development has been reorganized and the historical facts have been mined in depth; the major planning contents of the past ten years have been supplemented; an English edition has been supplemented; and some omissions in the 2009 edition have been corrected.

Based on the development of urban and rural planning, this book tells the history of Wuhan's urban construction and planning development since 1889. The text is divided into four chapters according to the development form of Wuhan's urban space pattern, which respectively describes Independent Construction of Three Towns, Planning Period for Integration of Three Towns, Cohesion of the Main City, and Planning during Regional Integration. It reflects the main process in which the three towns of Wuhan gradually merged after independent construction, and then realized rapid development and transformation.

This book is a rigorous study of history to bring out brilliant insights, hoping to provide a certain reference for urban and rural planners as a theoretical monograph of planning history.

前言
FOREWORD

两江交汇，三镇对峙以分；龟蛇锁江，七桥横空而架。百年以来，九省通衢的武汉凭借着其绝佳的自然禀赋崛起成为今日之宏伟江城。大城崛起，日新月异，一代又一代的规划工作者见证了这座城市的复兴之路。

许多人习惯将城市比作一本"石刻的史书"。历史是延续的，是发展的。我们研究历史的目的，是为了吸收、传承传统文化的精髓，古为今用；是为了延续历史文脉，传递人类文化的信息。本书的再版工作立足于对武汉百年规划文稿和图纸的再思考，希望在罗列的史料中勾勒出事件主线和碰撞思想火花。本书的再版工作距离上版已过去十年，十年间，我们见证了中国城镇化的巨大转型，也亲历了武汉城市规划和建设的华丽蝶变，城市经济实力进一步跃升，城市空间格局进一步优化，城市魅力特色进一步彰显。编者通过深挖史料，进一步补充完善了近十年的城乡规划资料，并对全书体系、内容进行了较大篇幅的调整，本书再版是对规划院四十年历史的精彩见证，也是对武汉规划历史的百年传承。

我院自1979年设立以来，不断着眼未来、意存高远、力求创新，一直把规划设计工作和理论研究工作统筹考虑。当前，我国城市发展进入新常态，国土空间规划改革已全面铺开，规划工作已从增量规划向存量规划转型发展，如何继续发挥好空间规划在城市发展中的战略引领作用，是每一位规划工作者的值得思考的问题和探索的方向。

知往鉴今，以启未来。做好规划，是任何一个城市发展的首要任务，而了解一座城市的规划发展历史，也是做好规划工作的必要条件。希望本书对广大规划工作者的探索实践具有一定的启发和指导意义，我们亦将再接再厉，不忘初心，继续发扬匠心精神，在长江经济带发展战略的指引下，为打造武汉城市样本而不懈努力！

Wuhan, located at the intersection of the two rivers, is divided into three towns; Turtle Mountain and Snake Mountain stand on both sides of the Yangtze River, over which there are seven bridges. Over the past century, Wuhan as the gateway to nine provinces has grown into a magnificent riverside city as it is today thanks to its excellent natural endowments. This great city rises and changes with each passing day, and generations of planners have witnessed its revival.

A city used to be compared to a "stone-carved history book". History is continuous and developing. Our purpose of studying history is to absorb and pass on the essence of traditional culture, to make the past serve the present, and to continue the historical context and convey the message of human culture. The second edition of this book is based on the rethinking of manuscripts and drawings of Planning Wuhan: 100 Years, hoping to outline the main storylines and spark trains of thought in the listed historical data. A decade has passed since the first edition of this book was printed. In the past decade, we have witnessed the tremendous transformation of urbanization in China and experienced the magnificent development of urban planning and construction in Wuhan. The city has seen its economic strength further strengthened, its urban spatial pattern further optimized, and its charm and characteristics further highlighted. The compilation group supplemented and improved the urban and rural planning data of the past ten years by digging deep into the historical data and made extensive adjustments to the system and content of the entire book. The second edition of this book is not only a wonderful testimony to the 40-year history of the planning & design institute, but also an inheritance of the 100-year planning history of Wuhan.

Since our establishment in 1979, we have been looking to the future, aiming for higher goals and striving for innovation. We have always taken planning & design and theoretical research into overall consideration. At present, China's urban development has entered a new normal, the reform of national land space planning has been comprehensively rolled out, and there is a transition from incremental planning to inventory planning. How to continue to give full play to the strategic leading role of space planning in urban development is the direction worth thinking and exploring for every planner.

The past serves as a reference for today and opens the way to the future. Excellent planning is the first priority for any city, while understanding the planning history of a city is a necessary condition for excellent planning. It is hoped that this book can enlighten and guide the vast number of planners in their exploration and practice. We will go the extra mile, stay true to the original aspiration, continue to carry forward the spirit of originality, and make unremitting efforts to build a sample city of Wuhan under the guidance of the development strategy of the Yangtze River Economic Belt!

武汉市规划研究院

Wuhan Planning & Design Institute

背景
BACKGROUND

武汉筑城起源于3500年前的盘龙城，它是"九省通衢"武汉之根。自三国始，武昌、汉阳因其地处长江、汉水要冲，其重要军事功能促进了两地的封建城堡建设，"双城重镇"的形态由此而生，开启了长达1000多年的"武汉双城"的历史。至明成化年间，汉水改道由龟山之麓入江，汉口从汉阳析出，其后汉口在近代水运和商业的带动下加速扩张，武汉自此由"双城"向"三镇"转变。至1927年，当时国民政府决定将武昌、汉阳和汉口三镇合并，但其后因时局动荡，三镇在建制上又经历了分离，直到1949年武汉解放，三镇合并，乃取三镇之首字命名为"武汉"。

汉阳城在三镇之中筑城最早。东汉末年，戴监军于鲁山（今龟山）北筑却月城。建安十三年（公元208年）孙权破城，同年刘琦于鲁山南之凤栖山筑鲁山城。唐武德四年(公元621年)扩鲁山城筑汉阳城，设八门。元末城败，明初大修，崇祯十六年（公元1643年）汉阳城又战毁，清顺治十八年（公元1661年）重建，1928年拆除。

武昌有城，始于三国时期东吴黄武二年（公元223年），孙权在江夏山（今蛇山）东北筑夏口城。南朝宋孝建元年（公元454年），孝武帝设郢州，修葺和扩建夏口为郢州城。唐敬宗宝历元年（公元825年），改土城为砖城，筑成鄂州城。明洪武四年（公元1371年），增拓武昌府城，设九门，1927~1929年拆除。

Built on Panlong City 3,500 years ago, Wuhan is known as "the gateway to nine provinces". Since the beginning of the Three Kingdoms Period, Wuchang and Hanyang as two transport hubs on the Yangtze River and the Hanjiang River had important military functions, which promoted the construction of feudal castles in the two places. The layout of "two important towns" was thus established, marking the formation of "Two Towns of Wuhan" that have existed for over 1,000 years. During the Chenghua period of the Ming Dynasty, the Hanjiang River changed its course, joining the Yangtze River at the foot of Turtle Mountain. Hankou was separated from Hanyang and then accelerated its expansion driven by modern water transportation and commerce. Since then, "Two Towns" of Wuhan transformed into "Three Towns". In 1927, the Kuomintang government decided to merge the three towns of Wuchang, Hanyang and Hankou. However, due to the turbulent situation, the three towns experienced an institutional separation. It was not until 1949 that Wuhan was liberated and the three towns merged. Wuhan gained its name by combining "Wu" from the first town and "Han" from the other two.

Hanyang was the first of the three towns to be built. In the late Eastern Han Dynasty, an army supervisor surnamed Dai built Queyue City in the north of Lu Mountain (today's Turtle Mountain). In the thirteenth year of the Jian'an period (208 AD), Sun Quan captured the city. In the same year, Liu Qi built Lushan City in Fengqi Mountain, south of Lu Mountain. In the fourth year of the Wude period of the Tang Dynasty (621 AD), Lushan City was expanded into Hanyang City with eight gates. The city was destroyed in the late Yuan Dynasty and repaired in the early Ming Dynasty. In the sixteenth year of the Chongzhen period (1643 AD), Hanyang was destroyed by war. It was rebuilt in the eighteenth year of the Shunzhi period of the Qing Dynasty (1661 AD) and demolished in 1928.

The construction of Wuchang can be traced back to Xiakou City built by Sun Quan in the second year of the Huangwu period of Eastern Wu in the Three Kingdoms Period (223 AD) in the northeast of Jiangxia Mountain (today's Snake Mountain). In the first year of the Xiaojian period of Song in the Southern Dynasties (454 AD), Emperor Xiaowu set up Yingzhou, and repaired and expanded Xiakou into Yingzhou City. In the first year of Emperor Jingzong of the Tang Dynasty (825 AD), the earth city was rebuilt into a brick city called Ezhou City. In the fourth year of the Hongwu period of the Ming Dynasty (1371 AD), Wuchang was built as a prefectural city with nine gates that was demolished between 1927 and 1929.

三镇之中汉口的历史最短，筑城最迟。明成化三年(公元1467年)，汉水改道，汉口、汉阳始分两地，至嘉靖年间设汉口镇巡检司，属汉阳县。其后汉口由于水运之便而迅速发展成为长江中游最大的物资集散地之一，曾一度与河南朱仙镇、广东佛山镇、江西景德镇并称为"天下四大镇"。

武汉"因水而兴，因商而立，因武而昌，因文而盛"。近百年来，以晚清洋务运动为肇始，武汉城市地位逐步提高，城市格局也基本形成。回顾武汉百年城市发展历程，大致可以分为三镇独立建设、三镇融合发展、主城内聚发展和区域一体化发展四个时期。百年城市发展历程也伴随着武汉城市规划的演变，成为具有代表性的中国城市规划历史样本。

Hankou with the shortest history among the three towns was the last to be built. In the third year of the Chenghua period of the Ming Dynasty (1467 AD), the Hanjiang River changed its course, and Hankou was separated from Hanyang. In the Jiajing period, Hankou Town Inspection Department was set up and put under the administration of Hanyang County. Later, with convenient water transport, Hankou rapidly developed into one of the largest distribution centers of materials in the middle reaches of the Yangtze River, and was once known as "Four Largest Towns in the World" together with Zhuxian Town in Henan, Foshan Town in Guangdong and Jingdezhen in Jiangxi.

Wuhan "thrives on water, commerce, martial arts and literature". In the past hundred years, since the Westernization Movement in the late Qing Dynasty, Wuhan's status as a city has gradually improved and its urban pattern has basically taken shape. Wuhan's urban development in the past 100 years can be roughly divided into four stages: independent construction of three towns, integration of three towns, cohesion of the main city and regional integration. The 100-year urban development process along with the evolution of Wuhan's urban planning has become the historic representative of China's urban planning.

1 2

1

武汉三镇浮雕图（1748年，现存于湖南芷江文物管理所）

Relief sculpture of three towns of Wuhan (1748, now stored in Hunan Zhijiang Cultural Relic Administration)

2

湖北舆地图（1862年，现存于北京国家图书馆）

Map of the Hubei region (1862, now stored in the National Library of China)

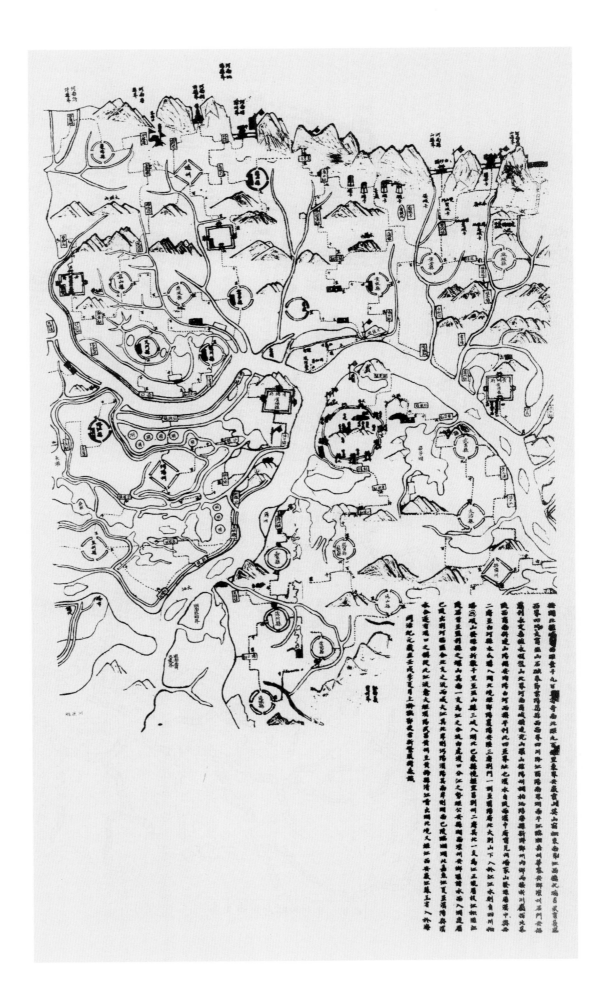

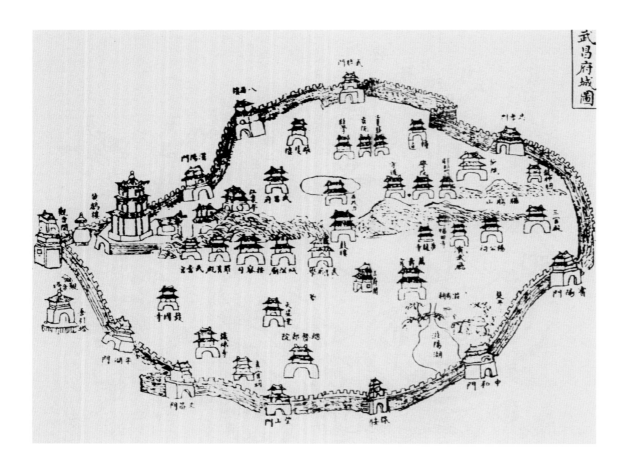

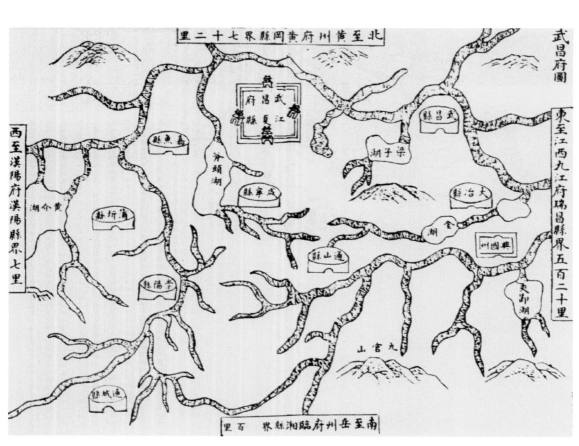

1	3
2	4

1

武昌府城图（1733年，现存于中国第一历史档案馆）

Wuchang Prefectural City Map (1733, now stored in the First Historical Archives of China)

2

武昌府图（1733年，现存于中国第一历史档案馆）

Wuchang Prefecture Map (1733, now stored in the First Historical Archives of China)

3

江夏县疆域图（1869年，现存于中国第一历史档案馆）

Jiangxia County Territory Map (1869, now stored in the First Historical Archives of China)

4

江夏县城池图（1869年，现存于中国第一历史档案馆）

Jiangxia County Map (1869, now stored in the First Historical Archives of China)

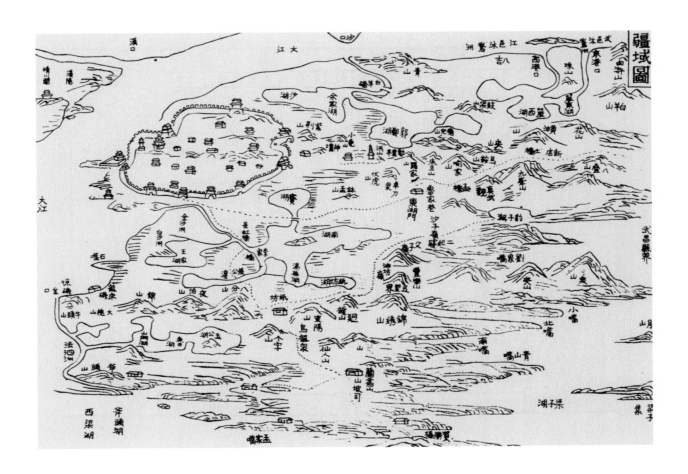

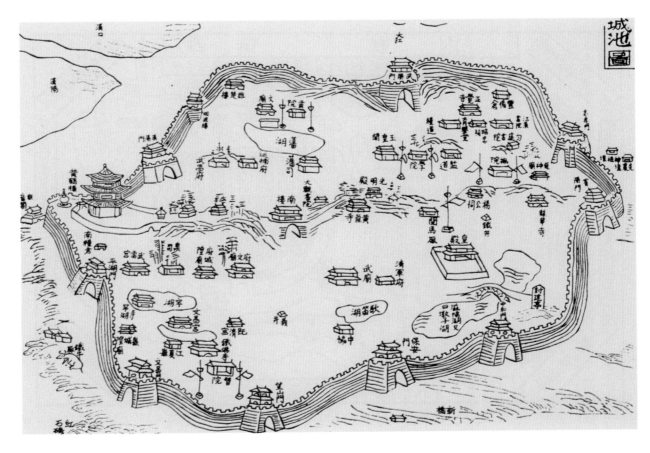

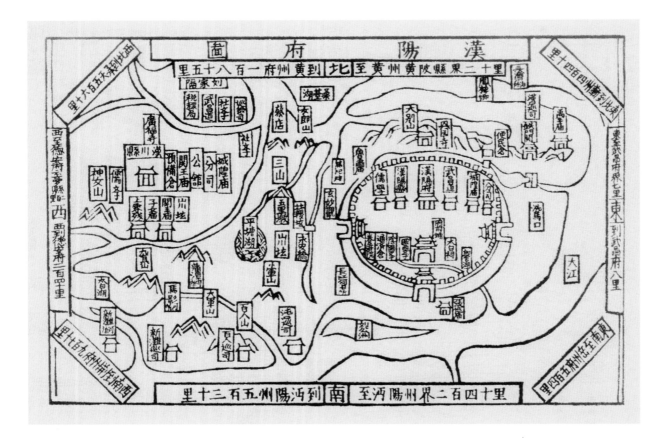

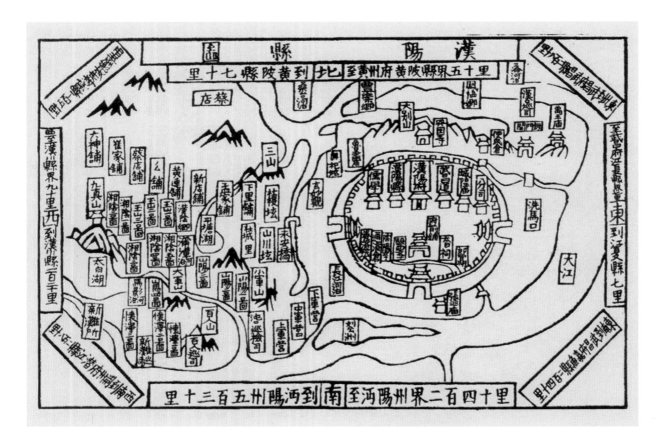

1	3
2	4

1

汉阳府图（1546年，现存于中国第一历史档案馆）

Hanyang Prefecture Map (1546, now stored in the First Historical Archives of China)

2

汉阳县图（1546年，现存于中国第一历史档案馆）

Hanyang County Map (1546, now stored in the First Historical Archives of China)

3

汉阳县境全图（1748年，现存于中国第一历史档案馆）

General Map of Hanyang County (1748, now stored in the First Historical Archives of China)

4

汉阳县治图（1748年，现存于中国第一历史档案馆）

Hanyang County Map (1748, now stored in the First Historical Archives of China)

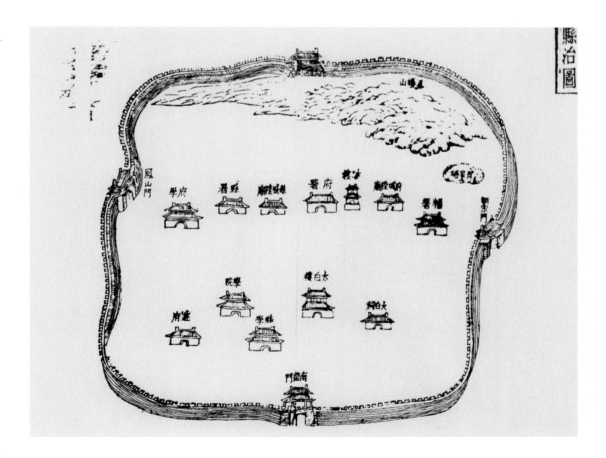

目录
CONTENTS

020	一、晚清时期的建设 Construction in the Late Qing Dynasty 1889 - 1910
046	二、民国初期的规划 Planning in the Early Republic of China 1911 - 1926

第壹章
三镇独立建设时期
CHAPTER ONE
Independent Construction of Three Towns
1889 - 1926

054	一、民国中期的规划 Planning in the Middle Period of the Republic of China 1927 - 1936
068	二、战争时期的规划 Planning During the War 1937 - 1949

第贰章
三镇融合规划时期
CHAPTER TWO
Planning Period for Integration of Three Towns
1927 - 1949

086	一、新中国成立初期的规划 Planning in the early days of People's Republic of China 1949 - 1977
114	二、改革时期的规划 Planning in Age of Reform 1978 - 1999

第叁章
主城内聚发展时期
CHAPTER THREE
Cohesion of the Main City
1949 - 1999

160	一、新世纪发展期的规划 Planning in the New Century 2000 - 2011
196	二、新时代转型期的规划 Planning During Transformation in the New Era 2012 - *present*

第肆章
区域一体化时期的规划
CHAPTER FOUR
Planning During Regional Integration
2000 - PRESENT

316	大事记 CHRONICLE OF EVENTS
317	参考文献 REFERENCES
322	后记 AFTERWORD

CHAPTER
第 壹 章
ONE
1889-1926

三镇独立建设时期
Independent Construction of Three Towns

1840年，西方列强敲开了古老封闭满清王朝的大门。伴随着近代商贸的加速发展，由历代帝都、省、府、州、县治所而逐渐形成的封建城市发展规律被打破，开始涌现出一批新兴的工商业、港口城市，于是由古老政治中心和新兴经济中心组成的"双城"格局应运而生，如北京与天津，济南与青岛等。汉口于1861年开埠之后，加速扩张为繁盛的近代商贸港口城市，而武昌、汉阳是古城，是府、县治所，三镇因此具有了政治中心和经济中心的双重功能。

1889年起，张之洞督鄂大力推行洋务运动，兴办近代教育、民族工业、水利工程和交通设施，促使汉口转向商贸重镇、武昌转向行政教育、汉阳转向近代工业等功能，武汉三镇开始由以内陆内部循环为主的传统商业市镇演变为面向海外市场、进行国际大循环的国际性通商口岸。

1911年辛亥革命后，武汉"敢为天下先"的精神为世界瞩目，孙中山以世界大都市的目标制订三镇发展计划，确定规划目标、城市的发展定位以及发展规模，并以模范市的标准提出汉口重建计划。

张之洞、孙中山等以政治家、实业家的视野和气魄，为武汉制订了宏大的发展规划，建设了自武汉辐射四方的"十字型"区域交通网络，奠定了武汉大都市的基本雏形和框架，从此启动了近半个世纪的区域性大都市建设。

In 1840, the Western powers knocked on the door of the ancient closed Qing Dynasty. With the accelerated development of modern commerce and trade, the development law of ancient cities gradually formed by the imperial capitals, provinces, prefectures, states and counties in past dynasties was broken and a number of new industrial, commercial and port cities emerged. Thus, the pattern of "two towns" consisting of an ancient political center and a new economic center has emerged, such as Beijing and Tianjin, Jinan and Qingdao. After its port opening in 1861, Hankou accelerated its expansion into a prosperous modern commercial port city. Wuchang and Hanyang were ancient cities and prefecture and county seats. Therefore, the three towns had dual functions as political centers and economic centers.

From 1889 onwards, Hubei governor Zhang Zhidong vigorously carry out the Westernization Movement and set up modern education, national industry, water conservancy projects and transport facilities, which turned Hankou into an important commercial city and promoted the administration and education in Wuchang and the modern industry and other functions in Hanyang. The three towns of Wuhan began to transform from traditional commercial towns of inland internal circulation into international commercial ports of international circulation which faced overseas markets.

After the 1911 Revolution, Wuhan's spirit of "daring to be the first" attracted the world's attention. Sun Yat-sen formulated the development plan for three towns with the goal of building a world metropolis, determined the planning goal, the city's development orientation and scale, and put forward the Hankou reconstruction plan according to the standard of a model city.

After the 1911 Revolution, Wuhan's spirit of "daring to be the first" attracted the world's attention. Sun Yat-sen formulated the development plan for three towns with the goal of building a world metropolis, determined the planning goal, the city's development orientation and scale, and put forward the Hankou reconstruction plan according to the standard of a model city.

一、晚清时期的建设
Construction in the Late Qing Dynasty

1889年,张之洞提出"开源利民",力举修筑卢汉铁路,促成了清政府修筑铁路的决议。由此被调任至湖广总督,开启了他长达18年督鄂的辉煌生涯……

张之洞任内实施"湖北新政",以洋务运动、实业发展推动三镇建设。他基于武汉在全国的区位优势,决定贯通武汉与四方的交通,规划了自武汉向外辐射的"十字型"区域性交通线路,同时为消除长江水患,扩大城市用地,主持修建了武昌南堤、北堤及汉口后湖长堤,又依据武汉三镇自然条件分设职能,其中汉阳以工业为主,武昌以教育为主,汉口以商业为主,据此规划布局了相关功能设施。

这一时期,武汉三镇的外贸、商务、工业、教育、交通等得到长足发展,一举改变了其在近代中国经济布局中的格局,开始从传统的政治中心和商业市镇向现代化的国际性工商业城市的转型,成为"驾乎津门、直逼沪上"的国际性大都市。

In 1889, Zhang Zhidong put forward the idea of "taping resources to benefit the people", and urged the construction of Lugouqiao-Hankou Railway, which prompted the resolution of the Qing Government to build the railway. He was thus transferred to the post of governor-general of Hu-Guang, and began his glorious 18-year career in Hubei.

During his tenure, Zhang Zhidong implemented "Hubei New Policy", promoting the construction of three towns with the Westernization Movement and industrial development. In consideration of the location advantage of Wuhan in the country, he decided to link Wuhan with its surrounding areas, and planned a "cross-shaped" regional traffic centered in Wuhan. To eliminate the Yangtze River flood and expand urban land, he presided over the construction of South Dike and North Dike in Wuchang and Houhu Dike in Hankou. He designated functions of the three towns according to their natural conditions. Hanyang focused on industry, Wuchang on education and Hankou on commerce. On this basis, he planned and deployed function-related facilities.

During this period, the three towns of Wuhan have made great progress in foreign trade, commerce, industry, education and transportation, changing their status in the economic pattern of modern China at one stroke. They began to transform from traditional political centers and commercial towns into modern international industrial and commercial cities, becoming international metropolises that "overtook Tianjin and caught up with Shanghai".

1889 – 1910

1 - 迈入铁路、水运 并进时代的区域交通规划

Regional Transportation Planning in the Simultaneous Development of Railway & Water Transportation

1861年汉口开埠，汉口商贸国际大门从此大开，武汉三镇内联腹地，外通海洋，其区位优势发挥得淋漓尽致。随着19世纪末、20世纪初分别以汉口和武昌为终点的京汉、粤汉铁路通车，南北向的铁路交通骨架已初步形成，与东西向的川汉铁路及沪汉长江航线共同构成了以武汉为核心的"十字型"主干交通线，在"车船时代"全面进入"铁路时代"的过程中，武汉"九省通衢"的交通优势发挥到了极致。凭借着铁铁中转、水铁联运、通江达海的垄断优势，武汉改变了其在近代中国经济布局中的格局，一度成为仅次于上海的中国第二大外贸口岸，被称为"东方芝加哥"。

1908年，张之洞又兼任鄂境内川汉铁路大臣，认为川汉铁路应由宜昌入当阳、荆门、仙桃、蔡甸以接汉阳至武汉。虽然此条线后因"保路运动"未能全部建成，但其作为横向贯通长江中上游地区的重要铁路走廊，仍具有重大战略意义。

1905年汉口至日本的直达水路航线开通，武汉开启了以轮运为主导的水路航运时代，汉口港自此成为国际港。至清末，由汉口驶向国外的轮船，已可直达德国、荷兰、埃及、法国、比利时、意大利等地，沪汉长江航线的开辟使长江航道变成黄金水道，武汉自古以来的水运优势跃上了一个新的台阶。

1	2	3
粤汉铁路施工现场（资料来源于武汉市档案信息馆）	粤汉铁路株韶段施工现场（资料来源于武汉市档案信息网）	长江上码头（资料来源于《武汉旧影》）
Construction site of the Guangzhou–Wuchang Railway (data source: Wuhan Archives Information Network)	Construction site of the Zhuzhou–Shaoguan Section of the Guangzhou–Wuchang Railway (data source: Wuhan Archives Information Network)	A wharf on the Yangtze River (data source: *Old Pictures of Wahan*)

Hankou opened its port in 1861, marking its opening to international commerce and trade. The three towns of Wuhan are linked to the hinterland and open to the sea, giving full play to its location advantages. With the opening of the Lugouqiao-Hankou Railway and Guangzhou-Wuchang Railway ending in Hankou and Wuchang respectively at the end of the 19th century and the beginning of the 20th century, the north-south railway traffic network has taken shape initially. Together with the east-west Chengdu-Hankou Railway and the Shanghai-Hankou Yangtze River route, it formed "cross-shaped" traffic trunk lines with Wuhan as the core. In the overall transformation of the "vehicle and ship era" into the "railway era", Wuhan's advantages as "the gateway to nine provinces" has been brought to its utmost. Relying on its monopolistic advantages in railway transfer, rail-water combined transportation, and connection to rivers and seas, Wuhan changed its status in the economic pattern of modern China and became China's second largest foreign trade port after Shanghai, known as "Oriental Chicago".

In 1908, Zhang Zhidong held the concurrent post of minister of Chengdu-Hankou Railway in Hubei, and believed that the Chengdu-Hankou Railway should enter Dangyang, Jingmen, Xiantao and Caidian from Yichang to link Hanyang with Wuhan. Although the line was not completed due to the Railway Project Crisis, it is still of great strategic significance as an important railway corridor running through the middle and upper reaches of the Yangtze River.

In 1905, the direct route from Hankou to Japan was opened, and Wuhan entered the era of water transportation by wheel. Hankou has since become an international port. By the end of the Qing Dynasty, ships sailing from Hankou to foreign countries had direct access to Germany, Holland, Egypt, France, Belgium, Italy and other places. The opening of the Shanghai-Hankou Yangtze River route has turned the Yangtze River waterway into a golden waterway, and Wuhan's advantage in water transportation has leapt to a new level since ancient times.

1
1906年，张之洞在京汉铁路通车庆典上（资料来源于《武汉通史》）

Zhang Zhidong at the opening ceremony of the Lugouqiao-Hankou Railway in 1906 (data source: *A General History of China*)

2
武昌黄鹄矶下汉阳门码头（资料来源于《武汉通史》）

Hanyang Gate Wharf under Huanghuji, Wuchang (data source: *A General History of China*)

3
长江上的船舶（资料来源于《武汉通史》）

Ships on the Yangtze River (data source: *A General History of China*)

4
汉口外滩鸟瞰（资料来源于武汉市档案信息网）

A bird's-eye view of the Hankou Bund (data source: Wuhan Archives Information Network)

2 – 开启近代
工业长廊的实业规划布局

Industrial Planning in the Modern Industrial Corridor

张之洞督鄂期间,主张先重后轻、"自相挹注"的重轻型发展战略,大力倡办实业,先后创办一批近代工厂,建立了包括冶金、矿业、军工、纺织和交通等行业门类较为齐全的近代大工业体系,推动了武汉早期工业化的快速发展,使武汉迅速成长为当时中国内地最大的工业城市之一,奠定了武汉近代工业发展的基础和格局。

1891年,汉阳铁厂于汉阳龟山北与汉水南岸之间的沿河地段动工兴建,时为亚洲最大的钢铁联合企业,共建有六大厂(炼生铁厂、转炉炼钢厂、平炉炼钢厂、造钢轨厂、造铁货厂、炼熟铁厂等)和四小厂(机器厂、铸铁厂、打铁厂、造鱼钩钉厂等),炼钢炉2座,工人3000名。1894年铁厂西侧加建的湖北枪炮厂(后易名为湖北兵工厂)建成,包括炮架、炮弹、枪弹三厂,该厂和炼铁厂组成了一个从冶炼到机器制造的冶金、机械工业区,成为武汉第一个近代工业区。

1892~1899年,湖北织布局、纺纱局、缫丝局和制麻局分别在武昌文昌门、望山门、平湖门外建成,武汉就此成为当时仅次于上海的中国第二大纺织工业中心。1907年分别在武昌建设制皮厂、毡呢厂、模范工厂、湖北银元局、铜钱局,在武昌城外白沙洲建设造纸厂,在汉阳赫山建设针钉厂,在赫山南麓建设官砖厂,在汉口硚口下首建设贫民工厂。

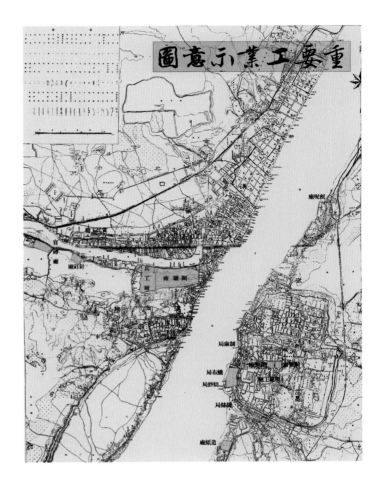

1 重要工业示意图(编者自绘,底图为1913年武汉三镇街市图)
Major industrial sites (drawn by the author; the base drawing is the street map of the three towns of Wuhan in 1913)

2 20世纪初的汉阳兵工厂(资料来源于《武汉通史》)
Hanyang Arsenal in the early 20th century (data source: *A General History of China*)

3 汉绣——平金夹绣(资料来源于《张之洞督鄂115周年》)
Gold thread embroidery of Han Embroidery (data source: *115th Anniversary of Zhang Zhidong's Governance in Hubei*)

During his tenure as Hubei governor, Zhang Zhidong advocated the development strategy of "developing heavy industry before light industry" and "mutual empowering". He strongly advocated the establishment of industry, successively set up a batch of modern factories and established a large modern industrial system covering a complete range of industries including metallurgy, mining, military industry, textile and transportation. This promoted the rapid development of Wuhan's early industrialization, made Wuhan rapidly grow into one of the largest industrial cities in mainland China at that time, and laid the foundation and pattern for Wuhan's modern industrial development.

In 1891, Hanyang Iron Works was built on the riverside section between the north of Turtle Mountain and the south bank of the Hanjiang River in Hanyang. It was then the largest combined iron and steel enterprise in Asia, with six large plants (pig iron plant, converter steel plant, open hearth steel plant, steel rail plant, iron goods plant, wrought iron plant, etc.), four small plants (machine plant, iron foundry, iron making plant, fishhook iron plant, etc.), two steel furnaces and 3,000 workers. In 1894, Hubei Firearms Factory (later renamed Hubei Arsenal) was built on the west side of the iron works, consisting of Gun Carriage Factory, Cannonball Factory and Bullet Factory. The factory and iron works formed a metallurgical and mechanical industrial area from smelting to machine manufacturing, which was Wuhan's first modern industrial area.

From 1892 to 1899, Hubei Textile Bureau, Spinning Bureau, Silk Reeling Bureau and Hemp Making Bureau were built respectively outside Wenchang Gate, Wangshan Gate and Pinghu Gate in Wuchang. Wuhan thus became the second largest textile industry center in China after Shanghai. In 1907, a leather factory, a felt factory, a model factory, Hubei Silver Dollar Bureau and Copper Coin Bureau were built respectively in Wuchang; a paper mill was built in Baishazhou outside Wuchang; a pin factory was built in He Mountain of Hanyang; an official brick factory was built at the southern foot of He Mountain; and a factory for poor was built under Qiaokou of Hankou.

3 – 探索三镇商埠发展的空间谋划

Spatial Planning of Commercial Ports in the Three Towns

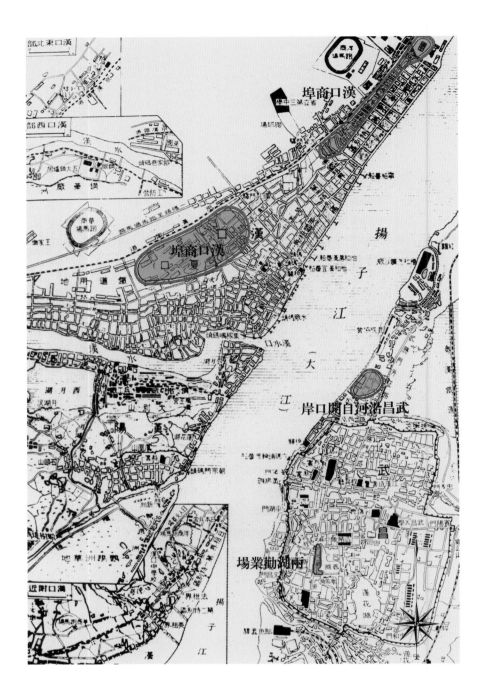

1

商埠口岸建设示意图（编者自绘，底图为1930年汉口、武昌、汉阳附近图）

Construction Plan of Commercial Ports (drawn by the author; the base drawing is the map of surrounding areas of Hankou, Wuchang and Hanyang in 1930)

2

武汉三镇堤防示意图（编者自绘，底图为1915年武汉地形图）

Dike Plan of the Tree Towns of Wuhan (drawn by the author; the base drawing is the Wuhan topographic map of 1915)

1894年，甲午战败，西方列强开始了又一轮的租界开辟和扩张。这一时期，武汉三镇轻重并举的工业格局已基本形成，为了限制外国租界进一步扩张，张之洞开始了自开商埠口岸的筹划，着力促成农工商一体化发展。自1899年开始，围绕着自主发展商埠口岸的战略目标，开始有计划实施"筑堤、拆城、修路、兴市"的策略，从此奠定了近百年来武汉城市发展的空间格局，大大加快了武汉近代城市化进程，实现了武汉城市发展史上一次大的飞跃。

一　以武昌口岸发展为目标的四大举措（1899～1903年）

1899～1903年，张之洞以实现武昌口岸自主发展为目标，在武昌先后实施了兴修南北堤、修筑省城马路、自开商埠口岸、创办劝业场的四大举措。通过"筑堤、扩城、修路"不仅有效缓解了武昌水患，大大拓展了陆地空间，还极大改善了原有的交通条件与市容市貌，为武昌商埠口岸的建设提供了有利条件和发展环境。

1899年，武昌北堤（武胜门外红关至青山）和南堤（白沙洲至金口）先后兴修，原旧堤面加高至5.7米，加宽至6.7米。考虑到内湖的水需设闸排泄，张之洞又在巡司河上、堤的北端分别建设武泰闸、武丰闸，于南段阁家河、袁家河一带修建3座石矶。南、北堤的建成拓展了20万亩良田，分别作为官办农场、畜牧场及农民租种使用。

鉴于原省城街道狭窄，年久失修，泥泞难行，张之洞分期建设马路，如阅马场至洪山、武胜门至沙湖边等马路，以及拆除障碍物辟通湘门，并下令武汉三镇的临街房屋，在修建时无论"铺面、住宅、公所，均应让出官街3尺以展宽马路"。

1900年，张之洞拟在武昌城北10里外的沿江地方（约在徐家棚和杨园一带）自开通商口岸，为此清查官荒土地和收买民地约3万亩，并雇请英国工程师斯美利来鄂，令其丈量土地，对建筑码头、填筑驳岸、兴修马路等工程进行详细勘估，绘制细图。后于1918年（民国7年）武昌商场局将武胜门至徐家棚一带规划了1:5000的武昌商埠全图，分甲、乙、丙、丁四区123块，并规划了驳岸、马路、沟渠等配套设施。

1902年，湖北近代最早的工业产品博览会——两湖劝业场在兰陵街（今武昌解放路）创办。场内分南北两场，每场有房79间，分一、二、三等，其陈列馆南北长约80米，东西宽约18米。内分3所，一是省内商品场，陈列湖北省内各种制造商品；二为省外商品场，陈列外省、外国的民用货物和机器；三是土产商品场，陈列湖北、湖南地区的各种土产，如各种天然矿产以及各种水果、茶叶、药材等物产。

一 围绕汉口新埠发展的四大举措（1904～1909年）

1904年，张之洞将目光开始转向汉口，他突破原有的"城墙"式的古典规划模式，开始了"筑后湖大堤、拆汉口城墙、建后城马路、兴汉口新埠"的系列谋划。通过修筑大堤首先保证汉口防洪功能，拓展铁路以北的城市发展空间，就此拆除阻碍城市扩张的汉口旧城墙，随后兴修一批近代化道路，并开始新埠的建设，以加强铁路与华界的联系。这一系列浩大工程促成了汉口道路网络的雏形，奠定了近代汉口城市规划与建设的基本格局。

1904～1906年，后湖大堤（又名张公堤）修筑，该堤东起堤角，西至舵落口，全长23.75公里。该堤的建成在防洪的同时又扩大了汉口的地理空间，使其与京汉铁路间的大片洼地变为可用的商居之地，市区面积因此扩展了7倍，计数10万亩。

张公堤的建成，使汉口堡失去了防洪功能，并成为城区向北扩展的障碍。故在修堤之时，张之洞已开始下令拆除汉口城墙，就原城墙基址改修成马路。1907年后城马路建成，时名后城马路（今中山大道），是汉口华界第一条近代化道路。这一"边建边拆"的做法不仅有利于市区交通的顺达，也使城垣拆除的砖石可用于修建堤防。

1905年，卢汉铁路汉口段已建成通车，由于租界面向大江，同时距离铁道较近，码头和道路等基础设施较华界旧市区更为完善，其市场发展在铁路通车后明显更具竞争优势。于是，张之洞提前谋划控制租界以北靠近大智门火车站、玉带门火车站的空间腹地；同年2月，汉口汉镇马路工程局设立，开始了汉口新埠建设，修筑市区靠近大智门车站一带的马路（大智门至玉带门），并拟扩充街、路直达江岸，以沟通铁路与旧市区，并开辟新市区，发展华界市场，展宽旧市区街道，控制市区房屋建造。

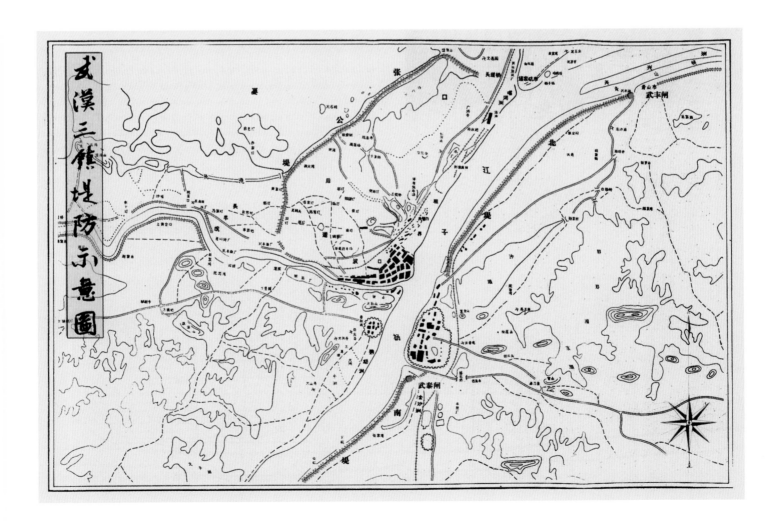

After China's defeat in the Sino-Japanese War of 1894, the Western powers began another round of concession establishment and expansion. During this period, the industrial pattern of developing light industry and heavy industry simultaneously has basically taken shape in the three towns of Wuhan. In order to restrict the further expansion of foreign concessions, Zhang Zhidong began to plan non-treaty ports and made great efforts to promote the integrated development of agriculture, industry and commerce. Since 1899, around the strategic goal of developing commercial ports independently, the strategy of "building dikes, demolishing cities, building roads and prospering cities" has been implemented in a planned way. The spatial pattern of Wuhan's urban development in the past 100 years has been established, the process of modern urbanization in Wuhan has been greatly accelerated, and a great leap in the history of Wuhan's urban development has been realized.

— Four measures to realize the independent port development in Wuchang (1899—1903)

From 1899 to 1903, to realize the independent port development in Wuchang, Zhang Zhidong implemented four measures successively in Wuchang, namely, building North Dike and South Dike, building roads in the provincial capital, setting up the non-treaty port, and establishing Quanyechang (a shopping center). "Building dikes, expanding cities and building roads" not only effectively

alleviated the flood in Wuchang and greatly expanded the land space, but also improved the traffic condition and the city image, providing favorable conditions and development environment for the commercial port construction in Wuchang.

In 1899, Wuchang North Dike (Hongguan outside Wusheng Gate to Qingshan) and South Dike (Baishazhou to Jinkou) were built. The old dike was raised to 5.7 meters and widened to 6.7 meters. To discharge the water in inner lakes, Zhang Zhidong built Wutai Sluice Gate and Wufeng Sluice Gate respectively on the Xunsi River and at the northern end of the dike and built three rocky ledges along the Kanjia River and Yuanjia River in the southern section. With the completion of South Dike and North Dike, 200,000 mu of fertile land was opened up for government-run farms, livestock farms and farmers rented land.

In view of the narrow streets of the provincial capital, which were in disrepair and muddy, Zhang Zhidong built roads in stages, such as the roads from Yuemachang to Hongshan and from Wusheng Gate to Sha Lake, and removed obstacles and built Tongxiang Gate. He also ordered that the houses with street frontages in the three towns of Wuhan "should be three *chi* off the government road to widen the road, whether they are shops, residential houses or government offices".

In 1900, Zhang Zhidong planned to open a non-treaty port at a place along the river about 10 *li* north of Wuchang (around Xujiapeng and Yangyuan). For this purpose, he checked waste government land and bought about 30,000 *mu* of non-government land, and hired British engineer Smiley to Hubei to measure the land, carry out a detailed survey and assessment, and draw detailed drawings for wharf construction, revetment filling, road construction and other projects. Later in 1918 (the 7th year of the Republic of China), the Wuchang Shopping Mall Bureau planned a 1:5000 general map of Wuchang's commercial ports from Wusheng Gate to Xujiapeng, dividing it into 123 blocks in the four districts of A, B, C and D, and planning supporting facilities such as revetment, roads and ditches.

In 1902, Hubei's earliest modern industrial products fair, Lianghu Quanyechang, was established in Lanling Street (now Jiefang Road in Wuchang). The venue was divided into northern and southern zones. Each zone had 79 rooms which were rated as grade one, grade two and grade three. The exhibition hall was about 80 meters long from north to south and 18 meters wide from east to west and was divided into three zones. The first zone displayed various goods manufactured

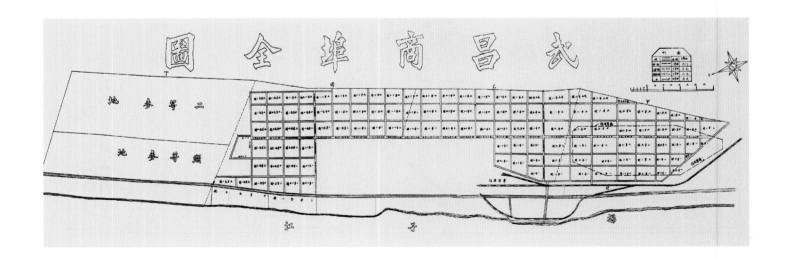

1	3
2	4

1

汉口市场（资料来源于武汉市档案信息网）

Hankou Market (data source: Wuhan Archives Information Network)

2

武昌商埠全图（1918年，现存于湖北省档案馆）

General Plan of Commercial Ports in Wuchang (1918, now stored in Hubei Provincial Archives)

3

湖北省城内外街道总图（1883年，现存于武汉市图书馆）

General Plan of Streets inside and outside the Provincial Capital of Hubei (1883, now stored in Wuhan Library)

4

汉口堡城址示意图（资料来源于《武汉市城市建设志》）

Site of Hankou Tower - (data source: *Wuhan Urban Construction Records*)

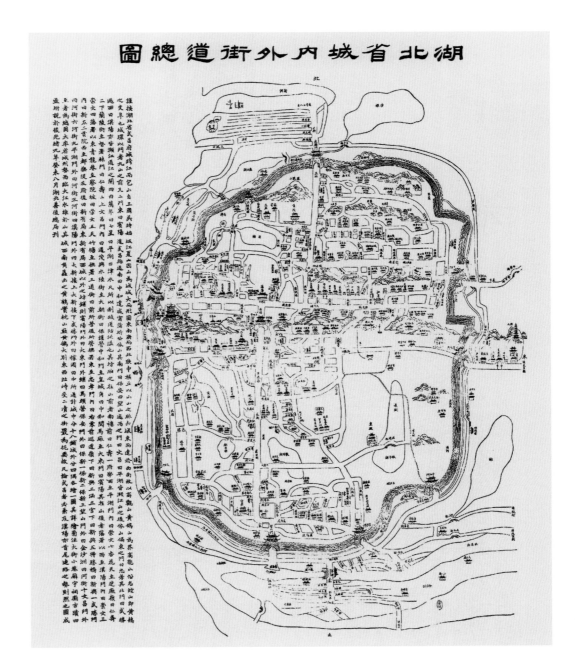

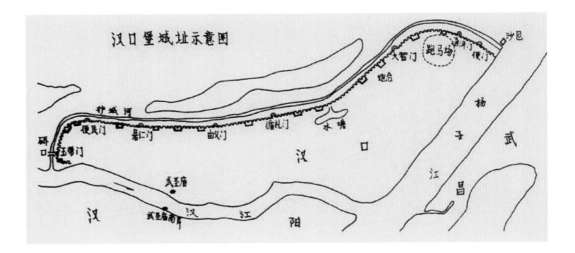

1

武昌省城最新街道图（1909年，现存于中国地图出版社）

Latest Plan of Wuchang Streets (1909, now stored in SinoMaps Press)

2

汉口自开市街用地示意图（编者自绘，底图为1908年汉口市与各国租界略图）

Land Plan for Non-treaty Market Streets in Hankou (drawn by the author; the base drawing is the sketch of Hankou and concessions of 1908)

in Hubei; the second zone displayed civilian goods and machines from other provinces and foreign countries; and the third zone displayed various native products of Hubei and Hunan, such as natural minerals, fruits, tea, medicinal materials and other products.

— Four measures for the new port development in Hankou (1904—1909)

In 1904, Zhang Zhidong turned his eyes to Hankou. He broke through the traditional "city wall" planning model and began a series of plans to "build Houhu Dike, dismantle city walls in Hankou, build Houcheng Road, and build a new port in Hankou". The construction of dikes ensured Hankou's flood control function. To expand the urban development space to the north of the railway, the old city walls of Hankou, which hindered urban expansion, were removed. Then a number of modernization roads were built and a new port was constructed to strengthen the connection between the railway and the Chinese community. These huge projects contributed to the prototype of Hankou's road network and laid the basic pattern of Hankou's modern urban planning and construction.

From 1904 to 1906, Houhu Dike (also known as Zhanggong Dike) was built, with a total length of 23.75 kilometers from the dike end in the east to Duoluokou in the west. The dike not only prevented flood, but also expanded the geographical space of Hankou, making a large depression between the dike and the Lugouqiao-Hankou Railway into a usable commercial and residential place, thus expanding the urban area by seven times to some 100,000 *mu*.

After the construction of Zhanggong Dike, Hankou Tower lost its flood control function and became an obstacle to the city's northward expansion. Therefore, when the dike was being built, Zhang Zhidong ordered the demolition of city walls in Hankou, and the former site of city walls was rebuilt into a road. In 1907, Houcheng Road was built. Houcheng Road (now Zhongshan Avenue) was the first modern road in Hankou's Chinese community. This practice of "building while demolishing" was beneficial to the urban traffic, and the bricks and stones removed from city walls could be used to build dikes.

In 1905, the Hankou section of Lugouqiao-Hankou Railway was completed and opened to traffic. As the concessions faced the river and were close to the railway, their infrastructure such as wharves and roads was more perfect than that of the old urban areas of the Chinese community, and they showed obvious competitive advantages in market development after the railway was opened to traffic. Then, Zhang Zhidong made plans ahead to control the hinterland near Dazhimen Railway Station and Yudaimen Railway Station north of concessions. In February of the same year, Hankou Town Road Works Bureau was set up; construction of the new port of Hankou began; and urban roads were built near Dazhimen Railway Station (Dazhimen to Yudaimen). Plans were made to expand streets and roads to the river bank to link the railway with the old urban area, open up new urban areas, develop markets in the Chinese community, broaden streets in the old urban area, and control the urban housing construction.

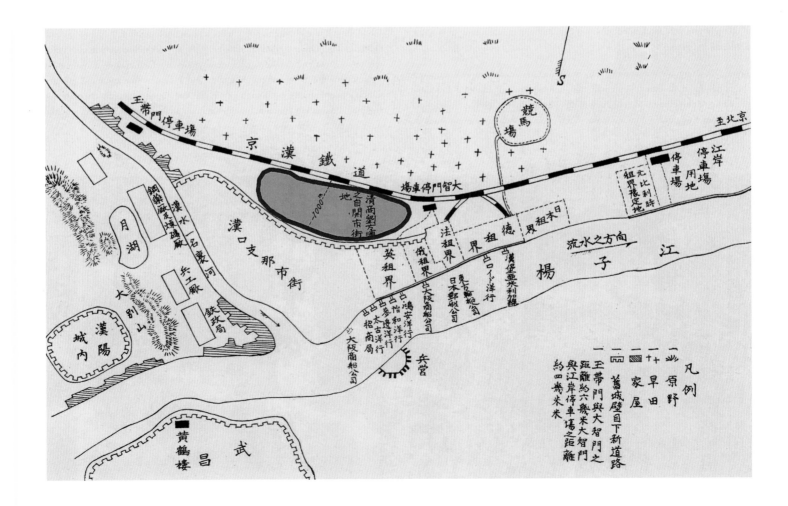

4 - 兴办新式教育的学堂规划布局

Planning of Schools with a New-Type Education System

张之洞认为"中国不贫于财而贫于人才"。他以"造真材，济时用"为宗旨，改书院、兴学堂、倡游学，武汉三镇就此形成了较为完备的近代教育体制，成为近代中国新式教育的中心之一。

张之洞推行书院改制，按照"中体西用"的指导思想调整江汉书院、经心书院和两湖书院课程。兴办新式学堂，先后创办了算学学堂(1891年)、矿务学堂(1892年)、自强学堂(1893年，今武汉大学)、湖北武备学堂(1897年)、湖北农务学堂(1898年，今华中农业大学)、湖北工艺学堂(1898年，今武汉科技大学)、湖北师范学堂(1902年)、两湖总师范学堂(1904年)、女子师范学堂(1906年)等，构建起传统教育与新式教育，普通教育与实业教育、军事教育相互配合、共同发展的近代教育体系，改变了此前私塾蒙馆、府县学宫和经学书院3个层次并立的传统教育结构。

Zhang Zhidong believed that "China doesn't lack money but talents". With the tenet of "training talents for the times", he reformed ancient academies, set up schools and advocated study tours. As a result, the three towns of Wuhan formed a relatively complete modern education system and became the center of modern education in China.

Zhang Zhidong carried out the reform of ancient academies and adjusted the curricula of Jianghan Academy, Jingxin Academy and Lianghu Academy according to the guiding idea of "absorbing Western science and technology on the basis of traditional Chinese ideology". He founded modern schools, such as Mathematics School (1891), Mining School (1892), Zi Qiang School (1893, now Wuhan University), Hubei Military School (1897), Hubei Farming School (1898, now Huazhong Agricultural University), Hubei Technological School (1898, now Wuhan University of Science and Technology), Hubei Normal School (1902), Hubei General Normal School (1904), Women's Normal School (1906), etc. He constructed a modern education system in which traditional education and modern education, general education and industrial education and military education cooperated with each other and developed together, changed the traditional education structure in which schools at three levels coexisted, namely, private schools, prefectural schools and Confucian classics academies.

1

两湖书院旧址（资料来源于《武汉通史》）

Former site of Lianghu Academy (data source: *A General History of China*)

2

方言学堂（资料来源于《武汉通史》）

Fangyan Academy (data source: *A General History of China*)

3

文华学院全景（资料来源于《武汉通史》）

A full view of Wenhua Academy (data source: *A General History of China*)

4

重要教育设施示意图（编者自绘，底图为1909年武昌省城最新街道图）

Plan of Key Educational Facilities (drawn by the author; the base drawing is the latest Wuchang street map of 1909)

5 – 引进西方近代城市建设形制的租界规划

Concession Planning by Introducing the Western Modern Urban Construction System

1861 年，汉口被迫开埠，先后划定为英、俄、法、德、日五国租界区。从全面历史辩证的角度来看，汉口租界的建立，引入了西方先进的城市建设与管理经验，租界当局分别制定租界规划，编制道路系统、下水道系统和给水、电力系统等规划，为武汉三镇建设提供了一个近代市政建设的模板。

自租界规划开始，武汉三镇正式引进了西方近代城市规划理论和思想。

— 租界设立情况

汉口旧租界区位于今沿江大道中段，即江汉路以北，麻阳街下码头以南，中山大道东南，其中法租界有部分越过中山大道，距京汉铁路东南约 200 米止，滨长江西北一带，沿江岸线共长约 3627 米。按地理位置由南向北顺序排列，分别为英、俄、法、德、日五国租界，共计面积 2968.61 亩。租界区共有八国设总领事馆，七国设领事馆，五国设代办处。各租界设立的时间、面积、收回时间，详见下表：

In 1861, Hankou was forced to open its port and was divided into five concessions of Britain, Russia, France, Germany and Japan. From the perspective of comprehensive historical dialectics, with the establishment of concessions, Hankou introduced advanced urban construction and management experience from the West. The concession authorities formulated concession plans, road systems, sewer systems, water supply and power systems and other plans, providing a template of modern urban construction for the three towns of Wuhan.

Since concession planning, the three towns of Wuhan have formally introduced Western modern urban planning theories and ideas.

— Concession Overview

The old concession district in Hankou is located in the middle section of present-day Yanjiang Road, namely, north of Jianghan Road, south of the lower wharf of Mayang Street, and southeast of Zhongshan Avenue. The French Concession, part of which crossed Zhongshan Avenue, was about 200 meters southeast of the Lugouqiao-Hankou Railway and extended along the northwest of the Yangtze River, with a total length of 3,627 meters along the river. The concessions of Britain, Russia, France, Germany and Japan were arranged from south to north geographically, with a total area of 2,968.61 *mu*. There are eight consulates general, seven consulates and 5 agencies in the concession district. See the table below for details of the establishment date, area and recovery date of each concession.

汉口原租界区基本情况一览表 表1

租界名称 Concession Name	设立时间、面积 Establishment Date and Area	扩展日期、面积 Expansion Date and Area	总面积 Total Area	收回时间 Recovery Date
英 Britain Concession	3/21 1861 《汉口租界条款》 Terms of Hankou Concession 458.33 亩 / *mu*	8/31 1898 《英国汉口新增租界条款》 Terms of Hankou New British Concession 337.05 亩 / *mu*	795.38 亩 / *mu*	3/15 1927 设立第三特别区 the third special district was established
俄 Russia Concession	6/2 1896 《汉口俄租界地条约》 Hankou Russian Concession Treaty 414.65 亩 / *mu*	—	414.65 亩 / *mu*	3/1 1925 设立第二特别区 the second special district was established
法 France Concession	6/2 1896 《汉口法国租界租约》 Hankou French Concession Lease 187 亩 / *mu*	11/12 1902 《汉口展拓法租界条款》170 亩，随后法国又不断向界外扩展 142 亩 Terms of Hankou French Concession Expansion 170 *mu* France extended another 142 *mu* beyond its concession	499 亩 / *mu*	8/14 1945 1943 年 2 月 23 日，法国维希政权曾向日伪交还租界权 August 14, 1945 (On February 23, 1943, the Vichy regime in France transferred the concession to Japan)
德 Germany Concession	10/3 1895 《中德汉口租界条约》 Sino-German Hankou Concession Treaty 600 亩 / *mu*	27/8 1898 清政府与德国修订德国汉口租界界地 36.83 亩 the Qing Government and Germany revised the Hankou German Concession of 36.83 *mu*	636.83 亩 / *mu*	3/14 1917 设立第一特别区 the first special district was established
日 Japanese Concession	7/16 日 1898 《汉口日本专管租界条约》 Hankou Japanese Concession Treaty 247.5 亩 / *mu*	2/7 1907 《日本添拓汉口租界条约》375.25 亩 Treaty of Hankou Japanese Concession Expansion 375.25 *mu*	622.75 亩 / *mu*	8/14 1945
合计 Total			2968.61 亩 / *mu*	

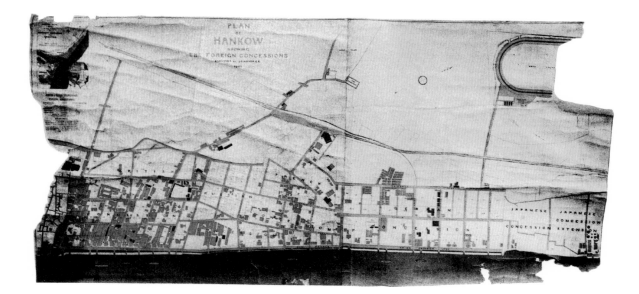

1
汉口租界图（1905年，现存于武汉市房地产管理局修志办公室）

Hankou Concession Plan (1905, now stored in the local chronicles office of Wuhan Housing Security and Management Bureau)

2
汉口租界分区图（编者自绘，底图为1918年汉口市街全图）

Hankou Concession Zoning Plan (drawn by the author; the base drawing is the Hankou street map of 1918)

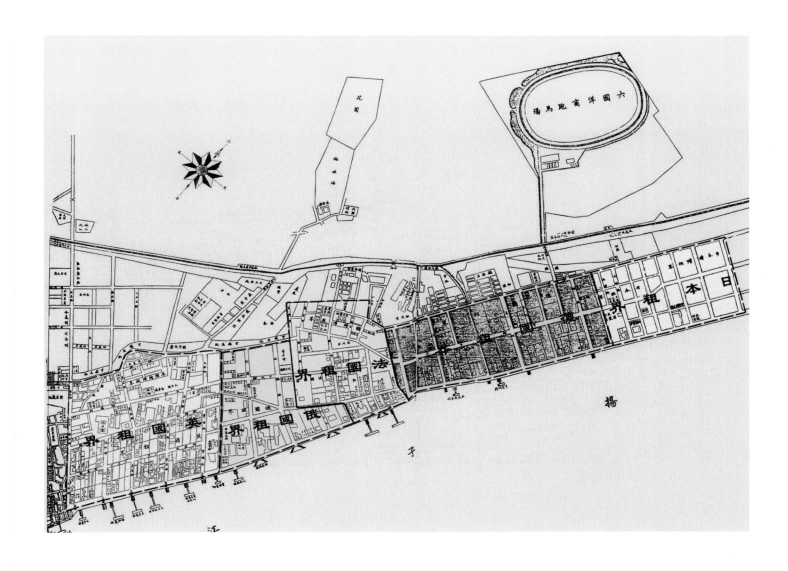

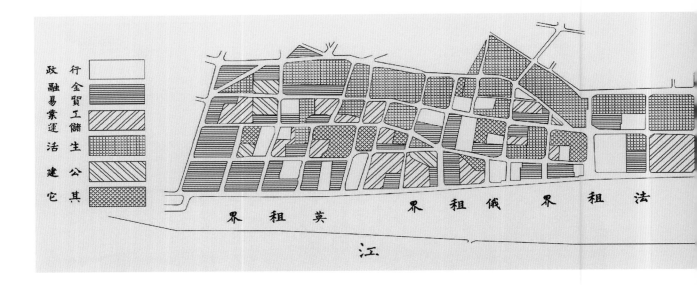

1

汉口租界用地分区图（资料来源于《武汉市城市规划志》）

Hankou Concession Land Zoning Plan (data source: *Wuhan Urban Planning Records*)

2

租界地区主要建筑分布图（资料来源于《武汉市城市规划志》）

Plan of Main Buildings in Concession District (data source: *Wuhan Urban Planning Records*)

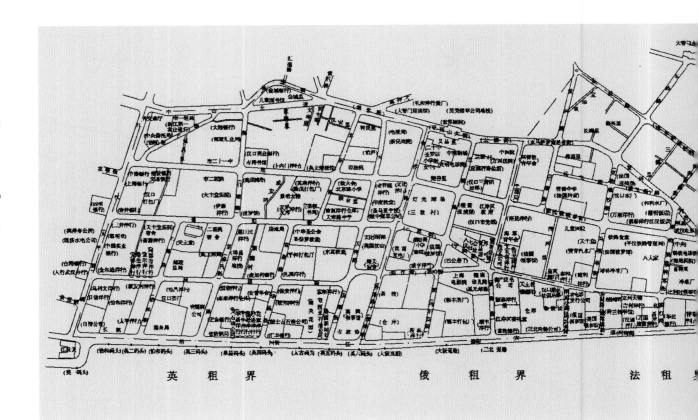

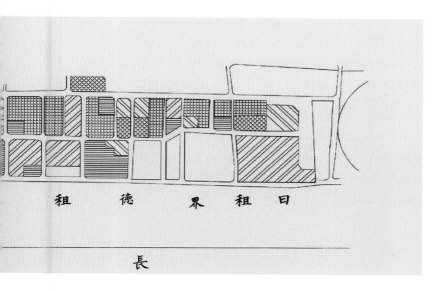

一　租界道路建设

五国租界建立时间虽有先后，但在修筑道路方面相互合作、相互照应，且考虑了老城区的格局，大多保留现状官路作为租界内及周边的主要通行道路，规划的纵向道路基本与老城区道路平行。租界的主要道路垂直于江面，通向码头。虽然各国租界分界处的横向道路存在局部错口，但临江主干道基本贯通相连，以联系各码头和租界各区域，满足商业贸易活动的需要，由此形成了较为规矩的方格路网体系。

此外，还存在局部的Y字形交叉道路，是由于英租界在1890年代前在汉口堡城墙内修建跑马场，为避让娱乐用地，通过通济门联系德租界，最终形成Y字形交叉道路。其后法租界扩张，拆除该跑马场，但在俄、法租界建设过程中依然保留这一路网，后期植入多段平行江岸的道路以完善路网系统。

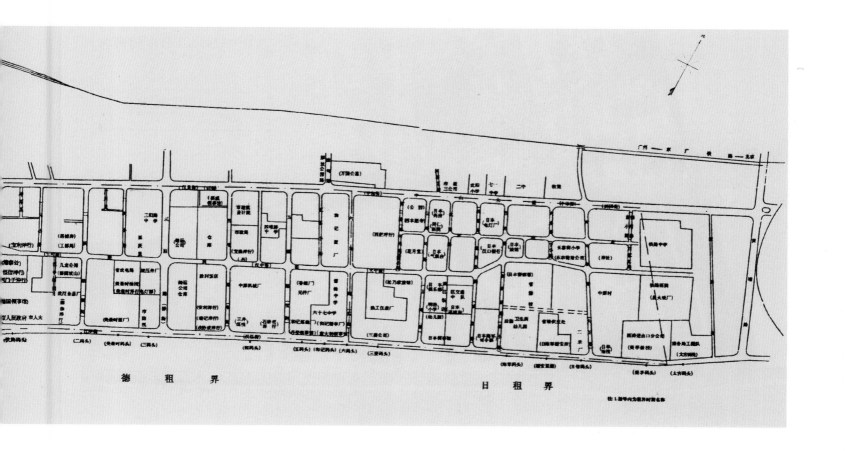

一 租界规划布局

为适应商业和航运的发展,各国租界均平行于长江横向布置,以求最大限度地获取临江岸线,有利于本国的港船贸易。其中英、德两国由于租界设立时间较早以及国力稍强,临江岸线明显多于其他三国,其码头数量也相应较多。

各租界地段内总体上统一了功能和布局,呈现"前商后居"的模式,从东至西按照商业和工业类用地、公共休闲类用地、居住类用地顺序排布。居住用地集中在法、德租界靠近京汉铁路的租界后方,环境优良,远离沿江闹市,沿江一带除开设少量领事馆用地外,多为商行、码头、仓库和货场等用地,以利于水上运输和集散。1863~1922年,租界区内共设有外国银行20家,到民国初年共设有外国银行114家,开设外国工厂42家,还有11家航运业轮船公司。

1906年京汉铁路的通车带来了汉口贸易的陆路交通时代,租界各国开始为争夺车站道路、占领车站周边地区而扩张各自的势力范围。各租界积极建设道路,将码头与之串联,或是直接引入支线,以打通水运、陆运交通。由于各国的利益争夺和势力扩展,租界和铁路之间因此形成了相对混乱的城市形态。

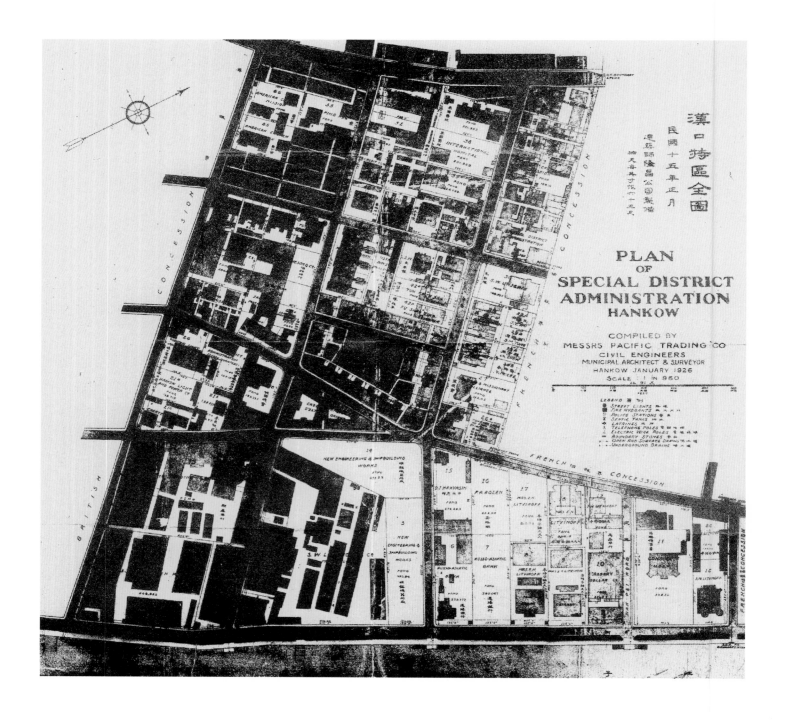

— Road Construction in Concessions

Although established at different times, the five concessions cooperated in road construction. Considering the pattern of the old city, they kept the existing government roads as the main roads in and around the concessions. The planned vertical roads were basically parallel to the roads in the old city. The main roads in concessions were perpendicular to the river and led to wharves. Although some horizontal roads at the boundary of concessions were staggered, main roads along the river were basically interconnected so as to connect the wharves and concessions and meet the needs of commercial and trade activities, thus forming a regular road network.

In addition, there were some Y-shaped cross roads. The British Concession built a racecourse within city walls of Hankou before the 1890s. To avoid the recreational land, a Y-shaped cross road was built, leading to the German Concession through Tongji Gate. Later, the racecourse was demolished due to expansion of the French Concession, but the road network was kept in the construction of Russian and French concessions. Several roads parallel to the river bank were built to improve the road network.

— Concession Planning

To adapt to the development of commerce and shipping, the five concessions were all arranged horizontally parallel to the Yangtze River, so as to maximize access to the riverfront and facilitate their port and ship trade. Due to the early establishment of concessions and the relatively strong national strength, Britain and Germany have larger riverfronts and more wharves than the other three countries.

On the whole, the functions and layout of each concession are unified, presenting the pattern of "commerce in the front and residence in the back". The commercial and industrial land, public and recreational land and residential land were arranged from east to west. Residential land was mainly in the back in French and German concessions close to the Lugouqiao-Hankou Railway. It had an excellent environment and was far away from the downtown along the river. In addition to a small number of consulates, most of the land along the river was used for commercial houses, wharves, warehouses and good yards, so as to facilitate water transportation and distribution. From 1863 to 1922, there were 20 foreign banks in the concession district. In the early years of the Republic of China, there were 114 foreign banks, 42 foreign factories and 11 shipping companies.

In 1906, the opening of the Lugouqiao-Hankou Railway brought Hankou trade into the land transportation era. The concessions began to expand their spheres of influence to fight for roads leading to the station and occupy areas around the station. They actively built roads to connect wharves with them or simply introduced branch lines to open up water and land transportation. Due to the interest contention and power expansion among concessions, a relatively chaotic urban form was formed between the concessions and the railway.

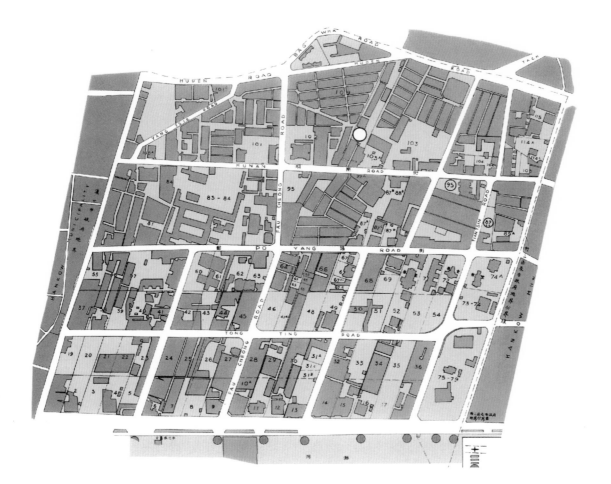

1　2

1

汉口特区（原俄租界）全图（1926年，绘制单位：建筑师隆昌公司）

Plan of Special District Administration, Hankou (1926), complied by Messrs Pacific Trading Company

2

汉口第三特别区（原英租界）图（1931年）（资料来源于《武汉历史地图集》）

Plan of Hankou Third Special District (formerly known as the British Concession) (1931) (data source: Wuhan Historical Atlas)

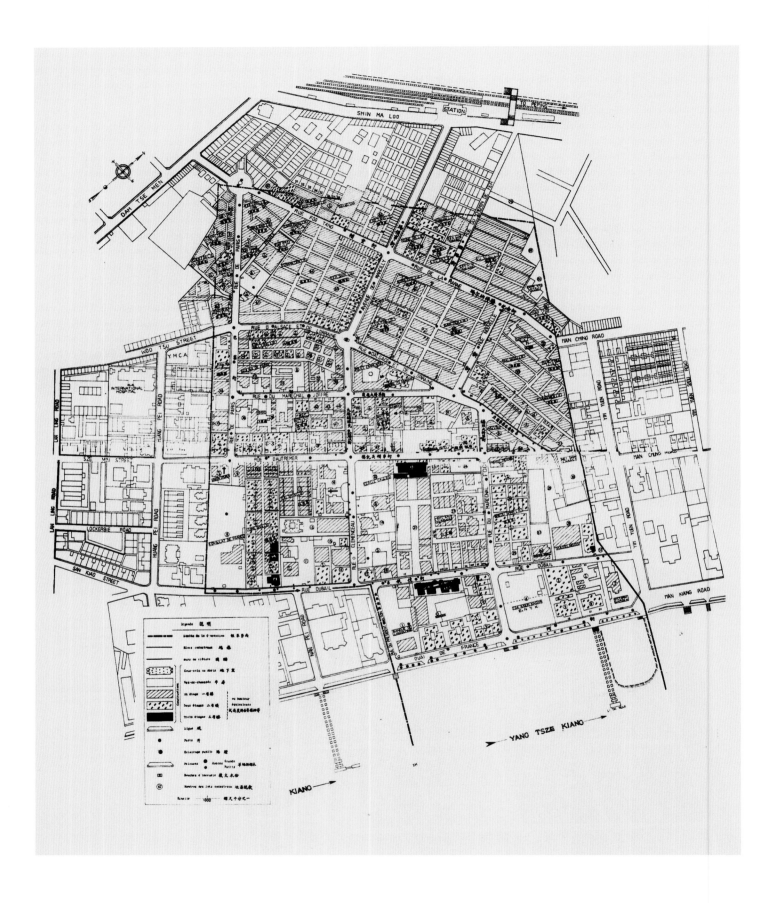

| 1 | 2 |

1
法租界图（1930年，现存于武汉市勘测设计研究院）

Plan of French Concession (1930, now stored in Wuhan Geotechnical Engineering and Surveying Institute)

2
汉口日本租界全图（1930年，现存于武汉市档案馆）

Plan of Japanese Concession, Hankou (1930, now stored in Wuhan Archives)

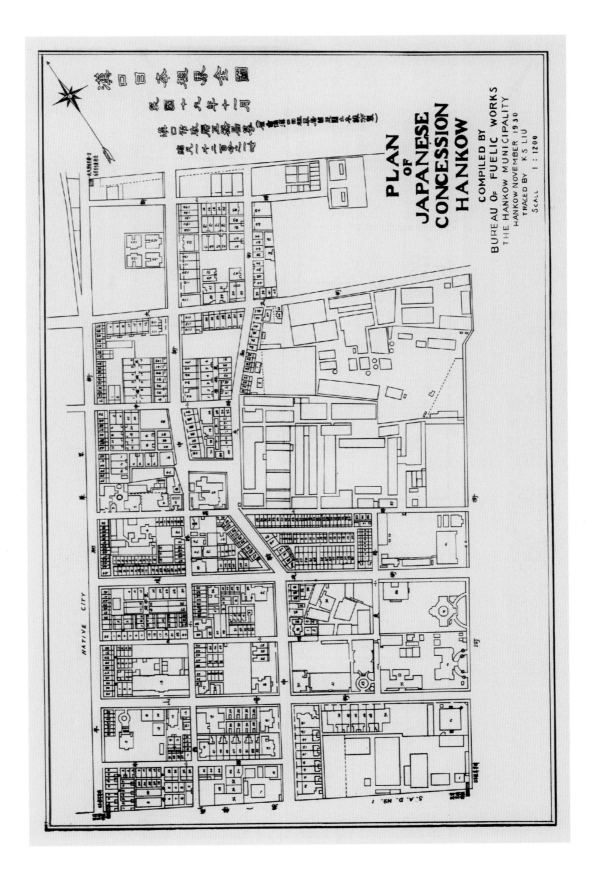

CHAPTER ONE • Independent Construction of Three Towns

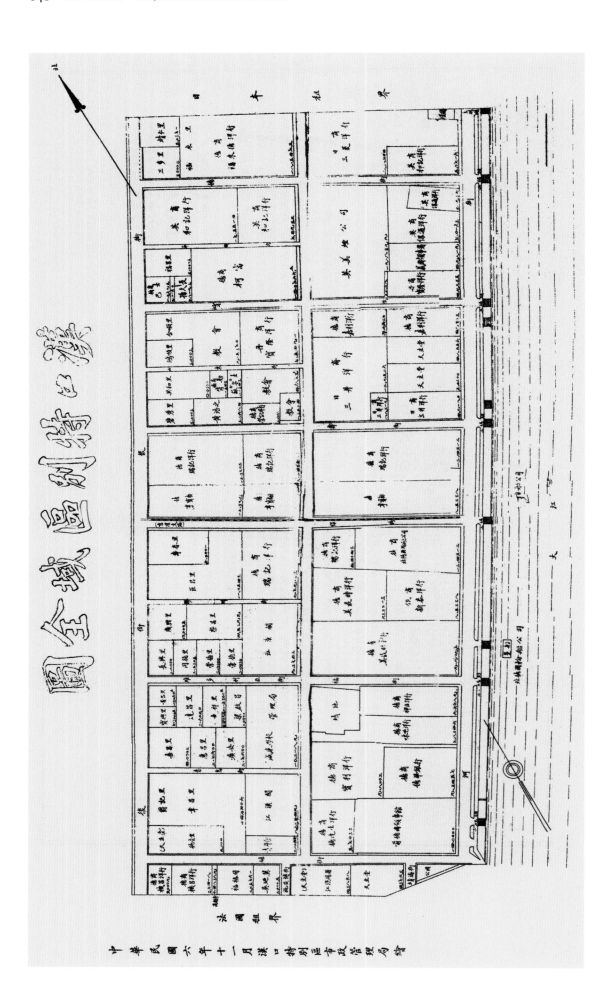

1
汉口特别区域（原德租界）全图（1917年，现存于武汉市档案馆）
Plan of Special District Administration (formerly known as the German Concession), Hankou (1917, now stored in Wuhan Archives)

2
福忠里、泰安里、保元里总平面图
General Layout Plan of Fuzhongli, Taianli, Baoyuanli Residential Buildings

3
海寿里、昌年里、海寿元里总平面图
General Layout Plan of Haishouli, Changnianli, Haishouyuanli Residential Buildings

一 新式里分住宅

伴随着租界及其附近地区的城市建设，使得西方建筑形式大量传入，产生了融合中国传统建筑形式的"新"建筑——里分住宅。19世纪末，汉口效法上海里弄的形式建造了一批里分住宅，是早期里分的代表，而与上海里弄行列式布局中入户门同向布置的方式不同，汉口里分多相向布置，延续了"临街两侧，对门而居"的街屋布局方式，形成了"门门相对，宽窄相间"的巷道空间形态。

里分住宅采用单元联排式布局，一般采取三间两厢或两间一厢的户型拼接方式，形成相向的入户门之间的错位关系。其在细部构造上仍保留武汉地方传统做法，在建设初期多无分户厨房、厕所，但由于建造周期短、造价低，又能通过商业发展提高地价，因此这种住宅形式得到了迅速发展。至1914年，汉口因模范市的建设再次掀起了建房热潮。这一时期的里分建造规模较大，多仿西式做法，多为2~3层砖木结构联排式房屋，街面是底层商店住宅，街后是联排式住宅，这种住宅形式节省用地，改变了传统街坊中"前店后寝"的布局。如长怡里、保和里、五常里等，其中以五常里为中心的中山大道两侧的店铺住宅房屋约200栋，为当时建房之冠。

从1900年里分建筑开始到1938年武汉沦陷，里分建筑基本结束，汉口的里分总数约为580条，这些里分纵横交织，构成大汉口的血管。直到新中国成立以后，武汉市人民政府为改善工厂职工的生活条件，在1950~1960年间建造了一批新型里分，后来随着单元式住宅的逐渐普及，里分住宅自此没有继续建造。

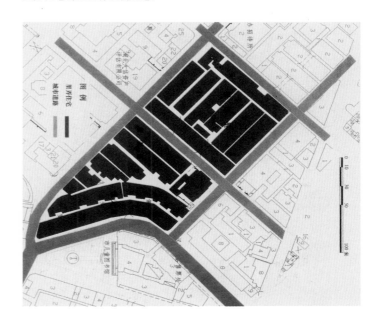

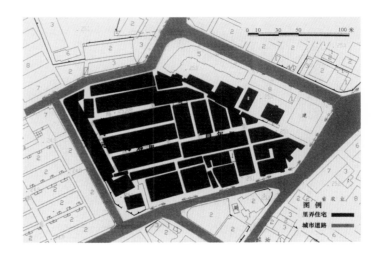

— New-type of *Li-Fen* Residential Building

The urban construction in concessions and its surrounding areas promoted large quantities of introduction of Western architectural forms, so a "new" type of architecture, the *Li-Fen* Residential Building, integrating the traditional Chinese architectural forms, came into being. At the end of the nineteenth century, a series of *Li-Fen* Residential Buildings in imitation of Linong houses in Shanghai constructed in Hankou were the representatives of *Li-Fen* Residential Buildings in early times. Different from methods for arrangement in same direction of entry doors in rowed layout of Linong houses in Shanghai, *Li-Fen* Residential Buildings in Hankou were arranged in opposite direction, which continued layouts of street houses "facing toward each other on both sides of street" and formed spatial forms of "door-to-door, narrow and wide" lanes.

Units in layouts of townhouses are adopted for *Li-Fen* Residential Buildings. In general, three rooms and two wings, or two rooms and one wing, are house types by splicing, forming misalignment relationship between entry doors in opposite direction. In terms of detailed structures, Wuhan local traditional methods are still adopted. At the initial stage of the construction, most of the households had no kitchens or water closets, but such residential forms were developed rapidly because of short construction cycle, low cost and increase in land price by commercial development. The construction boom was on the rise again because of building model city in Hankou by 1914. During this period, *Li-Fen* Residential Buildings were constructed at a relatively large scale with the imitation of Western ways. They were townhouses with two to three storey of brick and timber structures. In front of street were shops and residences on ground floor while behind the street were townhouses. Such forms of buildings not only saved land but also changed layouts of "bedroom after store" in traditional living quarters, such as the Changyili Residential Building, Baoheli Residential Building, Wuchangli Residential Building and others. There were about 200 stores and houses on both sides of Zhongshan Avenue centering on Wuchangli Residential Building, ranking the top of the buildings at that time.

Starting from 1990, the construction of *Li-Fen* Residential Buildings was stopped due to the Fall of Wuhan in 1938. About 580 *Li-Fen* Residential Buildings in total in Hankou were crisscrossed like blood vessels, spreading over the big city. After the founding of People's Republic of China, in order to improve the living conditions of factory workers, the Wuhan Municipal People's Government constructed a batch of brand-new *Li-Fen* Residential Buildings between 1950 and 1960. *Li-Fen* Residential Buildings were no longer constructed later due to the gradual popularity of apartment houses.

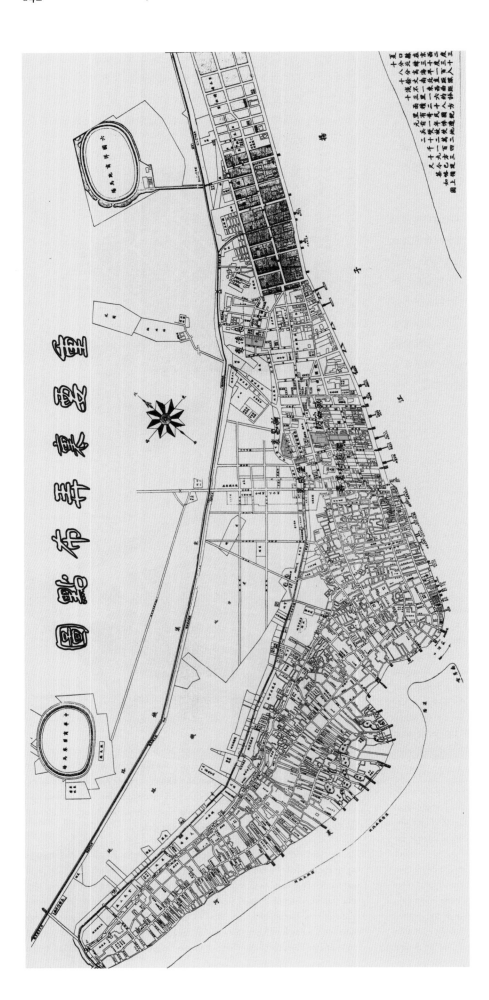

1

汉口里弄分布示意图（编者自绘，底图为1918年汉口街全图）

Plan of Linong in Hankou (drawn by the author; the base drawing is the Hankou street map of 1918)

2

汉口华商跑马场正厅图（资料来源于《夏口县志》）

Main hall of the Hankou Huashang Racecourse (data source: *Xiakou County Annals*)

3

跑马场建设示意图（编者自绘，底图为1922年武汉三镇街全图）

Schematic Diagram of Racecourse Construction (drawn by the author; the base drawing is the street map of the three towns of Wuhan in 1922)

4

西商跑马场马道（资料来源于《武汉通史》）

Track of the Xishang Racecourse (data source: *A General History of China*)

一 跑马场建设

1896年，英国人在今复兴街、昌年里一带修建的"马道子"和球场，为武汉最早的跑马场，"跑马之风"由此引入汉口。后由于法租界扩界，于是1905年英国人另辟跑马场所，建设西商跑马场，其位置以今解放公园为中心，但面积是解放公园的1.5倍以上。跑马场内有马道两圈，外圈宽30米，长1600余米，内圈是马球和橄榄球场，同时建有能容万名观众的阶梯式看台，以及写字间、酒吧、舞厅（在今中国人民解放军通信指挥学院宿舍范围内）。跑马场内还另辟了18洞的高尔夫球场（今解放公园南门附近），占地百余亩。

1906年，刘歆生与人合伙在今航空路、万松园路等大片土地上，修建"华商跑马场"。1926年，武汉商界王植夫、吴春生等人集资合股，又在唐家墩与姑嫂树一带，修建了万国跑马场。

跑马场建设示意图

— Racecourse Construction

The "riding track" and court constructed by the British in 1896 on Fuxing Street and Changnianli Residential Building nowadays are the earliest racecourse in Wuhan, thus introducing the "trend of horse race" into Hankou. Later, due to the expansion of the French Concession, in 1905, the British found another place for racecourse and constructed the Xishang Racecourse centering on the current Liberation Park with more than 50% area compared with it. There are two laps of riding tracks in the Racecourse. The outer ring is 30 meters in width and more than 1600 meters in length while the inner one comprises the polo and the rugby field. Meanwhile, there are also stepped grandstand with a capacity of 100, 000 audiences, office room, bar and ballroom (within the dorms of today's Communication College of People's Liberation Army). There is also the golf course with 18 holes in the Racecourse (nearing the south door of today's Liberation Park) covering over 100 *mu*.

In 1906, Liu Xinsheng cooperated with others to construct the "Huashang Racecourse" on the current Hangkong Road, Wansongyuan Road and other lands. In 1926, Wang Zhifu, Wu Chunsheng and others of Wuhan business circles constructed the Wanguo Racecourse around the Tangjiadun Street and Gusaoshu via fundraising partnership.

一　水、电设施建设

1884年汉口设电报局，1901年德商首先开办市内电话，1906年英商在合作路设立发电厂，同年既济水厂开始建设。随后1907年德商开办电灯公司，1913年日商开办电灯厂，此时租界区由三家外商电力公司供电。至1930年代，由华界商人宋炜臣等人创办的汉镇既济水电公司的业务开始深入租界区域。

既济水厂（今宗关水厂）建成之前，由于各租界皆担心经营自来水厂难以获利，故未铺设供水管道，用水直接从长江提取，所取之水经简单沉淀和过滤后饮用。1909年，既济水厂和水塔建成，是武汉自来水历史的开端标志。既济水厂所制之水，开始为汉口10万左右居民及租界区所用，同时位于中山大道的水塔是当时汉口城区（包括租界区）的最高建筑，高度约40米，其楼顶的钟楼也承担了汉口地区火灾报警的职责，以敲钟数字表明火灾区位。

租界除对防火有严格要求外，对于卫生和安全也特别重视。租界的下水管道以排放污水及粪水为主，直接排江；排水沟以排放雨水为主，法、

俄及德租界区域实行了早期的雨污分流。每到夏季，租界便定期对下水管道和排水沟进行消毒，大大改善了租界内的卫生状况。

为了保证租界防洪安全，各国挖后湖土以填高地面，又在租界区沿江岸一带修建红砂石驳岸码头，与马路平齐。1931年大水后，又整修了江汉关沿江以下的各租界段防水墙。1935年大水后，又以红砖浆砌高防水墙，高程为海拔29米，麻阳街以下加高为29.5米。

— Construction of Water and Electricity Facilities

In 1884, Hankou set up the Telegraph Office. In 1901, German businessmen took the lead to open the urban telephone system. In 1906, British businessmen established the power plant on the Hezuo Road and the Jiji Water Plant was being built in the same year. Later in 1907, German businessmen opened the electric light company while in 1913, Japanese businessmen founded the electric light factory. At that time, the Concession Area was supplied by three foreign electric power companies. By the 1930s, the Hanzhen Jiji Hydropower Company, founded by Chinese businessman Song Weichen and others, began to expand its businesses in the Concession Area.

Before the completion of the Jiji Water Plant (now Zongguan Water Plant), because Concessions worried about that it was hard to make profits from running waterworks, no water supply pipes were laid. The water was directly extracted from the Yangtze River, simply precipitated and filtered before drinking. In 1909, Jiji Water Plant and Water Tower were completed, marking the beginning of history of running water in Wuhan. Jiji Water Plant started to offer water for about 100,000 residents and Concession Areas of Hankou. Meanwhile, the Water Tower located on Zhongshan Avenue was the tallest building in Hankou city (Concession Areas included) at that time with a height of about 40 meters. The bell tower on the top of the Tower also took on the responsibility of fire alarm in Hankou area, indicating the location of the fire with numbers for ringing the bells.

In addition to strict requirements on fire prevention, the Concessions also attached special importance to the hygiene and safety. Sewer lines in the Concessions serve mainly the discharge of waste water and liquid dung into the river while the drainage ditches are mainly for storm-water drainage. Early diversion of rain and sewage water was conducted in the French, Russian and German Concession. Every summer, the Concessions regularly disinfected sewer lines and drainage ditches, greatly improving the sanitary conditions of it.

In order to ensure the safety of the Concessions in flood control, various countries dug the soils of Houhu to fill up the ground, and built Red Sandstone Revetment Wharf along the banks of the Concession Areas, which were parallel and level with the roads. After the Great Floods of 1931, waterproof walls of the Concessions below the banks of Jianghanguan River were renovated; after the Great Floods of 1935, red bricks and mortars were used for high waterproof walls with elevation of 29 meters above sea level. Areas below the Mayang Street were heightened to 29.5 meters.

	4
1	
2	3

1

1906年既济水电公司（资料来源于《武汉旧影》）

Jiji Hydroelectric Company in 1906 (data source: *Old Pictures of Wuhan*)

2

汉口电灯公司（资料来源于武汉市档案信息网）

Hankou Electric Light Company (data source: Wuhan Archives Information Network)

3

1913年水塔（资料来源于《武汉旧影》）

A water tower in 1913 (data source: *Old Pictures of Wuhan*)

4

市政公用设施示意图（编者自绘，底图为1931年武汉街市图）

Plan of Municipal Public Facilities (drawn by the author; the base drawing is the market street map of Wuhan in 1931)

二、民国初期的规划
Planning in the Early Republic of China

1911 – 1926

1911 年，武昌首义后，清军为击败革命军，不惜毁灭汉口，下令焚舍烧房，汉口 1/5 市区被焚毁，十里街道夷为焦土，三镇经历了近百年来的第一次浩劫。

辛亥革命后，孙中山非常看好"首义之地"的武汉，所著《建国方略》之《实业计划》模仿巴黎、伦敦规模规划武汉发展，明确了城市发展目标，提出了城市发展战略，制定了以武汉为中心、辐射全国的中央铁路系统和大港计划等。

1912 年修复已焚毁的华界市区被提上日程，孙中山指示内务部筹划重建汉口市区，并提出将汉口、武昌和汉阳作为一个整体进行规划的设想，随后分别对三镇进行修建计划，编制了《汉口全镇街道规划》、《武昌商埠全图》等规划方案，同时草拟《汉口市政建筑书》，进行了城市功能分区，量丈建设用地，开辟干道网络，布置过江通道，开展城市改良工作，意在与租界区媲美，形成模范市区。

这一阶段，三镇的城市空间得到了横向和纵向扩展。与此同时，租界区因在战火中幸免于难，使其成为当时武汉市区的相对安全区，一时间租界内人口增加，日益繁荣，汉阳、武昌的城市中心也逐渐向江边转移，大量工业项目部署在码头岸边。这一时期可视为晚清租界时期至民国时期一脉相承的城市规划发展过程。

In 1911, after the Wuchang Uprising, the troops of Qing Dynasty, in order to defeat the revolutionary army, wanted to destroy Hankou at all costs and ordered to burn houses. One fifth of Hankou's urban areas were burned down and ten miles of streets were burnt to ashes. The Three Towns experienced the first catastrophe within recent hundred years.

After the Revolution of 1911, Sun Yat-sen was very optimistic about Wuhan that was the "First Place of Uprising". His *The International Development of China: Material Reconstruction* imitated the scale of Paris and London to plan Wuhan's development, defined the urban development goals, put forward the guiding ideology of the urban strategic development, and formulated a central railway system and a large port plan centering on Wuhan and radiating across the whole country.

In 1912, the restoration of the burned-down Huajie urban area was put on the agenda. Sun Yat-sen instructed the Ministry of Internal Affairs to plan the reconstruction of Hankou urban area, and put forward the idea of planning Hankou, Wuchang and Hanyang as a whole. Subsequently, the Three Towns were built separately, and plans such as "*Hankou Town Street Plan*" and "*Wuchang Commercial Port Plan*" were drawn up. Meanwhile, the "*Hankou Municipal Architecture Book*" was drafted, which divided the urban functions, measured the construction land, opened up trunk road networks, arranged pathway crossing rivers, and carried out urban improvement work, aiming to be on a par with the Concession Areas and form a model downtown area.

At that stage, the urban spaces of the Three Towns were expanded horizontally and vertically. Meanwhile, the Concession Areas survived the war, making themselves relatively safe areas in Wuhan at that time. At that time, the population in the Concession Areas increased with increasingly prosperity. The urban centers of Hanyang and Wuchang also gradually shifted to the river banks, and a large number of industrial projects were deployed on the banks of wharfs. That period can be regarded as a continuous process of urban planning and development from the period in concessions of late Qing Dynasty to the period of the Republic of China.

1 - 对标纽约、伦敦的《建国方略》规划蓝图

Blueprint of The International Development of China modeling on New York and London

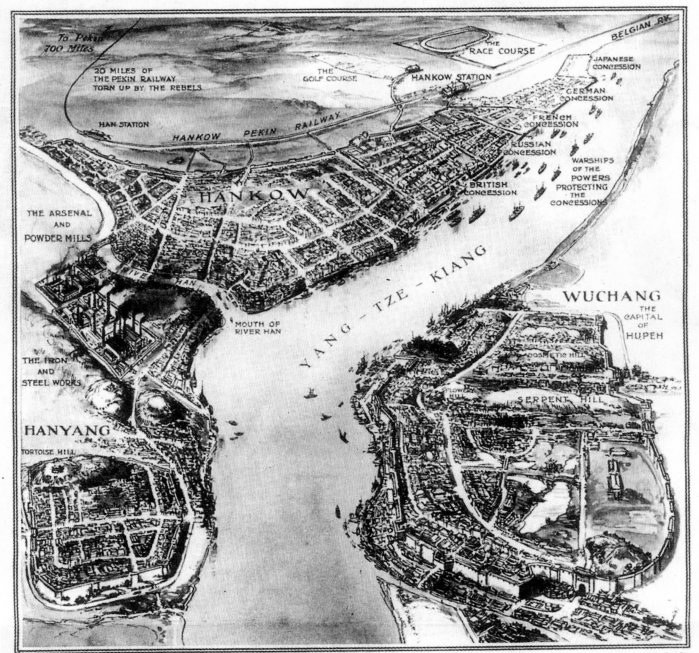

武汉三镇地理方位图（1911年，现存于武汉市博物馆）

Geographic Map of Three Towns of Wuhan (1911, now stored in Wuhan Museum)

孙中山在其所著《建国方略》中，对武汉的城市计划发展多有论述："武汉者，指武昌、汉阳、汉口三市而言。此点实吾人沟通大洋计划之顶水点，中国本部铁路系统之中心，而中国最重要之商业中心也"。此外，关于城市规模也有相应表述，"确为世界最大都市之一矣，所以为武汉将来计划，必须定一规模，略如纽约、伦敦之大"。

孙中山对武汉的规划思路是首先将武汉三镇连为一体，提出武汉城市桥梁隧道的选址设想，在京汉铁路于长江边第一转弯处设过江隧道，在汉水口设桥梁或隧道，以联络武昌、汉口、汉阳三城为一市，进而建立以武汉为中心的铁路系统，使其具有大城市的集聚效应，以此加强对腹地的联系与交流，辐射带动中西部地区的发展。

孙中山在其中央铁路系统和东南铁路系统计划中，以武汉为"中国本部铁路系统之中心"。除当时已经修建和计划修建的卢汉、粤汉及川汉线外，提出建设若干直达武汉的铁路线，还要修建八大铁路干线以接通汉口，包括南京汉口线、西安汉口线、北方大港汉口线、黄河港汉口线、芝罘汉口线、海州汉口线、新洋港汉口线和福州武昌线，这八大铁路干线总长度预计达5906公里。

为了保证武汉与外洋畅通无阻，孙中山主张长江两岸从汉口直达海滨，要修筑长堤；对长江水道由下而上要进行系统疏浚，以便大洋航船无论冬夏，皆可直达"汉口至于汉口以上"的水路，无论长江上游还是汉水，均要加以整治，以充分发挥航运之利。

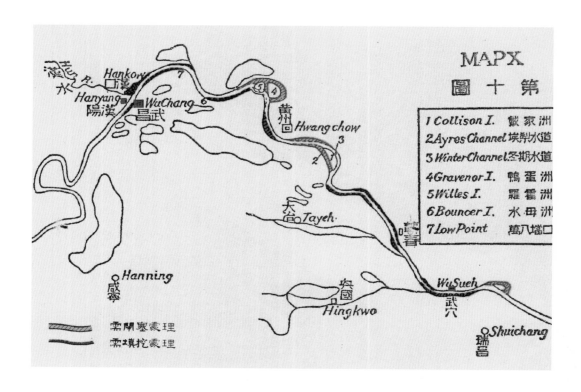

1

武穴至汉口河道整治建设示意图（编者自绘，底图为1998年版《建国方略》之二《实业计划》第十图）

River Regulation and Construction from Wuxue to Hankou (drawn by the author; the base drawing is the tenth illustration in The International Development of China: Material Reconstruction (1998 edition))

2

建筑汉口全镇街道全图（1912年，现存于武汉市图书馆）

Map of Streets and Buildings in Hankou Town (1912, now stored in Wuhan Library)

Sun Yat-sen in The International Development of China mentioned a lot about the planned urban development of Wuhan that, Wuhan comprised the Wuchang, Hanyang and Hankou, which was exactly the planned port for connection with foreigners, the center for local railway system and the most important business center in China. In addition, there were corresponding statements about the urban size that it would be the largest global cities and, for the future plans of Wuhan, it should be as large as the New York and London.

Sun Yat-sen's planning idea for Wuhan is to connect the three towns of Wuhan into a whole and to propose the location of Wuhan's urban bridges and tunnels. The Lugouqiao-Hankou Railway will be located at the first turning point along the Yangtze River, with a underwater tunnel. Bridges and tunnels are set at the mouth of Hanjiang River. The three cities of Wuchang, Hankou and Hanyang will be connected as a whole, and then a railway system centered on Wuhan will be set up to make it have the agglomeration effect of a big city, so as to strengthen the connection and communication with the hinterland and radiate the development of the central and western regions.

Sun Yat-sen took Wuhan as "center for local railway system" in his plans for the central railway system and the southeastern railway system. In addition to the Lugouqiao-Hankou Railway, the Guangzhou-Wuchang Railway and Chengdu-Hankou Railway that was built and to be built, several direct railway lines to Wuhan were proposed for construction. There were eight major trunk railways to be built for connection with Hankou, including Nanjing-Hankou Railway, Xi'an-Hankou Railway, Northern Dagang-Hankou Railway, Yellow River Port-Hankou Railway, Zhidai-Hankou Railway, Haizhou-Hankou Railway, Xinyang Port-Hankou Railway and Fuzhou-Wuchang Railway, totaling 5, 906 kilometers as estimated.

In order to ensure the convenience of Wuhan and foreign countries, Sun Yat-sen advocated building causeways on both sides of the Yangtze River from Hankou to the seashore. The Yangtze River waterway should be systematically dredged from bottom to top so that ocean ships can reach the waterway of "Hankou and above" in winter and summer. Both the upper reaches of the Yangtze River and the Hanjiang River should be regulated to give full play to the convenience of water transportation.

2 - 以模范市为目标
的辛亥革命战后重建规划

Reconstruction Planning after the Revolution of 1911 toward a Model City

— 汉口全镇街道规划（1912 年）

1911 年 10 月，武昌首义成功，英国《环球》画报以"中国革命中心"为题刊登武汉三镇地理方位图。该图描述如下，"长江南岸（图右方）为武昌，长江北岸（图左下）为汉阳，与汉阳隔江相望的是汉口。武昌是湖北政治中心，汉口有火车站、高尔夫球场、租界，而汉阳则拥有兵工厂和钢铁厂。这三个镇人口密集，三镇人口加起来有 125 万"（详建筑汉口全镇街道全图）。据了解，这些图片为辛亥武昌首义期间，外国记者用钢笔画好，寄回本国再印成报纸。

1912 年为重建汉口市区，孙中山指示内务部筹划修复事宜，务期"首义之区变成模范之市"，设立汉口建筑筹办处，任命李四光为"特派汉口建筑筹备员"，主要对街道建设进行了规划，后又仿巴黎、伦敦规模，规划汉口全镇道路 306 条，绘制汉口全镇街道图（不含租界区），以辟街道，建楼房。此时的道路系统规划追求理想的构图形式——"方格网十对角线"式构图，虽未考虑地形因素，最终并未实现，但反映了欧美近代功能主义的规划思想已经随着政治的大变革初步进入了中国。

— 汉口市政建筑计划书（1923 年）

民国 12 年（1923 年）汉口地亩清查局出版了孙武（1911 年辛亥革命起义时任军事总指挥参谋长，湖北军政府成立时任军务部部长）撰著的《汉口市政建筑计划书》，该书提出了"拓商耕农、交通为首、水陆并举……"的武汉城市发展主张，论述了汉口总体发展规划、铁路、干道、桥梁、排水等规划细节。

该《计划书》"以夏口全县区域为建筑汉口市政之范围"，面积约 70 平方公里，即汉口、汉阳、武昌三镇，划分为甲、乙、丙三部分。甲部为汉口旧市镇及张公堤至舵落口一段，辟做商场；乙部以张公堤至柏泉山一段，辟做工场；丙部以西湖巨龙岗至新沟一段，辟做农场。

《计划书》提出用铁路、公路、市内公共交通道路及江河道构成城市交通网络系统。市内交通系统规划采用全区域方格网道路系统，开辟东西向贯通干道 3 条，南北向分隔干道 8 条，并配套建设大马路（66.6 米宽）、中马路（40 米宽）、河边路（33.36 米宽）、小巷路（20 米宽），形成纵横交错的网状交通结构。

为使汉口充分发挥交通枢纽功能，计划提出在汉阳大别山麓（今龟山）与武昌黄鹤山（今蛇山）之间架设汉水桥，将京汉、川汉、粤汉三大铁路连贯一体，将大智门旧总车站移至硚口上段汉水边；在汉阳城东门（现汉阳大道沿江外）修建大型港口；在汉水两岸修建平行马路，河上架设 5 座能开启的铁桥，使汉口、汉阳搭臂相依，交通无阻；把汉口当时已收回的德租界辟为特别区，与市内交通干道连通形成市区交通控制中心。

孙武还十分重视城市排水防汛，计划修建河道来解决雨季排滞抗涝，在地势较低的汉口东端戴家山开挖深潭（水库）扩大城市蓄水防洪能力。

根据计划，孙武对汉口用地进行"鱼鳞量丈"，清理整治，并在刘歆生的支持下，倡导建设了汉口模范区，修建了吉庆街等十多条街道，意在其规划布局要"与租界区媲美"，是"汉口市区之模范"。

— Hankou Town Street Plan (1912)

In October 1911, the Wuchang Uprising was a success. The Britain's Globe pictorial published the geographic map of three towns of Wuhan under the title of "China's Revolutionary Center". The map was described as follows: "Wuchang is on the south bank of the Yangtze River (right in the map), Hanyang is on the north bank of the Yangtze River (lower left in the map), and Hankou faces Hanyang across the river. Wuchang is the political center of Hubei. Hankou has a railway station, golf course and concession, while Hanyang has an arsenal and a steel factory. These three towns are densely populated, totaling 1.25 million"; see details in the Map of Streets and Buildings in Hankou Town. According to information, these pictures were made by foreign journalists during the Wuchang Uprising of 1911. They were painted in pen and sent back to their home countries and printed into newspapers.

In 1912, in order to rebuild Hankou city, Sun Yat-sen instructed the Ministry of Internal Affairs to plan for the restoration so as to "transform the area of the Wuchang Uprising into a model city" and set up the Hankou Building Preparation Office, appointing Li Siguang as the "Special Hankou Building Preparation Official". The main task was to plan the construction of streets. Later, following the scale of Paris and London, he planned 306 roads and drew the street map of Hankou town (excluding Concession Areas) to build streets and buildings. The road system planning was in pursuit of the ideal composition form, namely the composition of "square grid and diagonal line" with no consideration of topographic factors, so it wasn't realized at last. but which reflected that the philosophy of modern functionalism in Europe and the United States has initially entered China with the great political changes.

— Hankou Municipal Construction Plan (1923)

In the 12th year of the Republic of China (1923), Hankou Land Inventory Bureau published Hankou Municipal Construction Plan written by Sun Wu (Chief of Staff of the Military Commander in Chief in 1911 Revolution Uprising and Minister of Military Affairs when Hubei Military Government was established). The Plan put forward Wuhan's urban development idea of "commercial development and husbandry, priority of transportation, simultaneous development of land and water communication", discussing the planning details of Hankou's overall development plan, and details, such as railways, trunk roads, bridges, and drainage.

The Plan took the entire Xiakou County as the scope for municipal construction in Hankou City. The area was about 70 square kilometers, covered Hankou, Hanyang and Wuchang Towns, and divided into the Part A, B, C. The Part A started from Jiushi Town and Zhanggong Bank and ended at the Duoluokou of Hankou and was constructed into the shopping mall. The Part B started from Zhanggong Bank and ended at the Baiquan Mountain and was constructed into the workshop. The Part C started from Xihu Julonggang and ended at the Xingou and was constructed into the farm.

The Plan proposed to use railways, highways, public traffic roads and river channels in the city to form the urban traffic network system. The municipal traffic system plan was adopted with the regional square grid road system, opening up three east-west breakthrough trunk roads and eight south-north division trunk roads with supporting construction of main roads (66.6 meters in width), middle roads (40 meters in width), riverside roads (33.36 meters in width) and alley roads (20 meters in width), and forming the criss-crossed grid traffic system.

In order for Hankou to give full play to its function as a transportation hub, the Plan proposed to set up a Hanjiang River Bridge between the foot of the Dabie Mountains in Hanyang (now Turtle Mountain) and the Huanghe Mountain in Wuchang (now Snake Mountain), to connect the major Lugouqiao-Hankou Railway, Chengdu-Hankou Railway and Guangzhou-Wuchang Railway, and to move the old Dazhimen main station to the Hanjiang River at Qiaokou. A large port at the east gate of Hanyang city (now outside Hanyang Avenue along the river) was built; on both sides of the Hanjiang River, parallel roads were built, and five iron bridges were erected on the river to connect Hankou and Hanyang and facilitate the traffic. The German Concession that Hankou had recovered at that time was turned into a special area, connecting with the main traffic roads in the city to form an urban traffic control center.

Sun Wu also attached great importance to municipal drainage and flood control, and planned to build river channels to solve the problem of stagnant water and water logging in rainy season, and to excavate deep pools (reservoirs) in Daijiashan at the eastern end of Hankou to expand the urban capacity of water storage and flood control.

According to the Plan, Sun Wu conducted the "Yulin Measurement" toward the construction lands in Hankou with cleanup and renovation. With the supports from Liu Xinsheng, he advocated the construction of Hankou Model Area, built Jiqing Street and other nine Streets and intended to develop it into a "match for the Concession district" and the "model of Hankou City" in terms of planning and layout.

1

夏口县地形及护堤全图（1923年）（资料来源于《武汉市城市规划志》）

General Plan of Terrain and Dikes in Xiakou County (1923) (data source: *Wuhan Urban Planning Record*)

2

汉口商场建筑全图（1923年）（资料来源于《武汉市城市规划志》）

General Plan of Business Buildings in Hankou (data source: *Wuhan Urban Planning Record*)

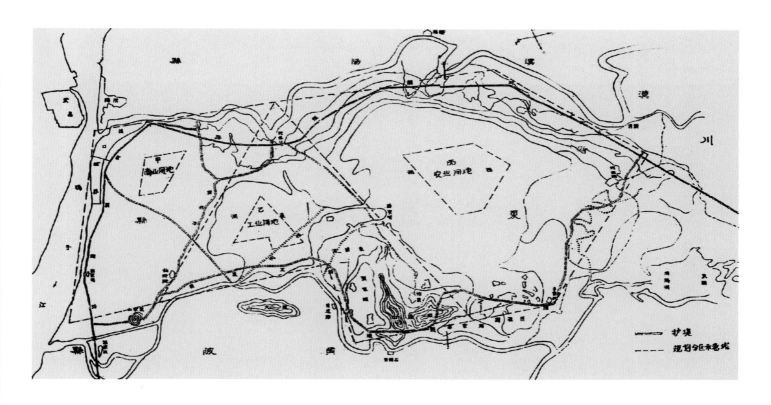

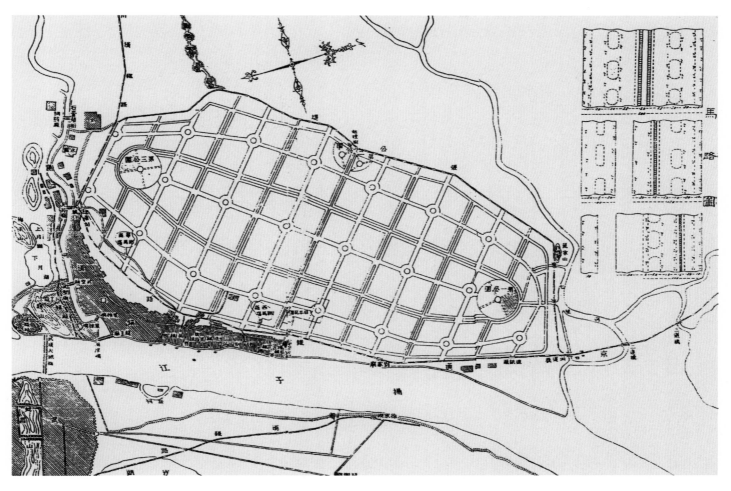

CHAPTER
第 贰 章
TWO
1927-1949

三镇融合规划时期
Planning Period for Integration of Three Towns

1927年，国民政府迁都武汉，三镇合而为一，完成了从县、厅到市的转变，武汉市成为中国当时第一个直辖市。这一时期，三镇的城市建制时分时合，社会发展并不稳定：前有辛亥革命时期清军在汉口纵火，继有北洋军阀统治时代的城市兵变，国民革命军北伐武昌时的围城战，抗战初期日军对武汉的轰炸，加之1931年水淹武汉，均影响了武汉城市发展的进程，但总体上朝着融为一体的城市近代化方向曲折迈进。

为打破传统三镇分割的局面，国民政府延续了孙中山《建国方略》的设想，将三镇作为一个整体进行规划建设，系统引进了西方城市规划理论，提出了三镇统一的功能分区，先后组织编制了《武汉特别市工务计划大纲》《武汉特别市之设计方针》《汉口都市计划书》等城市总体规划层面的都市计划。

1937年抗日战争全面爆发，次年武汉沦陷，导致了武汉近七年城市近代化进程几近停滞状态。至1943年抗日战争末期，三镇虽未真正融合，但区域规划进入了中国历史舞台，这一阶段首次提出了规划意义上的"大武汉"，先后发布《大武汉市建设计划草案》《武汉区域规划实施纲要》《武汉区域规划初步研究报告》和《武汉三镇土地使用与交通系统计划纲要》，开始由"功能分区+建设计划"向近代城市规划逐渐转变。

期间，国民政府还特别重视市政基础设施建设，组织编制了一系列的运河工程、抗涝排水、江滩改造、防空疏散等市政工程规划，以提高城市的安全性，扩展城市的发展空间。

In 1927, the Kuomintang government moved its capital to Wuhan, and the three towns merged into one, completing the transformation from county, administrative office to city. Wuhan became the first municipality directly under the central government in China at that time. During that period, the urban system of the three towns changed from time to time and the social development was not stable: The Qing army set fire to Hankou during the Revolution of 1911, followed by the city mutiny during the Northern Warlords, the siege of Wuchang during the Northern Expedition by the National Revolutionary Army, the Japanese bombing of Wuhan in the early Anti-Japanese War, and the flooding of Wuhan in 1931, all of which affected the process of Wuhan's urban development. However, Wuhan was on the whole moving towards the integrated modernization of the city.

In order to break the traditional division of the three towns, the Kuomintang government continued Sun Yat-sen's The International Development of China, planned and built the three towns as a whole, systematically introduced western urban planning theories, and proposed the unified functional division of the three towns. And it successively organized and compiled urban plans at the overall planning level, such as Outline of Public Works Programme for Special City in Wuhan, Design Guidelines for Special City in Wuhan, Hankou Urban Plan, etc.

The Anti-Japanese War broke out in 1937 and Wuhan fell in the following year, which led to the stagnation of Wuhan's city modernization for nearly seven years. By the end of the Anti-Japanese War in 1943, although the three towns were not truly integrated, the regional planning entered the stage of Chinese history. At this stage, the "Greater Wuhan" in the planning sense was put forward for the first time. The Draft Plan for the Construction of Greater Wuhan City, Implementation Outline of Wuhan Regional Planning, Preliminary Study Report on Wuhan Regional Planning and Outline for the Land Use and Transportation System of the Three Towns of Wuhan were successively released, gradually changing from "functional zoning + construction plan" to modern urban planning.

During this period, the Kuomintang government also paid special attention to the construction of municipal infrastructure, and organized and compiled a series of municipal engineering plans such as canal projects, waterlogging prevention and drainage, river beach renovation, air defense and evacuation, so as to improve the safety of the city and expand the development space of the city.

一、民国中期的规划
Planning in the Middle Period of the Republic of China

1927 - 1936

为指导三镇合一的武汉市建设，1929年4月，国民政府确定武汉为特别市，先后编制了《武汉特别市工务计划大纲》《武汉特别市之设计方针》，是武汉历史上第一次将三镇合为一体进行规划，从整体角度对三镇进行了功能分区，首次提出了工业区、商业区、行政区、住宅区等分区规划，明确城市发展规模。1936年编制了《汉口市都市计划书》，针对人口过于聚集的问题，提出对都市区用地进行功能分区、道路交通规划、城市绿化率覆盖计划等。

同时期还十分重视道路和堤防建设，先后编制了《武昌市政工程全部具体计划书》《汉口旧市区街道改良计划》《汉口道路建设计划》《湖北省会市政建设计划纲要》等系列规划，以指导市政基础设施建设。

这些规划很大程度上继承了孙中山《建国方略》实业计划的基本思想，结合当时西方国家城市建设经验和当时中国城市规划流行的做法，将城市按照功能分区实行重组计划，指导了武汉三镇合一初期的城市建设，标志着武汉近代城市规划的形成。

In order to guide the construction of Wuhan integrating three towns, in April 1929, the Kuomintang government designated Wuhan as a special city, and successively compiled the *Outline of Public Works Programme for Special City* in Wuhan and Design Guidelines for Special City in Wuhan. This is the first time in Wuhan's history that the three towns were planned as a whole and divided into different functional zones from an overall perspective. Zoning plans for industrial areas, commercial areas, administrative areas and residential areas were first proposed to clarify the scale of urban development. In 1936, the Kuomintang government compiled the Hankou Urban Plan, which proposed the functional zoning of urban land, road traffic planning, urban greening coverage plan, etc., to address the problem of over-concentration of population.

At the same time, great attention was also paid to road and embankment construction, and a series of plans such as *All Concrete Plans of Wuchang Municipal Project, Street Improvement Plan of Old Urban Area in Hankou, Hankou Road Construction Plan, and Outline of Municipal Construction Plan for Hubei Provincial Capital* were successively drawn up to guide municipal infrastructure construction.

These plans largely inherited the basic idea of Sun Yat-sen's *The industrial plan of the International Development of China*, combined with the urban construction experience of western countries at that time and the prevailing urban planning practices in China at that time, reorganized the city according to functional zoning, guided the urban construction at the initial stage of the unification of three towns in Wuhan, marking the formation of modern urban planning in Wuhan.

1 – 基于功能分区理念的都市计划

Urban Planning Based on Functional Zoning Concept

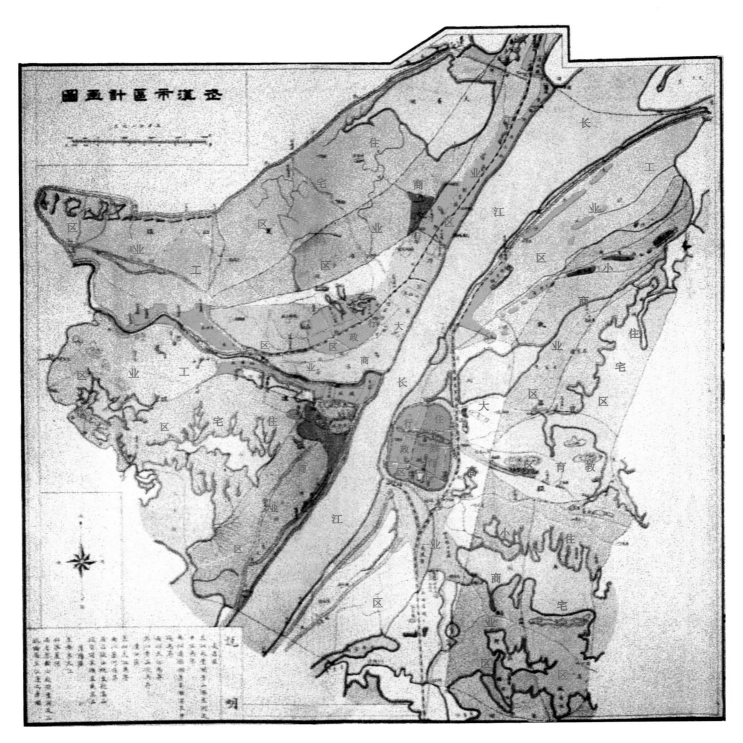

武汉市区计划图（1929 年，资料来源于《武汉市城市规划志》）

Wuhan Urban Planning Map (1929, data source: Wuhan Urban Planning Records)

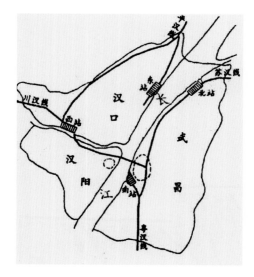

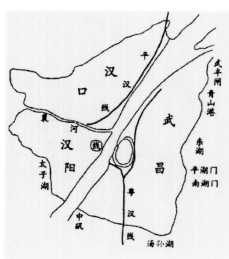
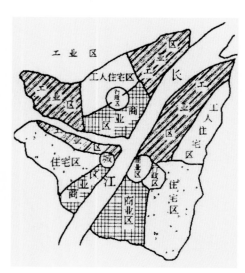

1
武汉特别市铁道车站图（资料来源于《武汉市城市规划志》）
Railway Station Plan of Wuhan Special City (data source: *Wuhan Urban Planning Records*)

2
武汉特别市公园系统图（资料来源于《武汉市城市规划志》）
Park System map of Wuhan Special City (data source: *Wuhan Urban Planning Records*)

3
武汉特别市市区图（资料来源于《武汉城市规划志》）
Area Map of Wuhan Special City (data source: *Wuhan Urban Planning Records*)

4
武汉特别市分区图（资料来源于《武汉城市规划志》）
District Plan of Wuhan Special City (data source: *Wuhan Urban Planning Records*)

5
汉口市分区计划图（1930年，现存于武汉市档案馆）
District Plan of Hankou City (1930, now stored in Wuhan Archives)

一 武汉特别市工务计划大纲（1929年）

1929年，时任市长刘文岛亲任市政管理组组长，由时任工务局长董修甲直接主持，参考欧美现代都市情形，组织制定了庞大的《武汉特别市工务计划大纲》，具体计划从分区、水陆交通、沟渠、公共建筑物、公共娱乐场、公用事业工程六大方面建设，并设定实施该计划时间表，成为其后汉口市各种建设计划的最初蓝本，对武汉工商业的合理分布产生了深远影响。

该《大纲》对武汉三镇按照行政、住宅、商业、工业、教育等功能进行分区；计划新建道路，对旧路进行翻修拓宽，提出展宽襄河出口和修浚扬子江，并利用河泥填高两岸；三镇街道的明沟改为阴沟；公共建筑物计划包括公署、市立图书馆与演讲厅、公共厕所、菜场、公墓和公共浴堂等方面；公共娱乐场所规划则包括公园、通气草地、博物院、游戏场与戏院、运动场与游泳池等；公共事业规划涉及武汉三镇水电、路灯、广告场、电气煤气事业等。

计划分三期完成：一期（1929~1933年）；二期（1934~1938年）；三期（1939年以后）。三期计划实施程序罗列较详细。虽后因水患和战乱，计划未能完全实现，但民国中期武汉市政建设大体遵循此大纲顺序进行。

一 武汉特别市之设计方针（1929年）

1929年4月，由时任武汉特别市政府总工程师张斐然撰写《武汉特别市之设计方针》，拟定全市工程计划的进行步骤及应对方针，"以从事科学之建设，谋产业之发达，并预防无秩序的人口膨胀"。其规划的武汉三镇行政区域范围是依据孙中山《建国方略》论述的精神，结合三镇原来的天然界线，预测其后60年间人口增加所需的城市建设空间及四周与中心之距离要求相等为原则加以划定，面积共401.76平方公里，其范围东至武昌的武丰闸、平湖门，南至汤逊湖、中矶及汉阳的老关，西至汉口的舵落口及汉阳的琴断口、太子湖，北至汉口张公堤。规划以60年为期限，按欧、美、日各国城市人口年递增比例3%计，规划人口从83万增至594万，人均用地面积可达60.3平方米。

基于欧洲的近代功能主义分区原则，规划将土地利用分区划分为工业区、商业区、住宅区（分工人住宅区与商人住宅区）、行政区。工业区位于长江下游的汉口、武昌两地，商业区位于汉口特区、旧市区、汉阳和武昌的长江上游临江地带，工人住宅区在汉口北、武昌下马庙，商人住宅区在武昌城内、汉阳城内、洪山狮子山一带，行政区在汉口循礼门、万松园、武昌博文书院附近。同时，按人均不少于9平方米的最低标准，在全市均衡布置了公园系统。按照总用地的20%～40%面积建设城市街道和交通系统。

规划建议修建跨长江的武汉大铁桥，预留汉口日租界至徐家棚的长江水下隧道，在汉口设东站（平汉）、西站（川汉），武昌设南站（粤汉）、北站（苏汉）四站。

一 汉口都市计划书（1936年）

《汉口都市计划书》系汉口市工务局于1936年5月编制，全书共分为：都市计划之必要、都市计划区域及分区、道路、公园及造林、桥梁及江底隧道、区段及宅地问题、市政府、中央车站及中央广场、下水及地下埋设物、结论等部分。

计划书针对人口过于聚集于城市的缘由、如何解决城市发展问题的方法等方面进行分析，提出对汉口进行了分区，包括工业区、商业区、住宅区，选择水陆交通便利的宗关一带为第一工业区，谌家矶沿江附近为第二工业区；在交通便利的江汉路以西旧市区及中正路（现称解放大道）一带划为商业区；在旧特区、日法租界及模范区一带为第一住宅区，张公堤以南为第二住宅区。

为便于将来发展区与市中心的联系，计划书选择格网式配以放射线的道路系统，以棋盘式街道为主，分设数个交通中心进行分流，以避免交通集中于一点，并与园林大道衔接，建设园林都市。

计划书在汉水的中码头、武圣庙、硚口、宗关、罗家墩等五处各建桥梁一座。接受美国桥梁专家华德尔的建议，认为"水底隧道可不碍船舶的通行，较之水上桥梁自为便利，自汉口日本租界以下地区至武昌徐家棚，建设江底隧道，使粤汉、平汉、川汉三线在汉口接轨"。此外，计划还对工业、商业和住宅等不同用途的城市土地区段的大小、建筑密度做了详细的规定。

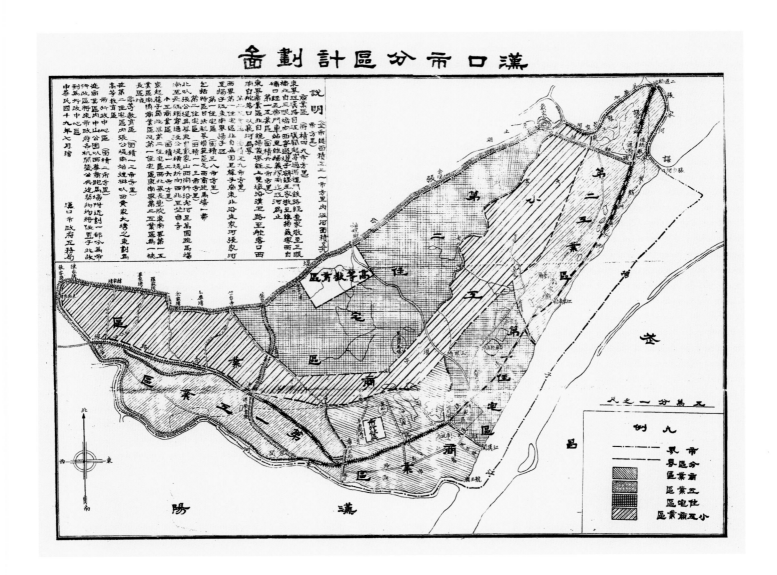

— Outline of Public Works Programme for Special City in Wuhan (1929)

In 1929, Mayor Liu Wendao was appointed head of the municipal management team, directly presided over by Director of Public Works Dong Xiujia. Referring to the situation of modern cities in Europe and America, he organized and formulated a huge *Outline of Public Works Programme for Special City in Wuhan*. Its specific plan included six major aspects of construction: zoning, land and water transportation, ditches, public buildings, public entertainment venues, and public utilities and also set the implementation schedule. This Outline became the original blueprint for various subsequent construction plans in Hankou City and exerted a profound impact on the rational distribution of Wuhan's industry and commerce.

The Outline divided the three towns in Wuhan into different districts according to their administrative, residential, commercial, industrial and educational functions. It was planned to build new roads, renovate and widen the old roads, propose to widen the Xianghe River outlet and repair and dredge the Yangtze River, and fill up the two banks with river mud. The open ditches in the streets of three towns were changed into sewers. The public building plan included government offices, municipal libraries and lecture halls, public toilets, food markets, cemeteries and public bathhouses. The planning of public entertainment places included parks, ventilated lawns, museums, playgrounds and theatres, sports grounds and swimming pools, etc. Public utility planning involved water and electricity, street lamps, advertising sites, electric and gas utilities in the three towns of Wuhan.

This plan was completed in three phases: Phase I (1929—1933), Phase II (1934—1938) and Phase III (after 1939). The implementation procedures of Phase III plan were detailed. Although this plan was not fully realized due to floods and wars, Wuhan municipal construction in the middle of the Republic of China generally followed this Outline sequence.

— Design Guidelines for Special City in Wuhan (1929)

In April 1929, Zhang Feiran, Chief Engineer of Wuhan Special Municipal Government, wrote the *Design Guidelines for Special City in Wuhan* to draw up the steps and guidelines for the city's project plan, "to engage in scientific construction, seek industrial development and prevent disorderly population expansion". Its planned administrative area of Wuhan's three towns was defined according to the spirit of Sun Yat-sen's *The International Development of China* and the principle of equal space for urban construction and distance between the surrounding areas and the center in anticipation of population growth in the next 60 years as well as original natural boundaries of the three towns. Covering an area of 401.76 square kilometers, it extended to Wufeng Sluice Gate and Pinghu Gate in Wuchang in the east, to Tangxun Lake, Zhongji and Laoguan in Hanyang in the south, to Duoluokou in Hankou and Qinduankou and Taizihu Lake in Hanyang in the west, and to Zhanggong Dike in Hankou in the north. It was scheduled to be completed within 60 years. According to the 3% annual increase in urban population in Europe, America and Japan, the planned population increased from 0.83 to 5.94 million, with a per capita land area of 60.3 square meters.

Based on the principle of modern functionalism zoning in Europe, the land use zoning was planned to be divided into: industrial zone, commercial zone, residential zone (divided into workers residential zone and merchants residential zone) and administrative zone. The industrial zone was located in Hankou and Wuchang in the lower reaches of the Yangtze River. The commercial zone was located in Hankou special zone, old urban area, Hanyang and the riverside area in the upper reaches of the Yangtze River in Wuchang. The workers' residential zone was located in North Hankou and in Xiamamiao Wuchang. The merchants' residential zone was located in Wuchang City, Hanyang City and Hongshan Lion Rock. The administrative zone was near Xunlimen, Wansongyuan in Hankou and Bowen Academy in Wuchang. At the same time, according to the minimum standard of not less than 9 square meters per capita, the park system was evenly distributed throughout this city. The city streets and transportation system were constructed within 20%-40% of the total land area.

This plan proposed to build the Wuhan Railway Bridge across the Yangtze River, reserve the tunnel under the Yangtze River from the Japanese concession in Hankou to Xujiapeng, and set up East Station (Lugouqiao-Hankou Railway) and West Station (Chengdu-Hankou Railway) in Hankou, and South Station (Guangzhou-Wuchang Railway) and North Station (Suzhou-Wuchang Railway).

— Hankou Urban Plan (1936)

The *Hankou Urban Plan* was compiled by the Hankou Works Bureau in May 1936. The book was divided into the following parts: the necessity of city plan, the city plan areas and divisions, roads, parks and afforestation, bridges and tunnels under the river, sections and homesteads, municipal government, central station and central square, underground water and buried objects, and conclusions.

This Plan analyzed the reason why the population was too concentrated in cities and how to solve the problems of urban development. It proposed to divide Hankou into industrial zone, commercial zone and residential zone. The Zongguan area with convenient land and water transportation was chosen as the first industrial zone and the vicinity of Shenjiaji along the Yangtze River as the second industrial zone. The old urban area west of Jianghan Road with convenient transportation and the Zhongzheng Road area (now Jiefang Avenue) were designated as a commercial zone. The old special zone, the Japanese-French concession and the model area were the first residential zone and south of Zhanggongdi was the second residential zone.

In order to facilitate the connection between the development area and the city center in the future, this Plan selected a grid-type road system with radiation, with checkerboard streets as its main part, and set up several traffic centers to divert traffic, so as to avoid traffic concentration at one point and connect with garden avenue to build a garden city.

This Plan was to build one bridge each at Hanjiang River's Zhongmatou, Wushengmiao, Qiaokou, Zongguan and Luojiadun. This Plan also took the advice of American bridge expert Wardell, "the underwater tunnel is more convenient than the water bridge since it will not hinder the passage of ships. The tunnel under river, extending from the area below the Japanese concession in Hankou to Xujiapeng in Wuchang, will be built to connect the three lines of Guangzhou-Wuchang Railway, Lugouqiao-Hankou Railway and Chengdu-Hankou Railway." In addition, this Plan stipulated in detail the size and building density of different urban land sections for industrial, commercial and residential purposes.

2 - 以道路、堤防为重点的市政建设计划
Municipal Construction Plans Focusing on Roads and Dikes

一 武昌市政工程全部具体计划书（1929 年）

1929 年 6 月武昌市工程处成立，次年 10 月改组为湖北省会工程处后，由时任工程处主任汤震龙署名编著《武昌市政工程全部具体计划书》。该《计划书》分总论、工程程序、全市区测量费之预算额、全部干线马路建筑费之总预算额、紧要马路建筑费之预算额以及结论等六章，主要包括武昌市区范围、用地分区、道路系统规划、公共建筑分布、市政公共设施计划及工程实施程序等内容。

该《计划书》确定了武昌市区的界限：自青山起，经郭郑湖、东湖达东湖门、南湖门，绕狮子山过汤逊湖、小黄家湖达中矶，再沿江直下经白沙洲、徐家棚抵青山港口。1935 年的《湖北省会市政建设计划纲要》中对省会区域的划定也基本沿袭此界，总面积约为 160 平方公里，约占武汉市区总面积的 40%。计划书还将市区土地划分为行政区、教育区、商业区、工业区、住宅区、军用区域等功能分区。

同时该《计划书》指出"旧有街道已成格子形，新辟区域的街道沿江上下或东西横贯亦可成格子形，再加数条斜街道，街道即可称便利"。在此基础之上，规划了城市道路干线 31 条，基本采用东西南北横直相交的形式，局部考虑天然地势采用斜线相交形式，由此形成了武昌城市方斜混合式交通系统，并区分了道路等级。同时，配合城墙的拆除，计划书提出了环城马路计划，并设想两处跨江铁桥：一由黄皓矶头至汉阳凤凰山；二为徐家棚附近跨江至汉口。两桥均可通车辆、电车及火车。

一 汉口旧市区街道改良计划（1930 年）

《汉口旧市区街道改良计划》刊于 1930 年《新汉口市政公报》第一卷第十二期上，由市工务局提出。

该《计划》针对汉口旧市区街道大都狭窄低湿、异常曲折、水陆运输不能联络、前后市区不相贯通等问题，分期实施街道改良计划。在《计划》中划分各街道等级，将有水陆客货运交通功能的沿江沿河马路以及中山路确定为主要交通干道，宽度为 30 ~ 40 米；将中山路与沿江河路之间的八条垂江道路确定为次要道路，宽度为 24 ~ 30 米；将黄陂街与直通硚口的正街以及其他小街都作为内街，宽度为 10 ~ 20 米不等。

一 汉口道路建设计划（1933 年）

民国初期，市政当局意识到道路发展直接影响城市现代化发展的趋势，故成立汉口市开辟马路委员会和武昌市开辟马路委员会，着手现代城市道路的开辟和改造。1927 年，汉口市政府制定《拓宽街道办法》，规定街道分四级，分别为 9 米、7 米、6 米和 4 米，后被武汉市政府推广。在此基础上，武汉特别市政府于 1929 年完成了城市道路系统特别是汉口道路体系规划，在法规的拓宽道路和建造限制章节中规定了市内道路和里巷的等级宽度。1930 年市政会议公布了汉口旧市区马路干线计划，拟定了旧城两条主干道、九条次干道，对民国中期以来的汉口道路建设具有实际的指导作用，产生了重大影响。此后不断完善，1933 年汉口市政府制定的道路系统规划，汉口新区有名字的道路 86 条，主要走向分为沿张公堤走向、沿汉水与长江流向、垂直于张公堤与江流水道等几类。

一 湖北省会市政建设计划纲要（1935 年）

1935 年 3 月时任湖北省会工程处主任方刚，因省会市政工程机关七八年来不断更迭，在市政建设方面却无完整切实宏远的计划可以遵循，于是起草了《湖北省会市政建设计划纲要》。

该《纲要》除了对省会区域、人口估计和城市职能分区做出了相应描述外，还提出再筑武泰闸以利于在巡司河口避风泊船，在汤逊湖与梁子湖湖汊距离最近处开凿运河，以沟通两湖，便于大量农产、水产直达武汉。此外，《纲要》还列出了若干市街工程和下水道工程等迫切需要的应办项目。

一 三镇堤防、水电设施建设

1931 年、1935 年两次大水给江城武汉造成极为惨重的灾难。1931 年大水，全市受灾人口 63 万余人，死亡 3619 人。此后，堤防建设成为市政建设的重要内容。民国中期，先后扩修武金堤、武昌城区堤、武青堤、武惠堤，汉阳的拦江堤、沿河堤、鹦鹉堤，汉口的沿河堤、沿江堤、张公堤等，基本上构成武汉三镇城区的防洪体系。

这一时期城区的水电设施发展较快。1934 年建成平湖门水厂，1935 年成立武昌水电厂（在此以前于 1931 年，武汉大学在珞珈山北侧创建日供水 800 吨的自备水厂）；1928 ~ 1935 年的 7 年间，扩建汉口宗关水厂，日供水能力由 2.7 万吨增至 9 万吨。民国中期先后扩建、新建的有既济水电公司、武昌电灯公司、汉阳电气公司、汉口电灯公司，至 1937 年三镇各厂电力总容量近 4.2 万千瓦，年发电量 1.47 亿度。

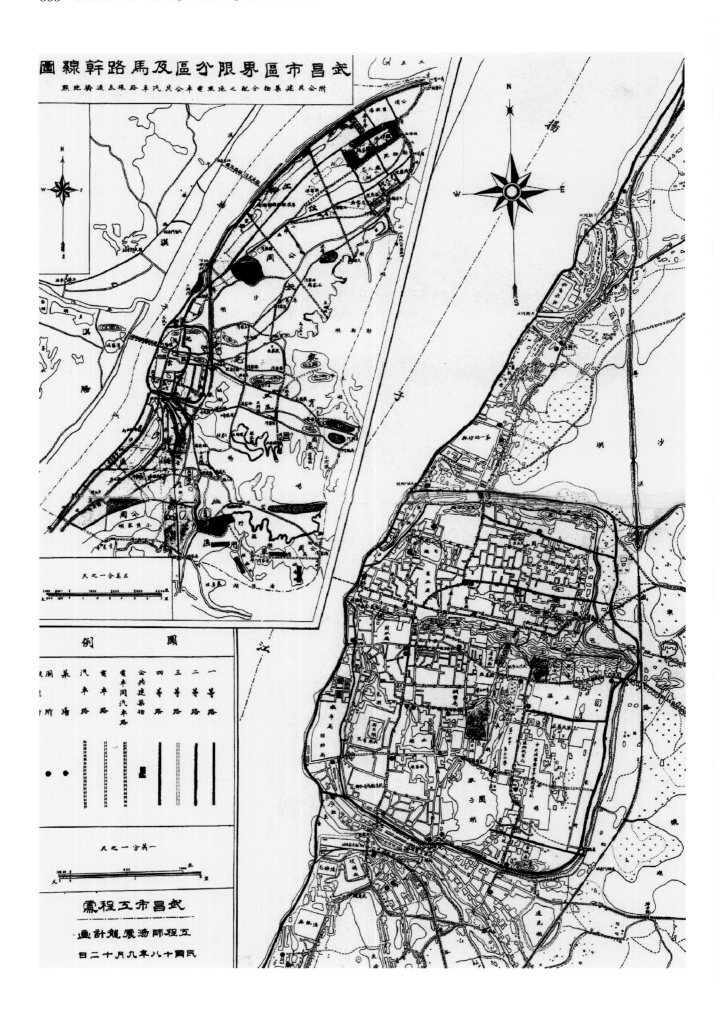

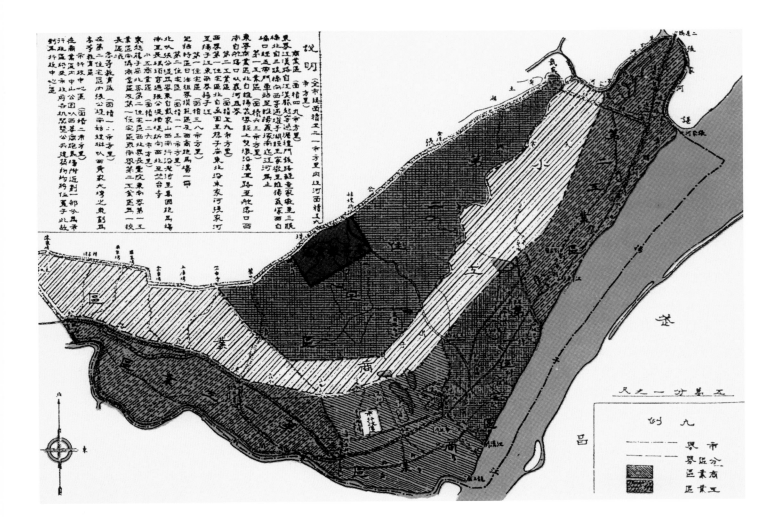

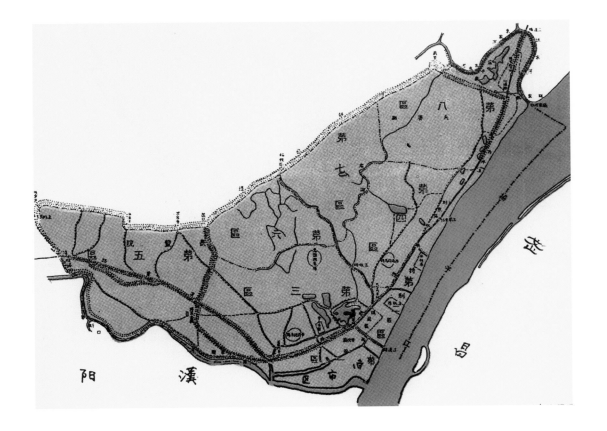

1

武昌城内马路干线及街道图（1929年，现存于湖北省档案馆）

Map of Main Roads and Streets in Wuchang City (1929, now stored in Hubei Provincial Archives)

2

汉口市分区计划图（1933年，现存于武汉市档案馆）

District Plan of Hankou City (1933, now stored in Wuhan Archives)

3

汉口市土地区划图（1933年，现存于武汉市档案馆）

Land Zoning Plan of Hankou City (1933, now stored in Wuhan Archives)

— All Specific Plans for Wuchan Municipal Engineering (1929)

In June 1929, the Wuchang Municipal Engineering Department was established, and in October of the following year it was reorganized into the Hubei Provincial Engineering Department. After that, Tang Zhenlong, Director of the Engineering Department, signed and compiled the *All Specific Plans for Wuchang Municipal Engineering*. This Plan was divided into six chapters: General Introduction, Project Procedures, Budget of the City's Survey Fees, Total Budget for Construction of All Trunk Roads, Budget for Construction of Critical Roads, and the Conclusion, mainly involving Wuchang urban area, land use zoning, road system planning, public building distribution, municipal public facilities planning, and project implementation procedures, etc.

This Plan defined the boundaries of Wuchang urban areas: starting from Qingshan, Guozheng Lake, East Lake to East Lake Gate and South Lake Gate, bypassing Lion Rock, Tangxunhu Lake and Xiaohuangjia Lake to Zhongji, and then reaching Qingshan Port along the river through Baishazhou and Xujiapeng. Division of provincial capital areas in *Outline of Municipal Construction Plan for Hubei Provincial Capital* in 1935 also basically followed this boundary, with a total area of about 160 square kilometers, accounting for 40% of the total area of Wuhan city. This Plan also divided urban land into administrative, educational, commercial, industrial, residential and military zones, etc.

This Plan pointed out that "the old streets have become latticed, and the streets in new area can also be latticed up and down along the river or across the east and west, plus several inclined streets, and the streets will be convenient". On this basis, 31 urban trunk roads were planned, basically in the form of horizontal and vertical intersection of east, west, north and south, and partial form of diagonal intersection in consideration of natural terrain, thus forming Wuchang city square and oblique mixed traffic system and distinguishing road grades. At the same time, with the demolition of the city wall, this Plan proposed a round-the-city road plan and envisaged two bridges across the river: one from Huanghao Jitou to Hanyang Phoenix Mountain, and the other from Xujiapeng to Hankou across the river. Both bridges were accessible to vehicles, trams and trains.

— Street Improvement Plan of Old Urban Area in Hankou (1930)

Street Improvement Plan of Old Urban Area in Hankou was published in the 1930 on *New Hankou Municipal Bulletin*, Volume 1, No.12, and was proposed by the Municipal Works Bureau.

This Plan was to implement the street improvement plan in stages, aiming at solving the problems of narrow streets with low humidity, unusual twists and turns, independent land and water transportation, and separate front and rear urban areas in Hankou's old urban areas. All streets are divided into different

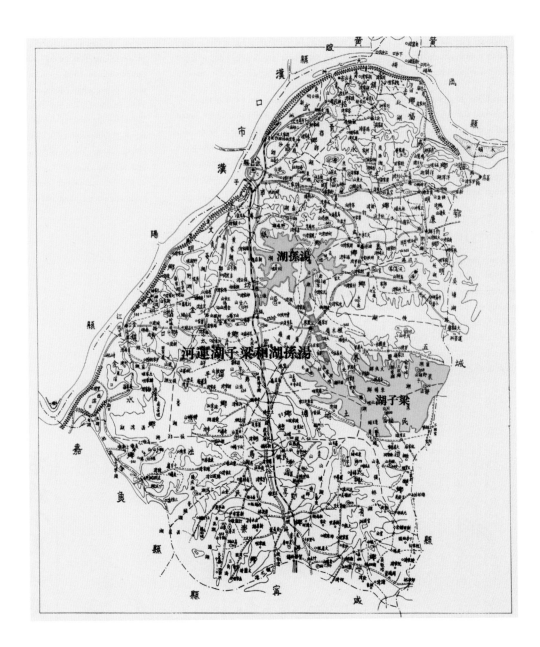

1

最近汉口市全图（1933年，现存于武汉市档案馆）

Latest Hankou City Map (1933, now stored in Wuhan Archives)

2

梁子湖、汤逊湖运河工程示意图（编者自绘，底图为1947年武昌县图）

Schematic Diagram of Canal Project between Liangzi Lake and Tangxun Lake (drawn by the author; the base drawing is Wuchang county map in 1947)

grades. The riverside roads enabling land and water passenger and freight transportation and Zhongshan Road were identified as the main traffic arteries with a width of 30-40 meters; eight roads perpendicular to the river between Zhongshan Road and riverside roads as secondary roads with a width of 24-30 meters; Huangpi Street, the main street leading straight to Qiaokou and other small streets as inner streets with widths ranging from 10 to 20 meters.

— Hankou Road Construction Plan (1933)

In the early days of the Republic of China, the municipal authorities realized that the development of roads directly affected the trend of urban modernization, so they set up the Hankou Road Opening Committee and the Wuchang Road Opening Committee to start the opening and renovation of modern urban roads. In 1927, Hankou municipal government formulated the *Measures for Widening Streets*, stipulating four levels of streets, namely 9 meters, 7 meters, 6 meters and 4 meters respectively, which were later promoted by Wuhan municipal government. On this basis, the Wuhan Special Municipal Government in 1929 completed the planning of urban road system, especially Hankou road system, and stipulated the grade width of urban roads and alleys in the sections of Widening Roads and Construction Restrictions under the laws and regulations. In 1930, the municipal council released the main road plan for Hankou's old urban areas, drawing up two main roads and nine sub-roads in the old city, which had a significant and practical guiding effect on Hankou's road construction since the middle of the Republic of China. Since then, this Plan had been continuously improved. In 1933, Hankou municipal government formulated a road system plan, and there were 86 named roads in Hankou New District, whose main trends are along the Zhanggong Dike, along the Hanjiang River and the Yangtze River, and perpendicular to the Zhanggong Dike and the river channels.

— Outline of Municipal Construction Plan for Hubei Provincial Capital (1935)

In March 1935, Fang Gang, Director of the Hubei Provincial Capital Engineering Department, drafted the *Outline of Municipal Construction Plan for Hubei*

Provincial Capital because there was no complete, practical and far-reaching plan to follow in municipal construction since the municipal engineering department in the provincial capital had been continuously changed for seven or eight years.

In addition to describing the provincial capital areas, population estimation and functional zoning of the city, this Outline also proposed to build the Wutai Sluice Gate again to facilitate the berthing of ships at the mouth of the Xunsi River, and to dig a canal at the nearest place between Tangxun Lake and Liangzi Lake to connect the two lakes, so as to facilitate direct access to Wuhan for a large number of agricultural products and aquatic products. Moreover, it listed several city street projects and sewer projects that were urgently needed.

— Construction of Dikes and Hydropower Facilities in Three Towns

The two floods in 1931 and 1935 caused great disasters to Wuhan along the River. In 1931, the flood affected over 630,000 people and killed 3,619. Since then, dike construction had become an important part of municipal construction. In the middle of the Republic of China, the Wujin Dike, Wuchang City Dike, Wuqing Dike, Wuhui Dike, and Lanjiang Dike, Yanhe Dike, Parrot Dike in Hanyang, and also Yanhe Dike, Yanjiang Dike, Zhanggong Dike in Hankou were successively expanded and repaired, which basically formed the flood control system of the three towns in Wuhan.

Urban hydropower facilities developed rapidly during this period. Pinghumen Water Plant was completed in 1934 and Wuchang Water Power Plant was established in 1935 (before that, Wuhan University established a self-provided water plant with a daily water supply of 800 tons on the north side of Luojia Hills in 1931); During the seven years from 1928 to 1935, Hankou Zongguan Waterworks was expanded with its daily water supply capacity increasing from 27,000 tons to 90,000 tons. In the middle of the Republic of China, Jiji Hydropower Company, Wuchang Electric Company, Hanyang Electric Company and Hankou Electric Company were successively expanded and newly built. By 1937, the total power capacity of each plant in the three towns was nearly 42,000 kilowatts with an annual generating of 147 million kWh.

1

市政公用设施示意图（编者自绘，底图为1931年武汉街市图）

Schematic Diagram of Municipal Public Facilities (drawn by the author; the base drawing is the Wuhan market street map in 1931)

2

民国时期堤防建设示意图（编者自绘，底图为1952年武汉市区全图）

Schematic Diagram of Dike Construction during the Republic of China (drawn by the author; the base drawing is the full map of Wuhan urban areas in 1952)

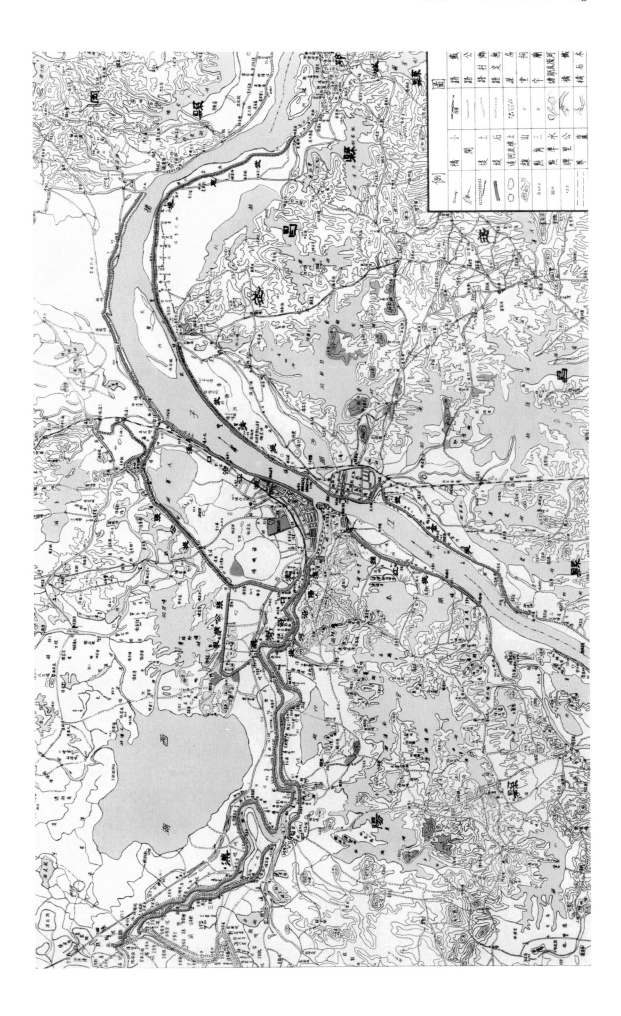

3 – 国立武汉大学早期建筑
Early Buildings of National Wuhan University

— 新校舍总平面设计

武汉大学前身是 1890 年的自强学堂，后相继改为武昌高师、武昌师范大学、武昌大学、武昌中山大学。1928 年 7 月，南京国民政府改组国立武昌中山大学，组建国立武汉大学。随后南京国民政府大学院（后改教育部）指派成立"武汉大学建筑设备委员会"（简称建委会），由李四光担任委员长。建委会最终选定在珞珈山麓、东湖之滨建设新校舍，并专赴上海聘请美国建筑设计师凯尔斯（F.H.Kales）、结构设计师莱文斯比尔（A.Levenspiel）、萨克瑟（R.Sachse）主持设计。1929 年破土动工，1932 年春武汉大学主体工程落成。1932~1987 年的 55 年间，武汉大学不断在原地扩建，成为享有盛誉的全国重点综合性大学之一。

— 武汉大学早期建筑

武汉大学早期建筑一共 15 处 26 栋，占地面积 3000 余亩，建筑面积约 5 万平方米，建委会成立之初便确定了武汉大学校园建筑群的建造原则："以宏伟坚牢适用为原则不求华美"，奠定了武汉大学早期建筑群的空间品质。整个建筑群采用轴线对称、主从有序、中央殿堂、四隅崇楼的形式，开创了我国建筑史上的新风尚。

1
2
3

1
李四光（资料来源于《武汉大学早期建筑》）
Li Siguang (data source: *Early Buildings of Wuhan University*)

2
亚伯拉罕·莱文斯比尔（资料来源于《武汉大学早期建筑》）
Abraham A. Levenspiel (data source: *Early Buildings of Wuhan University*)

3
国立武汉大学新校舍图设计平面总图（1929 年，现存于武汉大学档案馆）
General Layout of the New School Building Plan of the National Wuhan University (1929, now stored in Wuhan University Archives)

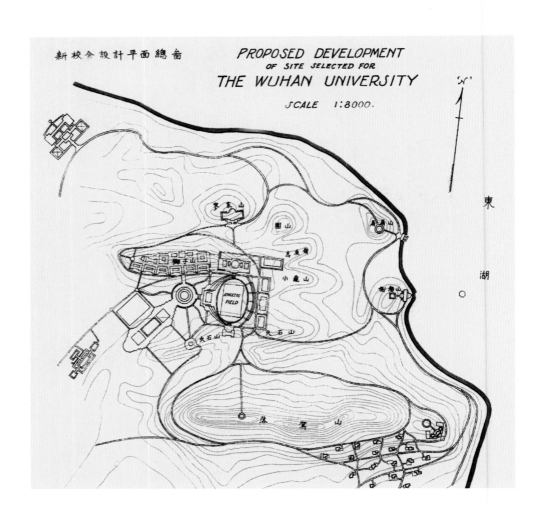

— General Layout of new School Buildings

Wuhan University was formerly known as Ziqiang School in 1890, and later subsequently changed into Wuchang Higher Normal University, Wuchang Normal University, Wuchang University and Wuchang Zhongshan University. In July 1928, the Nanjing Kuomintang government reorganized the National Wuchang Zhongshan University and established the National Wuhan University. Subsequently, the Nanjing National Government College (later renamed the Ministry of Education) appointed the "Wuhan University Building Equipment Committee" (the Building Committee), with Li Siguang as its chairman. At finally, the Building Committee chose to build new school buildings at the foot of Luojia Hills and the bank of East Lake, and went to Shanghai to hire American architects F. H. Kales, structural designers A. Levenspiel and R. Sachse to preside over the design. The construction began in 1929. In the spring of 1932, the main project of Wuhan University was completed. During the 55 years from 1932 to 1987, Wuhan University continued to expand in situ and became one of the key comprehensive universities with high reputation in China.

— Early Buildings of Wuhan University

The early buildings of Wuhan University had a total of 26 buildings in 15 places, covering an area of more than 3,000 *mu* and a construction area of about 50,000 square meters. At the very beginning of its establishment, the Building Committee decided that the construction principle of Wuhan University campus buildings was "to adhere to the principle of grandeur, firmness and application, and not to seek beauty", which laid the foundation for the space quality of early buildings in Wuhan University. The whole building complex adopted the form of symmetrical axis, master-slave structure, central palace and four-corner buildings, which has created a new fashion in China's architectural history.

1		
2 3 4		
5 6		

1　武汉大学装饰墙
The decorative wall of Wuhan University

2　街道口建立的国立武汉大学石牌坊（1933年）
The stone archway of National Wuhan University at the street entrance (1933)

3　武汉大学老图书馆飞檐（1935年）
The eaves of the old library of Wuhan University (1935)

4　街道口建立的国立武汉大学木牌坊（1931年）
The wooden archway of National Wuhan University at the street entrance (1931)

5　武汉大学校园全景（20世纪30年代）
A panoramic view of the campus of Wuhan University (1930s)

6　工学院龙凤柱外廊（1936年）
The veranda of dragon and phoenix columns in the School of Engineering (1936)

（资料来源于《武汉大学早期建筑》）
(Data source: *Early Buildings of Wuhan University*)

二、战争时期的规划
Planning During the War

1937 – 1949

1937~1945年抗日战争期间，武汉的各项城市建设和规划工作基本陷入停滞状态。至抗战胜利后，租界得以全面收回，武汉市具有了正常发展的可能。

1943年，湖北省当局开始筹划武汉战后重建事宜，次年元月《大武汉市建设计划草案》出台，首次提出了规划意义上的"大武汉"，明确大武汉的发展规模为3600平方公里、容纳人口1000万人。

1945年，武汉区域规划委员会成立，大武汉区域规划拉开序幕。分别于1945年12月、1946年4月和1947年7月发布《武汉区域规划实施纲要》《武汉区域规划初步研究报告》和《武汉三镇土地使用与交通系统计划纲要》。

这一时期规划提出延续孙中山《建国方略》设想，促进汉口大港地位的提升，以"武汉区域"为防洪单位，侧重于全域规划建设、前期规划研究、三镇交通、土地集约发展、武汉区域规划等方面，并贯彻"武汉区域规划"提出的将建设港埠作为首要任务的思想，编制了《汉口市运河工程计划书》《改变江滩计划》等。此外，为应对战争灾害，还组织编制了《武汉三镇防护实施纲要》《防空疏散计划》《应变措施》等，提高了城市的安全性。

During the Anti-Japanese War of 1937—1945, Wuhan's urban construction and planning basically stagnated. After China won the war, all the concessions were fully regained, providing Wuhan the possibility of normal development.

In 1943, the authorities of Hubei Province began to plan for the post-war reconstruction of Wuhan. The Draft Construction Plan of Greater Wuhan was issued in January of the next year, which first put forward "Greater Wuhan" in plan. The Draft made it clear that the development scale of "Greater Wuhan" was 360° square kilometers, with a population of 10 million.

In 1945, the Wuhan Regional Planning Commission was established. And the greater Wuhan regional plan began. The *Implementation Outline of Wuhan Regional Planning, the Preliminary Study Report on Wuhan Regional Plan,* and *the Outline for Land Use and Transportation Systems in Three Towns of Wuhan* were issued respectively in December 1945, April 1946 and July 1947.

In this period, the plan put forward the idea of continuing Sun Yat-sen's *The International Development of China* and promoting the status of Hankou Port. It took Wuhan Region as a flood control unit, focusing on the aspects as overall planning and construction, pre-planning and research, transportation of Three Towns, intensive land development and regional planning of Wuhan. It also carried out the idea in "Wuhan Regional Plan" that the construction of ports should be the government's primary task. *Hankou Canal Project Plan and River Beach Change Plan* were issued at that time. In addition, cope with the warfare hazards, Outline of Protection Implementation, Air Defense Evacuation Plan and Contingency Measures of Three Towns in Wuhan were organized and compiled to improve the security of the city.

1 – 基于区域
规划理念的都市计划

Urban Plans Based on the Regional Planning Concept

— 大武汉市建设计划草案（1944年）

1943年鄂西会战之后，抗日战争胜利的希望已经出现，时任湖北省政府主席陈诚约请专家研讨武汉战后重建事宜。几经酝酿，于1944年元月编印了《大武汉市建设计划草案》。该《草案》分为建设大武汉市的政策和规划两部分。在政策中分列大武汉的经济、土地、独占事业、社会、劳动、文化、住宅以及卫生等各个方面的政策；规划部分主要包括市区范围、各市区性质、交通规划、建筑分布、市政设施计划等。

《草案》确定的大武汉市区范围东起白虎山（今白浒山），通过左家岭、神塘湖西岸，沿五里界、纸坊、金口，渡江至大军山，经黄陵矶、大集厂抵蔡甸；渡襄河过西湖，经巨龙岗、白水湖达横店，武湖抵仓子埠，向南过龙口渡江达南岸之白虎山，面积约3600平方公里。假定以1%面积供建筑用，建筑物平均以3层计，则大武汉可容纳1000万人。

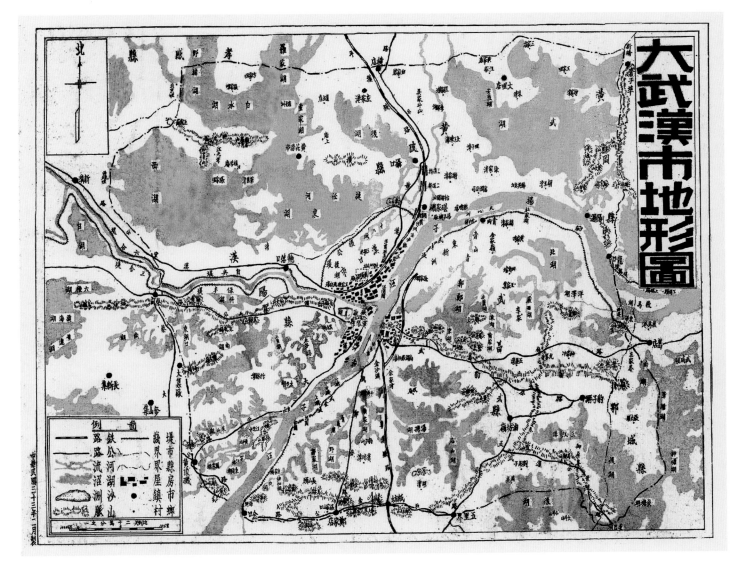

大武汉市地形图（1944年，现存于湖北省档案馆）

Topographic Map of Greater Wuhan City (1944, now stored in Hubei Provincial Archives)

《草案》将汉口市区沿江沿河地带划定为大武汉的主要商业区，将张公堤内旧市区的中间地带划为住宅区，新开运河北岸划为工业区；将汉阳市区沿襄河南岸达蔡甸的区域划为大武汉的主要工业区，墨水湖、太子湖等湖泊沿岸辟为住宅区，沌口西南划为文化区；将武昌市区的油坊岭（今流芳）定为中央行政区或省级行政区，油坊岭东南则划为大武汉的主要住宅区，旧武昌市区划为大武汉的市级行政区，青山南部和金口东北划为大武汉的文化区（含教育），青山至徐家棚的沿江地带划为工业区，卓刀泉至九峰山一带划为住宅区。

《草案》首次提出建设地铁以应对防空疏散和高速交通的需要，确定了"中心＋放射＋环路"的市区交通网。在市区边界各市镇建雄伟的郭门，以示尊严。利用三镇中心区放射道路，通过电车线路强化市中心区与各门之间的交通联系。

在江岸、徐家棚之间设置大型火车轮渡码头，建设汉阳晴川阁至武昌黄鹤楼、汉口民族路至汉阳集家咀以及硚口至汉阳工营场的三座铁桥，挖掘"舵落口—七里河故道—谌家矶"段长约20公里的运河，另辟"新沟—西湖—白水湖—后湖—武湖—阳逻"段长约70公里的河道，以利水运。扩大汉口王家墩及武昌南湖旧有机场，水上机场暂设于武昌东湖和汉口分金炉（今分金街一带）。

《草案》将武汉定位为国际都市，并多次设想将武汉定为国都的情况下，对经济性建筑物、行政性建筑物、儿童教育建筑物、高等教育建筑物、文化事业建筑物、社会性建筑物、卫生性建筑物等的设置和分布进行了详细的说明。

— 武汉区域规划实施纲要（1945年）

抗日战争胜利后，1945年11月，当时湖北省政府公布《武汉区域规划委员会组织章程》，我国第一个区域规划机构——武汉区域规划委员会应运而生，该委员会由朱皆平主持，由此开创了我国区域规划的先河。

《武汉区域规划实施纲要》在武汉区域规划委员会成立之后一个月发布，《纲要》用大量的篇幅详细阐述了"武汉区域规划"的工作范畴、组织形式、运行机制等问题。

该《纲要》将武汉定位为"国际政治经济文化枢纽之一"，明确大武汉市之建设应以防洪为其基础，以港埠为首要。强调"武汉区域"（约一万二千平方公里，包括武昌、汉阳、黄陂、鄂城、黄冈等五县之湖沼区域，以统筹治水防灾为主之建设单位）——"大武汉市区"（纵横六十公里）——"武汉市中心区"的三级划分形式，指出三级层次划分之间的关系：欲现有三镇之繁荣，必须以大武汉市区纵横六十公里为市政建设单位，同时欲大武汉市区之正常发展，必须以"武汉区域"为防洪单位。

在武汉市中心区规划中，武昌、汉口之间的交通联络问题主张采用江底隧道，建议将渡江铁桥改建于三镇上游，远离市区，以保持市容清洁；建议于汉水上多架桥梁，以缩短汉口与汉阳间距离；建议汉口铁路北迁，以增加市区面积，同时可将铁路基作为第一道堤防，提高防洪安全。

此外，《纲要》还提出疏散主义、卫星城等近代城市规划理论，以武汉市中心区为"母体都市"，以武汉区域内的各市镇为其"子体"，并围以绿色地带，即"多点式发展"。

— 武汉区域规划初步研究报告（1946年）

《武汉区域规划初步研究报告》于1946年4月发表，延续了《武汉区域规划实施纲要》中关于"武汉区域规划"工作分期的设想，分别为酝酿时期、推行时期、正常工作时期和工作开展时期，其中"正常工作时期"的主要工作为"综合各组工作结果，与对外技术上之联系，草拟计划报告"，而《武汉区域规划初步研究报告》即是当时设想的"计划报告"。

《报告》分为武汉三镇发展趋势、武汉市中心发展的物质基础、武汉市内外交通、卫生、城市美化、武汉市中心的规划原则等几个部分。

《报告》确定武汉的城市性质为"近代化工商业大城市"、"水陆交通的终点城市"。三镇就其个性发展，武昌为"政治文化城市"，以省政府机关、大专院校为主；汉口为工商业城市；汉阳为"园林住宅城市"。武汉市中心区人口规模限制在120万人，将规划区域内的大小市镇作为卫星城镇发展，城市应有绿色地带围绕，限制城市不致连片盲目扩大。

《报告》开篇即提出"武汉市政建设应列改进港埠为首要""其物质基础系于防洪工程"，为加强防洪工程和港埠建设，成立"港埠建设委员会"，并划定"港埠建设行政区"，上自鲇鱼套，下抵阳逻一段（江面两岸附近的陆地水道在内）。同时，整治汉口、武昌沿江淤积地带，实现孙中山《建国方略》中提出的疏浚长江中下游水道计划，江中泥沙挖出后可用来填高两岸土地。

市内外交通以将铁路、国道与市内交通完全分开为原则。长江铁路大桥拟建在青山，并在其附近建主要货运站和水陆联运码头，使青山形成"重工业卫星城"；道路系统要考虑划分市中心各条道路功能，设专用干道，不准在其沿街建筑房屋；将武汉市中心及周边地带划定为"卫生建设行政区"，统筹该区域的卫生建设，包括上下水道、垃圾处理、公厕、公园、运动场等。保护和恢复名胜古迹，使之成为"历史资本"。将三镇主要湖泊连通、通航，开辟"游道"，发展"旅游事业"。

三镇内部建设，规划采用"完整社区单位"，居住地与工作地点相距不远，并配以公益事业与社会活动场所。加强模范住宅、新村建筑和公园绿化系统建设。为确保规划的"动态性"，需要一面计划，一面执行计划，为此提出成立事业单位——武汉区域规划发展局。

武汉区域图（现存于武汉市档案馆）

Wuhan Area Map (1945, now stored in Wuhan Archives)

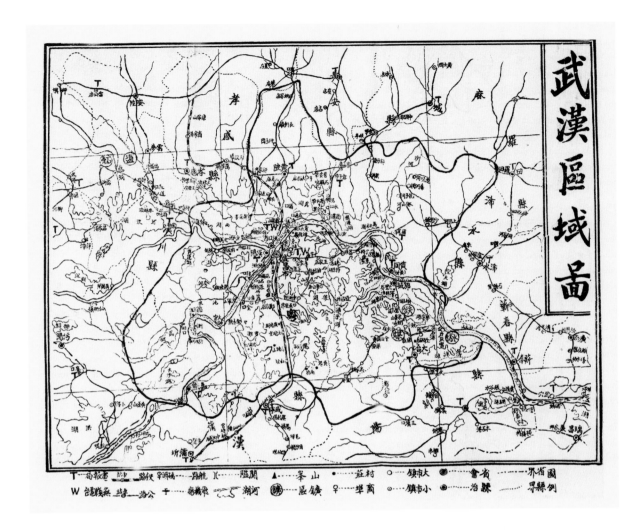

— Draft Construction Plan of Greater Wuhan City (1944)

After the Western Hubei Battle in 1943, the victory in the War of Resistance Against Japan had emerged. Chen Cheng, the chairman of the Hubei Provincial Government at that time, invited experts to discuss the post-war reconstruction of Wuhan. After a long period of deliberation, the Draft Construction Plan of Greater Wuhan was compiled in January 1944. The Draft is divided into two parts: policies and plans for the construction of Greater Wuhan city. The policy part lists the policies on economy, land, monopoly, society, labor, culture, housing, health and other aspects of Greater Wuhan are listed. The plan part mainly includes the urban scope, the nature of each urban area, traffic planning, building distribution, municipal facilities planning, etc.

The Draft defined the urban area of Greater Wuhan from White Tiger Mountain (now White Riverbank Mountain) in the east, to the Baihu Mountain on the South Bank of Longkou, crossing the Yangtze River to Dajunshan along the Wuli boundary, Zhifang and Jinkou along the West Bank of Zuojialing and Shentanghu Lake, then Huanglingji and Dajichang to Caidian, covering the Xianghe River to the West Lake, alone with Julonggang, Baishuihu Lake, Hengdian, Wuhu Lake and Cangzi Port. The Greater Wuhan covered an area of about 3600 square kilometers. Assuming that 1% of the area used for housing construction, three-floor per house, Greater Wuhan was able to hold 10 million people.

The Draft designated Hankou urban area along the river as the main commercial area of Greater Wuhan, the old urban area inside the Zhang gongdi as residential area, the North Bank of the new canal as industrial area. Hanyang City along the Bank of Xianghe River to Caidian was designated as the main industrial zone of Greater Wuhan. The bank of lakes as Moshui Lake and Taizi Lake was to be residential areas, and the southwest of Zhuankou cultural area. Youfangling (now Liufang) was designated as the central or provincial administrative region, southeast of Youfangling as the main residential area of Greater Wuhan, and the old city of Wuchang as the municipal administrative area. The southern part of Qingshan and northeast of Jinkou were to be the cultural area (including education), the area along the river from Qingshan to Xujiapeng the industrial area, and the area from Zhuodaoquan to Jiufeng Mountain the residential area.

The Draft proposed to construct subway to meet the needs of air defense evacuation and high-speed traffic for the first time, and determined the urban traffic network to be "center + radiation + ring road". It planned to build magnificent city gate on the border of cities and towns to show dignity. The Draft proposed to take the advantage of the radial roads in the central districts of the Three Towns of Wuhan to strengthen the traffic links between the central districts and the various departments through the bus lines.

A large railway ferry terminal was set up between the riverside and Xujiapeng. And three railway bridges were constructed from Qingchuan Pavilion in Hanyang to Yellow Crane Tower in Wuchang, from Hankou Minzu Road to Jijiazui in Hanyang, and from Qiaokou to Hanyang Gongyingchang. A 20-kilometer-long canal was excavated from Duoluokou to Shenjiaji, connecting by the old course of the Qili River. And another 70-kilometer-long canal was excavated from Xingou to Yangluo, coursing through West Lake Baishuihu Lake, Houhu Lake and Wuhu Lake. These canals facilitated water transport. The old airport of Wangjiadun in Hankou and Nanhu in Wuchang were expanded. The water airport was temporarily located in Donghu in Wuchang and Fenjinlu in Hankou (now Fenjin Street).

The Draft positioned Wuhan as an international metropolis and envisaged for many times that Wuhan would be the capital of the nation. It explained in detail the set and distribution of economic buildings, administrative buildings, children's education buildings, higher education buildings, cultural buildings, social buildings and sanitary buildings.

— Implementation Outline of Wuhan Regional Planning (1945)

After China won the Anti-Japanese War, the Hubei Province Government promulgated *The Constitution of the Wuhan Regional Planning Commission* in November 1945. Chaired by Zhu Junping at that time, it is the first regional planning organization in China, pioneered the regional planning throughout the country.

The Implementation Outline of Wuhan Regional Planning was issued one month after the establishment of the Wuhan Regional Planning Committee. The Outline elaborated in great length the working, organizational form and operational mechanism of Wuhan Regional Planning.

The Outline positioned Wuhan as "one of the international political, economic and cultural metropolis". It made it clear that the construction of Greater Wuhan should be based on flood control, with port construction as the first priority. The Outline emphasized on the three-level division: "Wuhan Area" (about 12,000 square kilometers, including the lake and marsh areas of Wuchang, Hanyang, Huangpi, Echeng, Huanggang and other five counties, with overall flood control and disaster prevention as the main construction unit), "Greater Wuhan City" (60 kilometers in length and breadth) , "Central District of Wuhan City". The relationship between the three-level division is as follow: For the prosperity of the Three Towns, it is necessary to take sixty kilometers across the city in Greater Wuhan as the main municipal construction unit. And the "Wuhan Area" must be taken as the flood control unit to guarantee its urban area's normal development.

For the planning of the central district of Wuhan and the traffic between Wuhan and Hankou, the Outline advocated to adopt tunnels under the river. The Outline suggested that the railway bridge across the river should be re-designated to the upstream of the Three Towns, away from the urban area, so as to keep the city clean. The Outline proposed to build more bridges on the Hankou River to make it more convenient for people travel between Hankou and Hanyang. It also proposed to move Hankou Railway northward to enlarge the urban area. And in that way, the railway foundation can be used as the first embankment to enhance flood control.

In addition, the Outline puts forward modern urban planning theories such as decentralism and satellite city. It takes the central district of Wuhan as the "mother", with the cities and towns in the Wuhan region as its "offspring", and surrounding the "mother" with plants, namely "multi-points development".

— Preliminary Study Report on Wuhan Regional Planning (1946)

The Preliminary Study Report of Wuhan Regional Planning was published in April 1946. It is a continuation of the idea of working carried out in stages of "Wuhan Regional Planning" in the *Implementation Outline of Wuhan Regional Planning*. The stages of work are divided into the preparation period, the promotion period, the normal working period and the implementation period. The main work of "normal working period" is as follows: "Drafting the plan report by synthesizing the results of each group's work and adopting foreign technologies." The Preliminary Study Report of Wuhan Regional Planning is the plan report conceived at that time.

The Report is divided into the following parts: the development trend of the Three Towns in Wuhan, the material basis for the development of Wuhan city center, the transportation inside and outside Wuhan, hygiene, urban beautification, and the planning principles of Wuhan city center, etc.

The Report identifies Wuhan as a "modern chemical commercial metropolis" and a "terminus city of land and water transport". In terms of the Three Towns' individual development, Wuchang shall be a "political and cultural city", with provincial government organs and colleges as the main body. Hankou shall be a commercial and industrial city. And Hanyang shall be a "city of garden and residence". The population of Wuhan downtown shall be limited to 1.2 million people. The cities and towns in the planned area shall be satellite towns. Cities shall be surrounded by plants to prevent the blindly expansion.

To begin with, the Report points out that "the improvement of harbors and ports shall be the priority in Wuhan municipal construction" and "the material basis lies in flood control works". To strengthen flood control and to accelerate port construction, the "Harbor and Port Construction Committee" was established. And the "Harbor and Port Construction Administrative Region" was delimited from Nianyutao on the upper side to Yangluo on the lower side (including land waterways near both sides of the river). At the same time, the siltation along the Hankou and Wuchang rivers shall be handled to realize the dredging plan for the middle and lower part of the Yangtze River proposed in The International Development of China by Sun Yat-sen. The silt can be used to fill up the land on both sides of the Yangtze River.

The principle of traffic inside and outside the city is to completely separate railway, national highway with urban traffic. The Yangtze River Railway Bridge is planned to be built in Qingshan, with major freight stations and land-water intermodal terminals nearby. It enables Qingshan to form a "satellite city of heavy industry". The road system shall clarify the functions of each road in the city center, deciding special main lines. No houses shall be allowed to be built along the street. The downtown and nearby areas of Wuhan shall be designated as a "hygiene construction administrative district" to coordinate the work of hygiene, including sewers, garbage disposal, public toilets, parks, stadiums and so on. It shall restore scenic spots and historic sites, making them "capital of history". It shall connect the main lakes of the Three Towns together through roads, rivers and canals, create up "tourist routes" and develop "tourism".

The inner construction of the Three Towns is planned to adopt a "complete community unit" scheme, which working places are not far from residences, and surrounded residences with places for public welfare and social activities. It lays emphasis on model residential buildings, new village construction and park greening system. To ensure the "dynamic" of the scheme, it is necessary to keep planning as well as carry these plans out. Therefore, the Wuhan Regional Planning and Development Bureau is established.

一 武汉三镇土地使用交通系统计划纲要（1947年）

1946年8月，武汉区域规划委员会重新改组后，由时任湖北省主席万耀煌作序，时任武汉区域规划委员会设计处长兼委员会秘书长鲍鼎主持编制了《武汉三镇交通系统土地使用计划纲要》。该《纲要》延续了此前的《区域规划实施纲要和初步研究报告》的思路、观点，不同之处在于此《纲要》的研究对象仅是"武汉市中心"，而不是"武汉区域"的全部范围，是由于当时的经费紧张，故缩小了研究对象范围，以将工作做得更为深入。文中对"武汉市中心"的人口、用地、交通等方面均做了较为细致的布置与设定，标志着"武汉区域规划"已经进入到较为具体的阶段。

《纲要》认为，从地理地位分析，武汉较芝加哥更为优越。武汉位于长江、汉水两大河流交汇处，可直航入海，流域腹地有1亿以上人口，资源丰富，工业化以后可形成内陆最大的都市。在分析现状人口和工业化后区域内的城镇发展趋势后，预测武汉未来区域总人口规模达400~500万人，并强调三镇人口必须平衡发展，将汉口密集人口向武昌、汉阳扩散。三镇的人口密度分为三级：市中心区居住人口不超过50000人/平方公里；城市边缘不超过25000人/平方公里；郊区不超过10000人/平方公里。

区域土地使用分为住宅区、商业区、仓库区、工业区、公共建筑地带以及永久农田与绿色地带等。近郊一带规划为第一住宅区，原市区的居住地带为第二住宅区，要形成不同规模的社区单位。工商业用地要考虑动力供应、雇工便利、交通方便等因素，并提出在建工厂的同时需建工人住宅。关于园林绿化，《纲要》提出扩大和改善中山公园，龟山、蛇山、洪山皆辟为公园。市区外围用绿化带围绕，以控制城市无限扩大，并与市内林荫路衔接。

《纲要》提出要疏浚长江、汉水河流，汉水船运要常年可达襄樊（今襄阳）。张公堤外开辟运河入江，减轻汉水船运拥挤状况。港埠建设方面，计划在刘家庙、徐家棚、龟山下游一带建货运码头，在王家巷下游一带建客运码头，并利用江汉关以下淤积滩地建码头和进行开发。计划将汉口市区原有铁路北移2公里，以免妨碍市区发展，在其中部设客运总站，刘家庙、徐家棚设铁路货站。过境货运公路计划绕市区外围经过，并与市内干道联系起来。汉口军用机场改为民用机场。

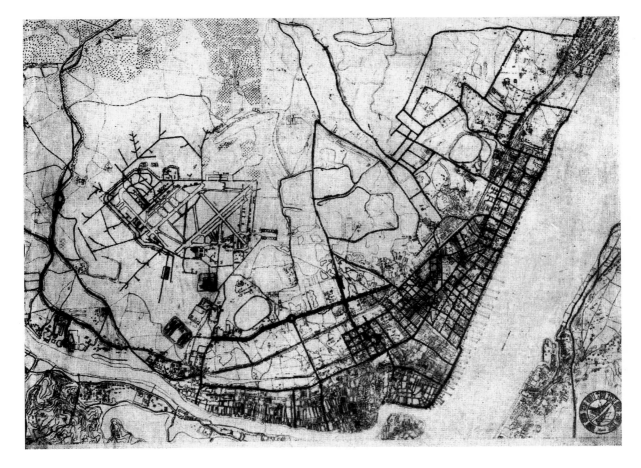

汉口现状图

Current State Map of Hankou

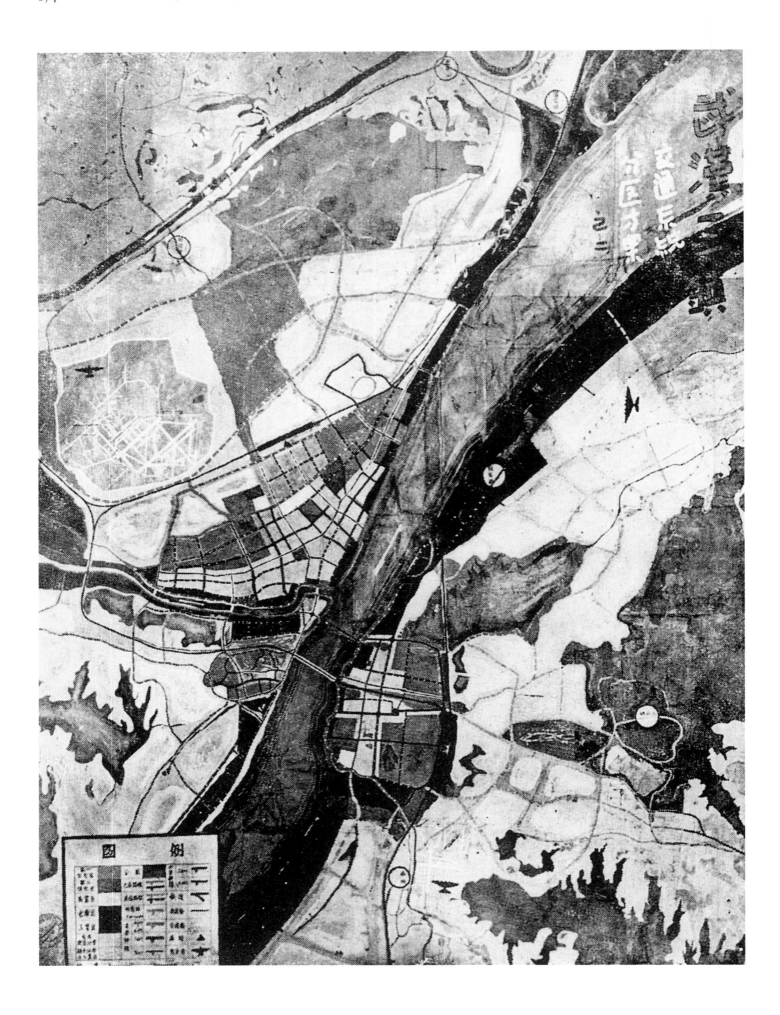

— Outline for Land Use and Transportation Systems in Three Towns of Wuhan (1947)

In August 1946, after the reorganization of the Wuhan Regional Planning Commission, Bao Ding, then Director and Secretary-General of the Wuhan Regional Planning Commission, compiled *Outline for Land Use and Transportation Systems in Three Towns of Wuhan*, and Wan Yaohuang, then Chairman of Hubei Province, wrote a preface for it. The Outline continues to hold the most of thoughts in the previous *Preliminary Study Report on Wuhan Regional Planning*. The difference lies in that the research object of the Outline is only "the downtown of Wuhan", not the whole "Wuhan Region". Because of the shortage of funds at that time, the scope of the research object had to be narrowed in order to deepen the work. In the Outline, the population, land use, transportation and other aspects of the downtown of Wuhan are laid out and set up in details, indicating that the "Wuhan Regional Planning" has entered a more specific stage.

According to the Outline, Wuhan is superior to Chicago in terms of its geographical position. Wuhan is located at the confluence of the Yangtze River and Hanjiang River, which allows ships to sail straight into the sea. With more than 100 million people and rich resources in the hinterland of the basin, it can become into the largest inland city after industrialization. After analyzing the current population and the development trend of cities and towns in the industrialized region, it is predicted that the total population of Wuhan will reach 4-5 million in the future. It emphasizes a balanced development of population of the Three Towns. The dense population in Hankou must be distributed to Wuchang and Hanyang. The population density of the Three Towns is divided into three levels: no more than 50,000 inhabitants per square kilometer in the downtown area; no more than 25,000 inhabitants per square kilometer in the edge of the city; and no more than 10,000 inhabitants per square kilometer in the suburbs.

Regional land use is divided into residential area, commercial area, warehouse area, industrial area, public building area, permanent farmland and green area, etc. The suburb is designated as the major residential area, while the original urban area as the minor residential area to form communities in different size. Industrial and commercial land shall pay attention to factors as power supply, convenience in employment and transportation. It proposes to build workers' residences while building factories. For landscaping, the Outline puts forward the idea of expanding and improving Zhongshan Park, making Turtle Mountain, Snake Mountain and Hongshan Mountain parks. Greenbelts shall be around the periphery of the city to control the unlimited expansion of the city and to connect with the city's shady avenue.

The Outline proposes to dredge the Yangtze River and Hanjiang River, allowing ships from Hanjiang River to reach Xiangfan (now Xiangyang) at all time. A canal to the river shall be excavated outside the Zhang gongdi, reducing the congestion of Hanjiang River shipping. For port construction, the Outline plans to build freight terminals in the downstream areas of Liujiamiao, Xujiapeng and Turtle Mountain and passenger terminals in the downstream areas of Wangjiaxiang, using silted beaches. It plans to move the original railway in Hankou 2 kilometers up to north, preventing it from hindering the development of the city. A passenger terminal will be set up in the middle of the railway, and a railway freight station in Liu jiamiao and Xujiapeng. Transit freight roads are planned to pass around the boarder of the city and be linked to the main lines inside. Hankou Military Airport was converted to civilian use.

1 2

1
武汉三镇交通系统土地使用
计划纲要总图（1947年）
General Land Use Plan for Transportation System of Three Towns in Wuhan (1947)

2
武汉三镇近郊形势图
Map of the Suburbs of Three Towns in Wuhan

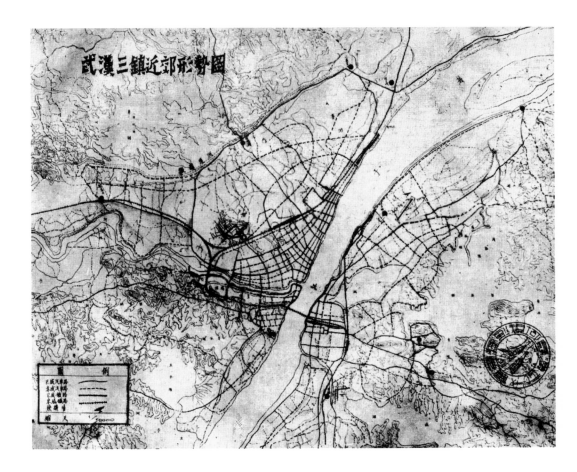

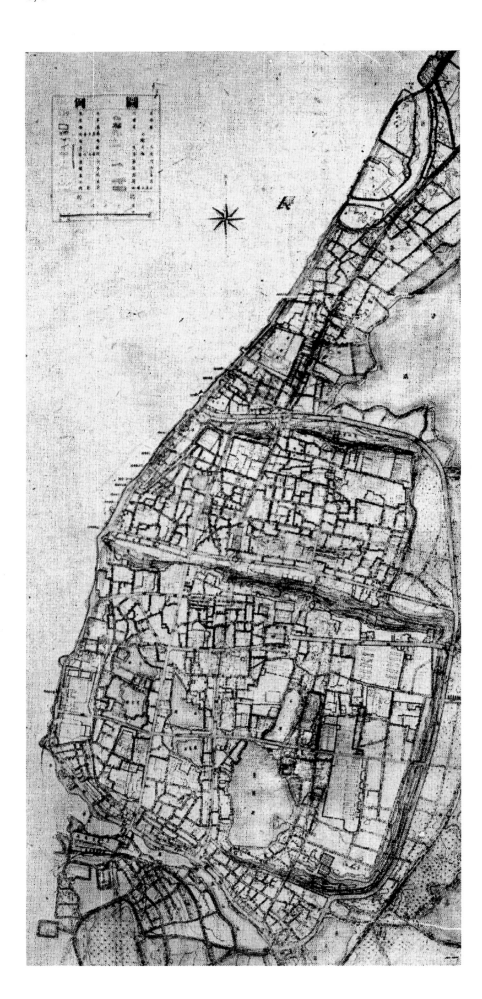

1
武昌道路系统图
Wuchang Road System Map

2
新汉口市计划图（1947年，现存于武汉市档案馆）
New Hankou City Plan (1947, now stored in Wuhan Archives)

2 - 以防洪和通航为重点的工程建设计划

Construction Plans Focusing on Flood Control and Navigation

一 新汉口市建设计划（1947年）

新汉口市建设计划由汉口市政府下属的工务科依据当时的"武汉区域规划"中关于汉口部分的内容而编制，是由于国民政府内政部根据联合国秘书长需征集各国乡村及都市房屋与城市建设计划资料的要求而做出的，主要包括市区范围及人口、用地分区、铁路改线、修建新港埠码头、开辟运河、桥梁建设、整治下水道系统、机场等内容。

该《计划》依据孙中山《实业计划》的论述，提出汉口市的城市性质为中部、南部之贸易中心，国内重要工商业、交通城市。市行政区范围延续1930年划定的界限，即东、南侧分别以长江及汉水的中心线为界，西从舵落口起，沿张公堤至岱家山，并以二道桥为起点，向南垂直以达江岸，总面积为115.5平方公里，共计人口约98万人，其中80%集中在市中心区。

该《计划》的城市用地功能分区，除基本沿用武汉区域规划提法外，只对市政工程建设方面提出了一些具体计划设想：填高江汉关至芦沟桥（亦称卢沟桥）路江滩，并疏通岸边水道，修建新式码头和仓库。利用张公堤外府河故道，修辟上通汉水下出长江的运河，如此既可减轻汛期洪水对城市的威胁，又可引航汉水船舶入运河，沿河建工厂码头，并利用挖运河的土方，填高张公堤内的低洼土地。长江大桥桥址仍选在龟山、蛇山之间，并规划汉阳过汤湖至汉口阮家台的铁路桥，同时将老兴巷对南岸嘴、硚口至汉阳、武胜路至汉阳作为公路桥址。计划在中山大道建总下水道至堤角谌家矶，并建污水处理厂。城市内部排水分江汉路上下两区，江汉路以上汇集单洞门穿中正大道（今解放大道）向东北流至三眼桥；江汉路以下则集中大智门，再向北排至三眼桥，与江汉路以上污水汇合，由岱家山闸出长江。计划将汉口军用机场改为民用，军用机场迁至青山与徐家棚之间。

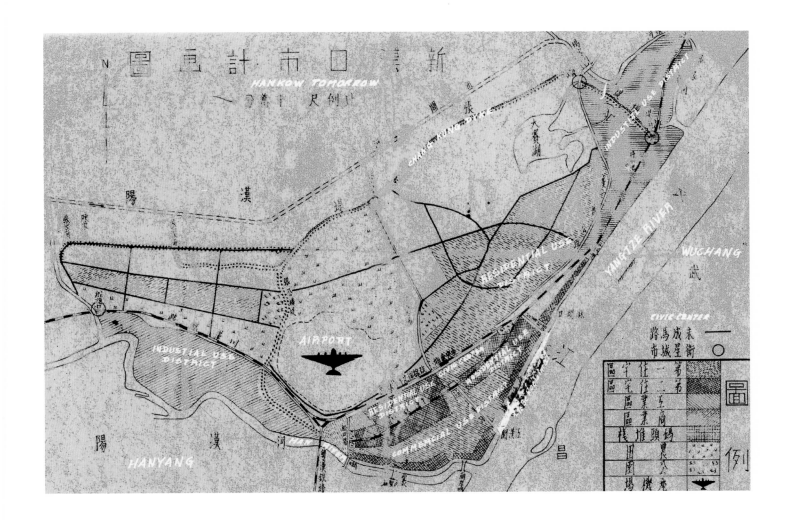

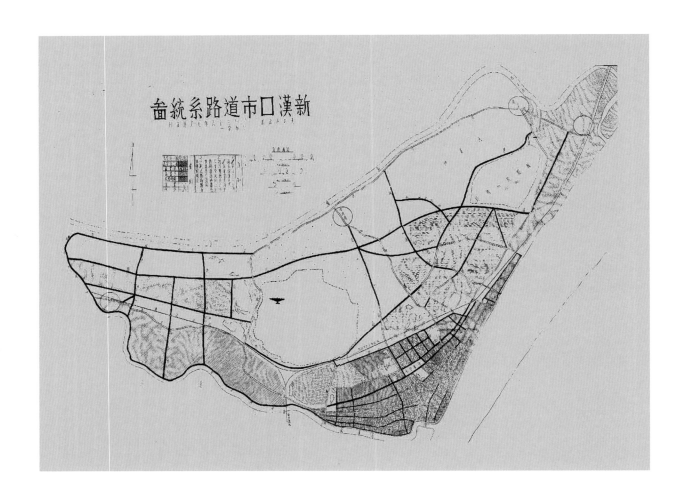

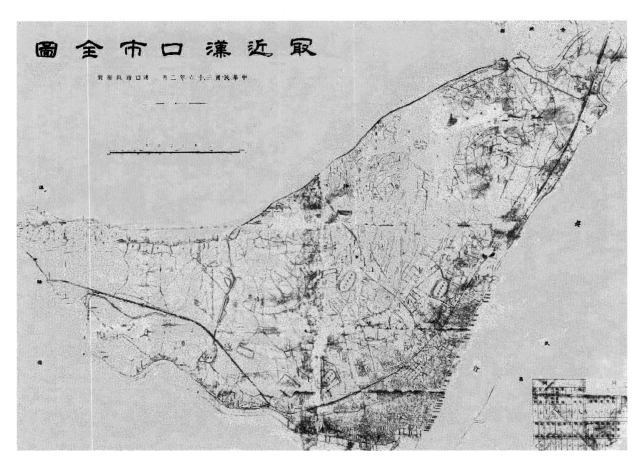

1

新汉口市道路系统图（1947年，现存于武汉市档案馆）

Road System Map of New Hankou City (1947, now stored in Wuhan Archives)

2

最近汉口市全图（1947年，现存于武汉市档案馆）

Latest Hankou City Map (1947, now stored in Wuhan Archives)

3

汉口市下水道系统图（1947年，现存于武汉市档案馆）

Hankou Sewerage System (1947, now stored in Wuhan Archives)

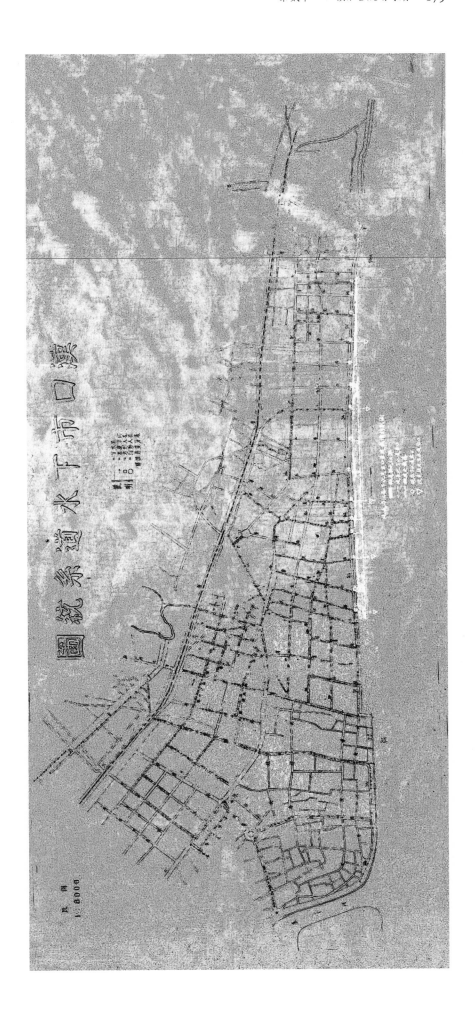

一　改良沿江下水道出口计划（1947 年）

该《计划》于《武汉区域规划初步研究报告》发布后一年所作，是为配合该《报告》中提出的港埠建设而制定的计划。由于自江汉关至芦沟桥路一带，前租界区的污水和雨水均直接排入长江，出水口多至 23 处，难于防水，因此确有改良之必要。

《改良沿江下水道出口计划》从市政原理、工程时间和费用等多方面，比较分析了改造下水道系统、延长沿江各下水道出口和延长沿江下水道出口并加建截沟等三种改良方式，认为第一种方式合乎市政原理，但工期长，费用较高；第二种方式较为简单，但出口太多，难于防水，且后期排水费用较高；第三种方式构造较为简单，在防水和排水方面可获得便利，建议先期按第三种方式操作。

一　汉口市运河工程计划书（1948 年）

《汉口市运河工程计划书》为进一步促进港埠建设发展所作，源起孙中山在《建国方略》中的设想"汉江可支配甚大之分水区域，汉口为航船之终点"。汉口位于江汉汇流之区，北抵张公堤，地形略似三角形，且市区湖沼错杂，故考虑连贯水利，在 1930 年、1945 年分别提出了沿张公堤外另开运河、开凿新沟至阳逻运河的设想，既可沟通航运，也可减轻汉口洪水压力，鼓励市区向北发展。后由于经费关系，以上设想均未付诸实施。

1947 年，由江汉工程局派人员作江汉运河工程测量，从杨家河起沿张公堤外 100 米直至长丰北垸朝家上湾，测量长度为 17100 米，作为开辟运河之路线。同年 11 月成立汉口市运河工程筹备处，由时任徐会之市长兼任处长。计划将运河线路定为由汉水左岸舵落口起，沿张公堤外经姑嫂树绕金银潭达谌家矶入江，全长 31 公里，除原有府河 11 公里可以利用、仅加疏浚外，新挖部分共 20 公里，并与旧河道衔接，以利货物转运。

一　《改良汉口江滩初步计划》（1948 年）

《改良汉口江滩初步计划》属于改良港埠计划之一。因自抗战以来，汉口江边码头废置，垃圾随意倾倒，导致长江部分区域淤积甚厚，尤其在冬季涸水时期，轮船须停泊在一百余米以外，上下轮船依赖长距离的栈桥，客货运十分不便。1948 年，由汉口市政府工务科拟定《改良汉口江滩初步计划》，在不妨碍泄洪的原则下，采取将江滩填高的方式，从事港埠改良。

新驳岸从江汉关起至山海关路，建议平均宽度控制在 100 米左右，维持 27 米的高度，以节省土方量，防水问题通过建筑防水墙或将江滩人行道抬高的方式解决。江滩改良后，不仅有利于轮船停泊，还可产生 348 亩土地，其中 35.99% 的土地用作仓库，10.81% 的土地用作公园，53.2% 的土地用作道路。

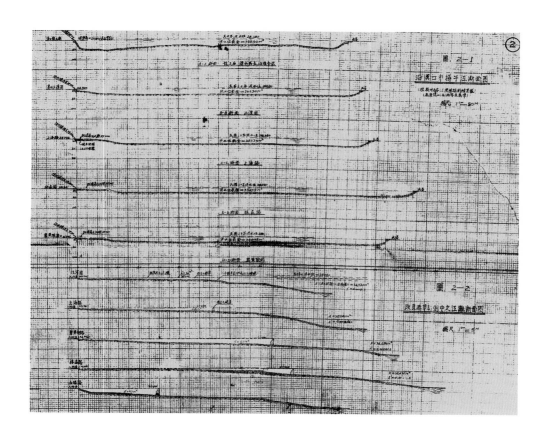

1

改良港埠计划之江滩断面图（1948 年，现存于武汉市档案馆）

Beach Section Map of the Port Improvement Project (1948, now stored in Wuhan Archives)

2

汉口港埠形势图（1948 年，现存于武汉市档案馆）

Hankou Port Situation Map (1948, now stored in Wuhan Archives)

— New Hankou City Construction Plan (1947)

The New Hankou City Construction Plan is compiled by the Engineering Department under the Hankou Municipal Government on the basis of the content of the Hankou section in the "Wuhan Regional Planning" at that time. It is made by the Ministry of Internal Affairs of the National Government in accordance with the request of the Secretary-General of the United Nations to collect information on rural and urban housing and urban construction plans in various countries. It mainly includes information like the urban area and population, land use, railway rerouting, construction of new ports and wharfs, canals, bridges, sewerage systems, and airports, etc.

According to The International Development of China of Sun Yat-sen, the Plan takes Hankou as a trade center in the central and southern parts of China and an important industrial, commercial and transportation city. The administrative area of the city remains mostly the same with the boundary demarcated in 1930, that is, the east and south sides separately bounded by the central line of the Yangtze River and Hanjiang River, the west starting from Duoluokou along Zhanggongdi to Daijiashan, and the south part starting from Erdaoqiao, reaching the river bank vertically southward. The area covers a total of 115.5 square kilometers. The whole population is 980,000, 80% of which are in the central area of the city.

The functional zoning of urban land use in The Plan has stuck to stick to the most part of Wuhan regional planning. The Plan only brings up some concrete plans and assumptions for municipal construction. The government shall fill up the riverbank between Jianghanguan and Lugou Bridge Road (or Lugouqiao Road) and dredge the waterway along the bank. The government shall build new wharfs and warehouses. The government shall excavate a canal to connect Hanjiang River and the Yangtze River using the old waterway of Zhanggongdi Waifu River. The new canal not only can reduce the threat of flood to the city in flood season, but also can allow ships from Hanjiang River into the Yangtze River. The government shall set up factories and wharfs along the river and fill low-lying land in the Zhanggongdi with the slit excavated from the canal. The site of the Yangtze River Bridge was selected between Turtle Mountain and Snake Mountain, and the railway bridge from Hanyang Tanghu to Hankou Ruanjiatai was planned. At the same time, road bridges were planned from Laoxing Lane to Nananzui, Qiaokou to Hanyang and Wusheng Road to Hanyang. It was planned to build a total sewer to Shenjiaji in Dijiao and a sewage treatment plant on Zhongshan Avenue. The internal drainage of the city was divided into upper and lower areas by Jianghan Road. For the area above Jianghan Road, sewage converged at Dandong Gate and flew northeast to Sanyan Bridge through Zhongzheng Avenue (now Jiefang Avenue). For the area below Jianghan Road, sewage gathered at Dazhimen, which was then drained northward to Sanyan Bridge, where it joined the sewage above Jianghan Road and flew to the Yangtze River from Daijiashan Sluice Gate. It was planned to the convert Hankou military airport to a civilian one and move the military airport to a place between Qingshan and Xujiapeng.

— Improvement Plan for Sewerage Outlets along the Yangtze River (1947)

The Plan, made one year after the publication of the Preliminary Study Report on Wuhan Regional Planning, is to coordinate with the port construction proposed in the Report. Since the sewage and rainwater in the former concession area from Jianghanguan to Lugouqiao Road are directly discharged into the Yangtze River, and the number of outlets is up to 23, difficult for flood control, it is necessary to improve the sewer outlets.

The Plan compares and analyses three solutions of improving sewer system including improve the sewer system, extending the outlets of sewers along the Yangtze River, and extending the outlets of sewers along the Yangtze River and adding ditches, from the aspects of municipal principle, construction time and cost. It finds that the first solution is in accordance with the municipal principle but takes longer construction time and higher cost. The second is relatively simple but left too many outlets to control and drainage cost in later period is very high. The third solution is simple in construction, convenient in flood control and drainage. So the Plan suggests to adopt the third solution in earlier stage.

— Hankou Canal Project Plan (1948)

The Plan is made to further promote port construction and development. It is originated from the idea that "Hankou is the terminal point for ships in a region where the Hanjiang River has a large waters at its disposal" in The International Development of China by Sun Yat-sen. Hankou is located in the confluence area of Hanjiang River and the Yangtze River, reaching Zhanggongdi in the north. Its terrain is slightly triangular and the location of its urban lakes and marshes are rather complicating. To connect rivers and lakes, the idea of excavating another canal along Zhanggongdi and digging a new ditch to Yangluo canal was put forward in 1930 and 1945. These canals can make it more convenient for shipping and reducing flood threat in Hankou, encouraging the urban development northward. Due to the financial shortage, the above ideas have been failed to put into practice.

In 1947, the Jianghan Canal Project was surveyed by the Jianghan Engineering Bureau. From Yangjia River, along 100 meters away from Zhanggongdi, to Chaojia Shangwan on Changfeng North Dike, the measured length is 17,100 meters, which is the waterway of the canal to be excavate. In November of the same year, the Canal Project Preparatory Office of Hankou was established. The mayor, Xu Huizhi was appointed Director. The Plan sets the canal to start from Duoluokou on the left side of Hanjiang River, coursing alone Zhanggong, through Gusaoshu and Jinyin tan to Shenjiaji and finally to the Yangtze River.

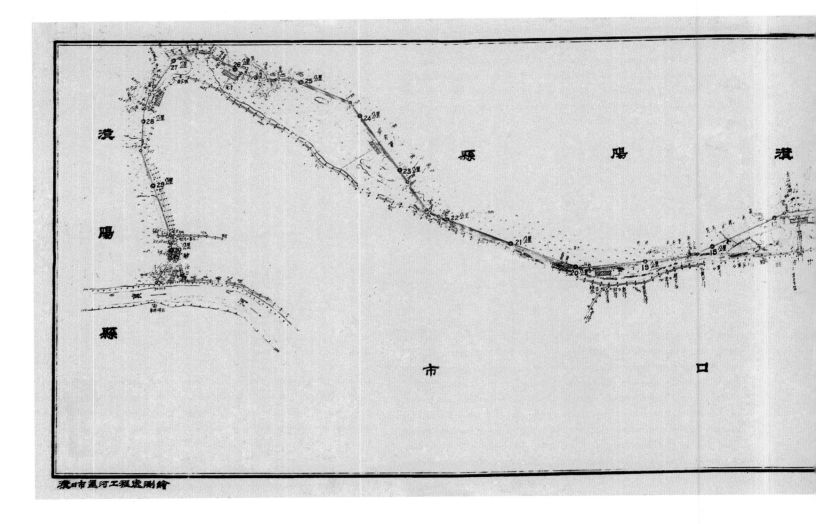

汉口市运河工程地形图（1948年，现存于武汉市档案馆）

Topographic Map of Hankou Canal Project (1948, now stored in Wuhan Archives)

The total length of the canal will be 31 kilometers. In addition to the original 11-kilometer Fuhe River that only needs to be dredged, the newly excavated part is in total 20 kilometer. The rest part is only dredged and connected with the old river waterways to facilitate the transportation of goods.

— Preliminary Plan for River Beach Improvement in Hankou (1948)

The Preliminary Plan for River Beach Improvement in Hankou is one of the plans for the improvement of ports. Since the War of Resistance Against Japan, the abandoned wharfs along the Hankou River and the dumping of garbage at will have resulted in thick siltation in some areas of the Yangtze River. Especially in the period of low water in winter, ships had to berth more than 100 meters away. Upper and lower ships rely on long-distance trestles, very inconvenient to passengers and freight transport. In 1948, the *Preliminary Plan for River Beach Improvement in Hankou* was drawn up by the Engineering Section of the Hankou Municipal Government. Without violation to the principle of flood discharge, the actions will be taken to improve the port by making the beach higher.

The new revetment shall be built from Jianghanguan to Shanhaiguan Road. The suggested average width is about 100 meters and the height is 27 meters to reduce earthwork. The flood control is done by building waterproof walls or lifting up the height of the sidewalks on the river beach. After the improvement, it is not only beneficial for ships to berth, but also increases 348 *mu* of land. The 35.99% is used for warehouse, 10.81% for park and 53.2% for road.

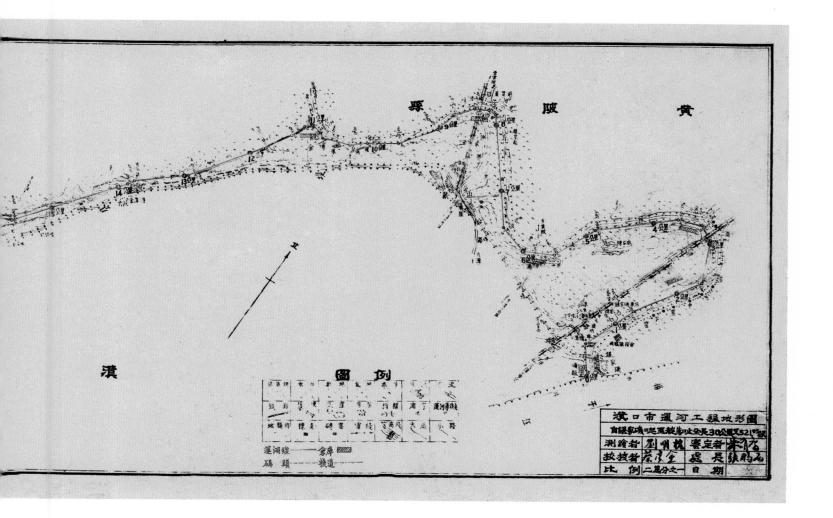

CHAPTER
第叁章
THREE
1949-1999

主城内聚发展时期
Cohesion of the Main City

武汉因水而生，因水而兴，自古以来一直是遵循沿江发展的规律。"一五""二五"时期，武汉被列为国家重点建设城市，受苏联规划思想和理论的影响，分别于1954年、1959年编制了城市总体规划和城市建设规划，远离市区布置了一批国家、地方工业项目，逐步拉开了城市骨架，引导了城市空间的跳跃式拓展。

"文革"十年，城市规划和建设工作基本停滞。改革之初，武汉市提出了"两通起飞"战略，在1973年城市总体规划的基础上，编制了1982年版的《武汉市城市总体规划（1982~2000年）》，启动了机场、港口、车站等大型交通设施选址和规划，确定了江南、江北独立成市的商贸流通体系，为大武汉的再次腾飞夯实了"两通"基础。

1984年，武汉市被确定为国家综合改革试点城市，于1988年编制了《武汉市城市总体规划修订方案》，启动了武汉开发区、东湖开发区等城市重要功能区以及花桥新村小区、钢花新村、常青花园等大型住宅区的规划建设，主城区的功能结构、用地空间逐步得到了丰富和完善。总体而言，这一时期的城市发展主要集聚在三环以内。

进入1990年代，武汉相继被批准为对外开放城市、开放港口，标志着武汉进入了新的重要发展阶段，城市活力进一步释放，部分发展较快的远城区开始出现贴近主城、与主城融为一体的发展趋势。

Wuhan, a city built along rivers, has been taking the advantage of natural endowment since ancient times. During the first and second "Five-Year Plan" periods, this river town was listed as a national key construction city. In 1954 and 1959, the city drew up its urban master plan and urban construction plan, and arranged a number of industrial projects of national and local levels far from the downtown, owing to the ideological influence of the Soviet Union. These efforts have gradually determined the city layout, and guided the expansion of urban spaces.

During the decade of the "Cultural Revolution", urban planning and construction work basically stalled. At the beginning of the reform and opening up, Wuhan launched the Liangtong Qifei strategy (literally "take-off on the strength of traffic and circulation") and in 1982 developed the Urban Master Plan of Wuhan based on the framework of the planning in 1973. Thus, Wuhan started the planning of large transportation facilities such as airport, port and station, which determined the city-based commercial circulation system in Jiangnan and Jiangbei, and laid a solid foundation for the re-take-off of Wuhan.

In 1984, Wuhan was listed as a pilot city for national comprehensive reform. In 1988, it revised the Urban Master Plan, and initiated in the main quarters the construction of important functional areas such as Wuhan Development Zone and East Lake Development Zone, as well as large residential areas, including Huaqiao Xincun Community, Ganghua New Community and Changqing Garden. Therefore, Wuhan has gradually improved in division of areas and land in its main urban areas. During this period, the city mainly saw development within the third ring road.

In the 1990s, Wuhan was approved as an open-up city and port, making a new important stage of development. Since then, the city has been more dynamic with some of its sub-districts catching up and integrating with the growth of the main quarters.

一、新中国成立初期的规划
Planning in the early days of People's Republic of China

1949 - 1977

时至今日，中国的城市化进程已经迈过了第 70 个年头。前 30 年，中国城市化的最大特征就是启动与徘徊。新中国成立后，受苏联实施中央计划经济和重化工业优先战略并在短期内取得巨大成就的影响，中国选择了变"消费城市"为"工业城市"的赶超战略，以期摘掉贫穷落后农业国的帽子，早日建设成为现代化的社会主义强国。在此期间，中国城市化的水平有了一定提高，但总体速度仍非常缓慢，滞后于发展中国家的平均水平，发展中国家由 16.2% 上升到 30.5%，中国内地仅由 11.2% 上升到 19.4%。与此同时，前三十年工业总产值增长了 38.18 倍，工业总产值在工农业总产值中的比重，由 30% 提高到 72.2%，中国城市化的水平还大大滞后于工业化发展水平。

1953 年，为落实国家"一五"计划，武汉市编制了《武汉市城市规划草图》。1954 年，为落实国家"156 项工程"中落户武汉的重大工业项目，借鉴苏联规划经验，编制了《武汉市城市总体规划》，该规划布局完全转移到为大型工业项目建设服务上来，建设了武钢、武锅、武船、肉联等大型企业，开辟了青山、中北路、石牌岭、白沙洲、堤角、易家墩、庙山等七个工业区，承载了全国 1/4 的重要工业项目，引导了城市空间的跳跃式拓展，迎来了武汉历史上第一轮空间扩展高峰。

"二五"时期，在"大跃进"的形势下，武汉提出了号称"200 项"的工业建设计划，并于 1959 年重新编制了《武汉市城市建设规划》，增加了关山、余家头、七里庙、唐家墩、鹦鹉洲等五处工业区和葛店化工区，选择了一些中小城镇和卫星城镇作为工业基地。此后的十余年中，城市规划工作停止。

China is now in its 70th year of urbanization. In the first 30 years, the biggest feature of China's urbanization was startup and wandering. After the founding of the People's Republic of China, influenced by the Soviet Union's priority strategy of developing a centrally planned economy and heavy chemical industry and its great achievements in a short period of time, China adopted the catch-up strategy of changing "consumer cities" into "industrial cities" in order to remove the label of poor and backward agricultural country and build itself into a socialist power with modern industry at an early date. During this period, the level of urbanization in China has improved to a certain extent, but the overall pace was still very slow, lagging behind the average level of developing countries, in which the level rose from 16.2% to 30.5%. In mainland China, the level rose from 11.2% to 19.4%. In the first 30 years, the total industrial output value increased by 38.18 times, and the proportion of the total industrial output value in the total industrial and agricultural output value increased from 30% to 72.2%. At the same time, the level of urbanization in China lagged far behind the level of industrial development.

In 1953, to implement the national "First Five-Year Plan", Wuhan compiled *Wuhan City Plan Draft*. In 1954, to implement the major industrial projects settled in Wuhan in the national "156 Projects", Wuhan compiled **Wuhan City Master Plan** drawing on the Soviet Union's experience in planning. The planning layout was completely shifted to serving the construction of large-scale industrial projects. Large enterprises such as Wuhan Iron and Steel, Wuhan Boiler Works, Wuchang Shipbuilding and Meat Processing Factory were built. Seven industrial zones were opened up such as Qingshan, Zhongbei Road, Shipailing, Baishazhou, Dijiao, Yijiadun and Miaoshan, carrying 1/4 of the country's major industrial projects, guiding the leapfrog expansion of urban space and ushering in the first round of space expansion peak in Wuhan's history.

During the "Second Five-Year Plan" period, in the context of the "Great Leap Forward", Wuhan put forward an industrial construction plan called "200 Projects". In 1959, **Wuhan Urban Construction Plan** was revised. Five industrial zones including Guanshan, Yujiatou, Qilimiao, Tangjiadun and Yingwuzhou were added and Gedian Chemical Industrial Zone was established; some medium and small towns and satellite towns were selected as industrial bases. In the following ten years or so, urban planning stopped.

1 – 变"消费城市"为"生产城市"引导下的两版城市总体规划

Two Versions of City Overall Plan Under the Guidance of "Changing Consumer Cities into Production Cities"

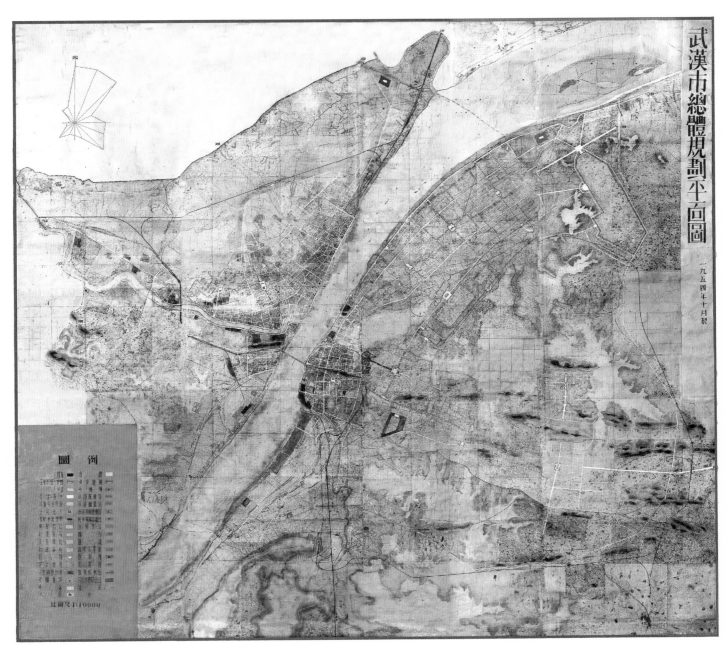

武汉市总体规划图（1954年）　　Wuhan City Master Plan (1954)

一 落实国家重大工业项目的《武汉市城市总体规划》
（1953 年、1954 年）

1953 年"一五"计划开始时，中国开启了以重点工业项目为主导的城市建设计划，武汉被列为重大工业项目的重点落户城市，为适应国家建设需要，武汉市城市建设委员会在国务院城建总局顾问——苏联专家巴拉金的指导下编制了《武汉市城市规划草图》。草图内容分为工业区、居住区、高等文教区、中南行政区以及湖北省和武汉市的行政中心、铁路运输用地、港务码头仓库用地、绿地（含风景区、疗养区）和郊区（含菜园苗圃）等。

随着国家"156 项工程"中武钢、青山热电厂一期工程、武重、肉联、武锅、武船等重大工业项目落户武汉，城市建设的重点转移到为大型工业建设服务，以致《武汉市城市规划草图》（1953 年）已不能很好地满足需要。1954 年底，武汉市城市建设委员会对《武汉市城市规划草图》（1953 年）进行修订，编制了《武汉市城市总体规划》，以实现工业化城市和三镇整体化的构想，同年 12 月上报国家计划委员会。

1954 年版《武汉市城市总体规划》确定 1972 年人口规模为 198 万人，全市规划使用土地共 203.5 平方公里。该规划将武昌旧城区以东作为市区未来扩展地区，东南、东北地区作为重型工业选址对象；汉阳旧城区向西、向南略作扩展；汉口增扩长丰南垸作为居住区，并将工业区向沿河地带转移。规划关山、黄土山、长丰垸、十里铺、武昌青山、答王庙、钵盂山等重工业区。高等文教区规划在珞珈山、喻家山、磨山一带。同时，该规划还提出沿东湖建设风景区。

1954 年版《武汉市城市总体规划》在武昌洪山中心开辟垂直于长江的道路一条，遥对汉阳南岸嘴，接转汉口解放大道人民广场，两旁各加辅助干道一条，将武汉三镇联系成一整体。这样，在城市景观上便形成了"中山公园前面广场—汉水边的集家嘴—汉阳南岸嘴—武昌洪山广场"的三镇主轴线，并在轴线上设立中南行政区、湖北省、武汉市三级行政中心，中南级中心在东湖、沙湖、洪山之间的三皇殿，省级中心在阅马场，市级中心在中山公园前人民广场。

同时，该规划还提出长江上计划修建三座桥，近期修建龟、蛇山之间的公铁两用大桥，远期修建分金炉至徐家棚公路大桥、天兴洲至谌家矶公铁两用大桥；汉江上则计划修建四座公路桥。道路系统以长江为主轴，对称设置，通过过江桥梁连接三镇主干道。同时，规划着重进行了铁路枢纽的选址定点。

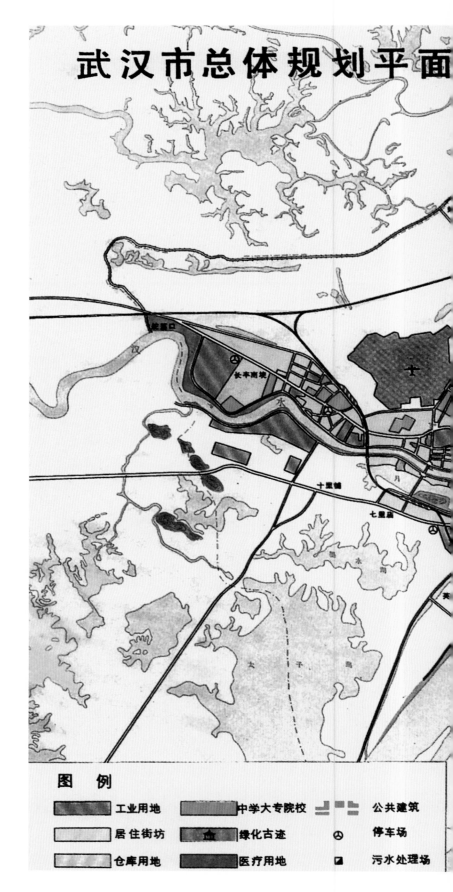

武汉市总体规划图
（1954 年）
Wuhan City Master Plan
(1954)

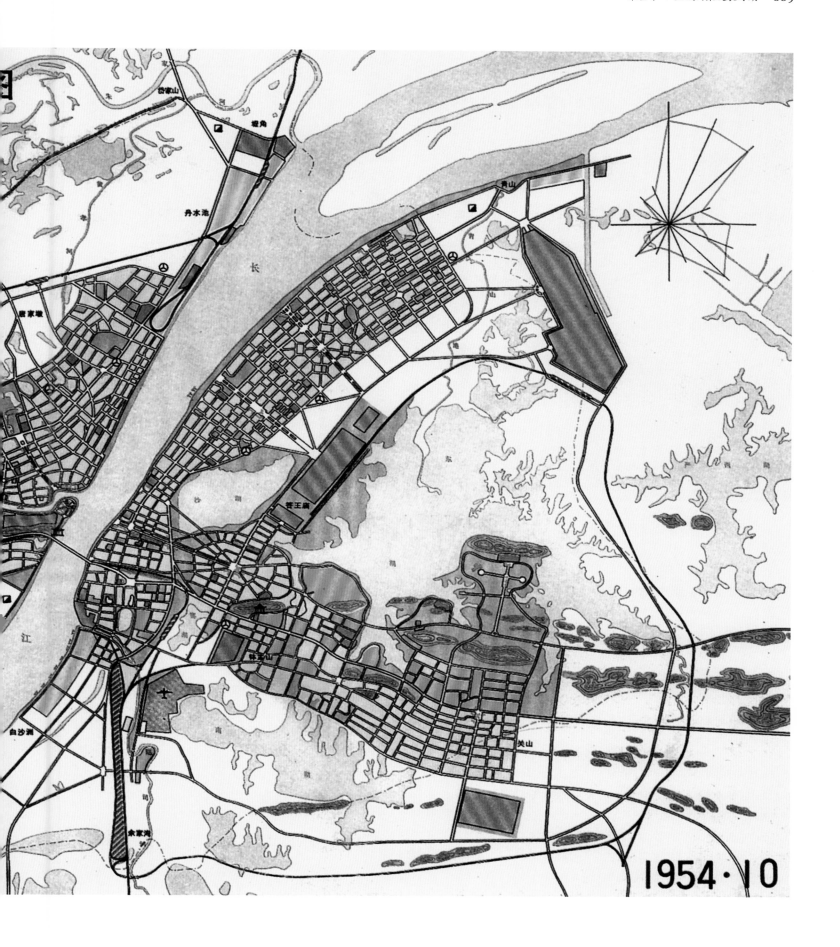

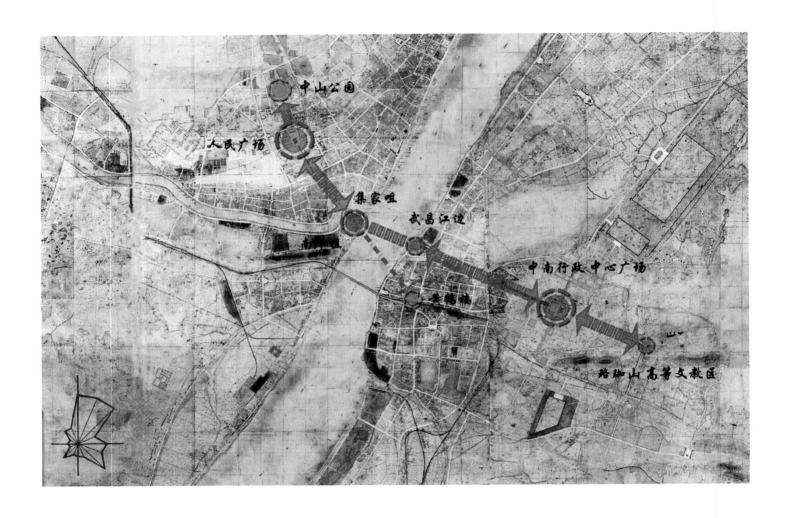

一 大、中、小工业并举的《武汉市城市建设规划（修正草案）》（1959年）

新中国成立初期，国家提出城市建设应集中力量，确保国家中心项目所在重点工业城市的建设，1955年提出"没有特殊原因，不建设大城市"。1958年"二五"开始，在"大跃进"的形势下，武汉市提出"200项"工业项目建设计划。在"大、中、小工业并举"的方针指引下，1959年按调整后的工业建设计划，武汉市基本建设委员会编制了《武汉市城市建设规划（修正草案）》。

该规划修正草案对城市人口和城市规模进行有效控制，1967年以前城市总用地167平方公里左右，人口规模240万人左右，提出城市性质及发展方向是"将要建成为钢铁、机械、化学等工业的基地，又是科学技术、文化教育的基地和交通枢纽，对于湖北省和华中协作区工农业的发展，担负着技术支援和经济协作的重大任务"。该规划修正草案进一步明确"一五"计划时期七个工业区，进一步配套建设居住区，形成综合工业组团。外迁某些不宜在市区的工业，选择一些有便利铁路、公路或水运交通，接近电源、靠近资源产地、水源及排水都能解决的中小城镇，并开辟卫星城镇（例如葛店）作为工业基地。同时规划重视旧城区改造，确定了省、市、区中心广场位置。

一 城市建设停滞期的城市总体规划

"文革"期间，武汉市人口数量下降，城市建设基本停滞，除建成江汉二桥、北湖码头等大型交通设施外，仅建设了白玉山小区、汉阳桥头居住区等配套住宅区。

为了配合1976~1985年的十年计划，武汉市城市规划设计院仍然编制了1973~1985年的城市总体规划。该规划未获批准，这一期间的城市建设基本以1959年的《武汉市城市建设规划（修正草案）》为依据。

1　武汉市总体规划城市中轴线示意图（1954年）（编者自绘，底图为1954年武汉市总体规划图）

The axis in Wuhan City Master Plan (1954) (drawn by the author; the base drawing is from Wuhan City Master Plan (1954)

2　武汉市中心规划示意图

Wuhan Center Plan

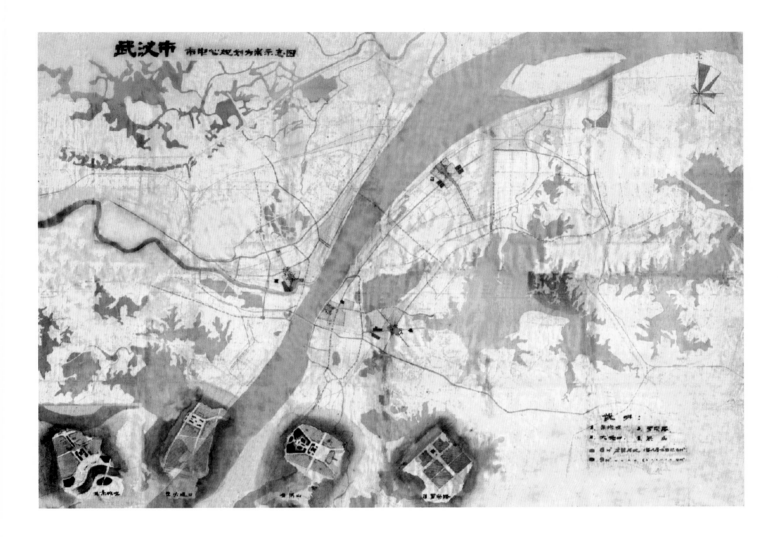

— Wuhan City Master Plan (1953, 1954) implementing major national industrial projects

At the beginning of the "First Five-Year Plan" in 1953, China started a city construction plan dominated by major industrial projects. Wuhan was listed as a key city for major industrial projects. In order to meet the needs of national construction, the Wuhan Urban Construction Committee, under the guidance of the Soviet Union expert Balakin, a consultant to the State Council's Urban Construction Administration, prepared the *Wuhan City Plan Draft*, which included industrial zones, residential areas, higher education and culture areas, administrative centers of Central South, Hubei Province and Wuhan City, land for railway transportation, land for port administrative affairs and wharf warehouses, green land (including scenic areas and convalescent areas) and suburbs (including vegetable and nursery gardens), etc.

As Wuhan Iron and Steel, Qingshan Thermal Power Plant Phase I Project, Wuhan Heavy Duty Machine Tool, Meat Processing Factory, Wuhan Boiler Works, Wuchang Shipbuilding and other major industrial projects in the national "156 Projects" settled in Wuhan, the focus of urban construction shifted to serving large-scale industrial construction; as a result, the draft plan could no longer meet the needs well. At the end of 1954, the Wuhan Urban Construction Committee revised the draft plan and compiled *Wuhan City Master Plan* to realize the idea of integrating the industrialized city and the three towns, which was submitted to the State Planning Commission in December of the same year.

According to the plan, the population size in 1972 was 1.98 million and the planned land area in the city was 203.5km^2. The east of the old urban area of Wuchang was used for the future expansion of the urban area, and the southeast and northeast areas were used as sites of heavy industries. The old urban area of Hanyang expanded slightly to the west and south. Hankou expanded Changfeng Nanyuan as a residential area and transferred the industrial zone to the riverside area. Heavy industrial zones were planned, including Guanshan, Huangtushan, Changfengyuan, Shilipu, Wuchang Qingshan, Dawangmiao and Boyushan. Higher education and culture areas were planned in Luojiashan, Yujiashan and Moshan. Meanwhile, the plan proposed to build scenic spots along East Lake.

It was planned to open up a road perpendicular to the Yangtze River in the center of Hongshan in Wuchang, facing Nan'anzui of Hanyang, and connecting to the People's Square on Jiefang Avenue in Hankou, with one auxiliary road on each side to connect the three towns of Wuhan into a whole. In this way, the main axis of the three town of "Square in front of Zhongshan Park — Jijiazui by the Hanjiang River — Hanyang Nan'anzui — Wuchang Hongshan Square" was formed in the urban landscape, and three administrative centers of Central South, Hubei Province and Wuhan City have been set up on the axis; the center of Central South was the Sanhuang Hall between East Lake, Shahu Lake and Hongshan, the provincial center is in Yuemachang, and the municipal center is the People's Square in front of Zhongshan Park.

According to the plan, three bridges would be built on the Yangtze River, a highway-railway bridge between Turtle Mountain and Snake Mountain would be

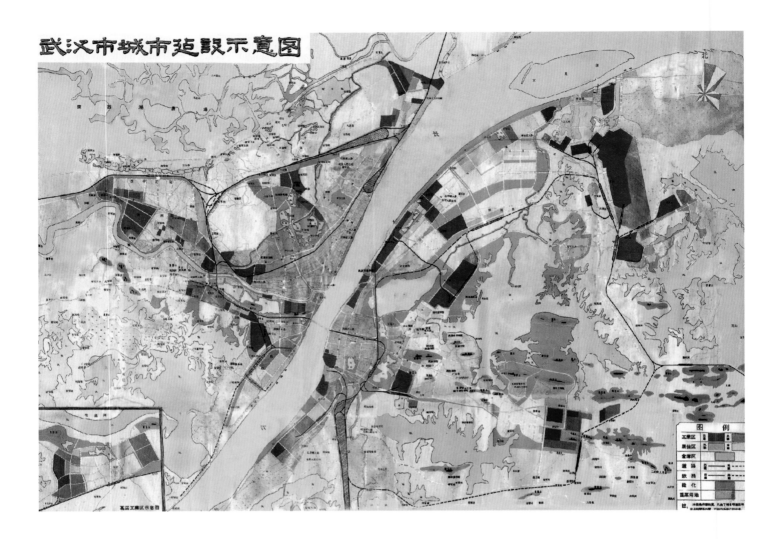

built in the near future, and a highway bridge between Fenjinlu and Xujiapeng and a highway-railway bridge between Tianxingzhou and Chenjiaji would be built in the long term; four road bridges were planned over the Hanjiang River. At the same time, the sites of railway hubs were determined. The road system was set symmetrically with the Yangtze River as the main axis, and the main roads of the three towns were connected through bridges across the river.

— Wuhan Urban Construction Plan (Amended Draft) (1959), developing large, medium and small industries simultaneously

In the early days of People's Republic of China, the state proposed that urban construction should focus on ensuring the construction of key industrial cities where national major projects were located. In 1955, it was proposed that "no big cities should be built without special reasons". From the "Second Five-Year Plan" in 1958, in the context of the "Great Leap Forward", Wuhan put forward an industrial project construction plan of "200 Projects". Under the guidance of the policy of "developing large, medium and small industries simultaneously", in 1959, the Wuhan Capital Construction Commission compiled *Wuhan Urban Construction Plan (Amended Draft)* according to the adjusted industrial construction plan.

The plan effectively controlled the urban population and urban size. Before 1967, the total urban land area was about 167km^2, with a population of 2.4 million. The plan specified the function and development direction of the city that "it will be the base of iron and steel, machinery, chemistry and other industries, as well as the base of science and technology, culture and education and the transportation hub, which shoulders important tasks of technical support and economic cooperation". The plan further defined the seven industrial zones during the "First Five-Year Plan" period, supporting the construction of residential areas to form comprehensive industrial groups. Some industries that were not suitable for urban areas were relocated; medium and small cities and towns that had convenient railway, highway or water transportation and were close to power supply, origins of resources, water sources and convenient drainage, and new satellite towns (such as Gedian) were selected as industrial bases. The plan attached great importance to the renovation of the old urban area and determined the location of the provincial, municipal and district central squares.

— Urban overall planning during urban construction stagnation

During the "Cultural Revolution", Wuhan saw a population decline and urban construction basically stopped. Apart from the large transportation facilities such as the Second Wuhan Yangtze River Bridge and North Lake Wharf, only supporting residential areas such as Baiyushan and Hanyang Qiaotou were built.

In line with the ten-year plan for 1976—1985, the urban master plan for 1973—1985 was prepared. The plan was not approved, and urban construction during the period was based largely on the 1959 master plan.

1
2
3

1
武汉市城市建设示意图
（1958年）
Wuhan Urban Construction Plan (1958)

2
武汉市城市建设十年规划示意图（1976~1985年）
Ten-Year Plan for Wuhan Urban Construction (1976—1985)

3
武汉市城市建设示意图
（1973~1985年规划方案）
Wuhan Urban Construction Plan (1973—1985 Plan)

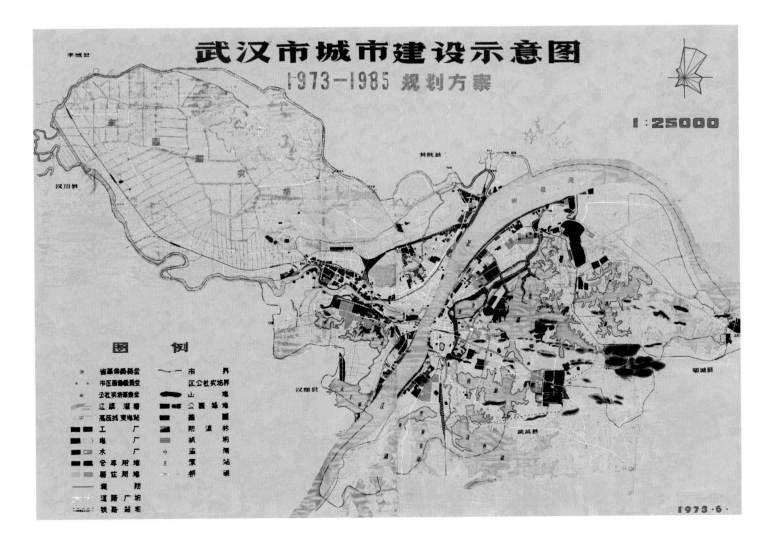

2 – 十三个工业区规划
Planning of Thirteen Industrial Zones

新中国成立初期，中国经济整体上仍以农业经济为主，1952 年农业产值比重为 57.7%，工业仅为 19.5%。在仿效苏联模式"重工业优先赶超战略"的指引下，我国迅速建立起了比较完整的工业体系，实现了从农业国向工业国的转型。

历史上，武汉因为是传统的商贸城市，所以新中国成立初期的第三产业比重很高，超过 50%，呈现"三、一、二"特征。随着重工业优先和抑制消费战略的实施，1956 年工业首次超过服务业，三产比例调整为 13.9：45.5：40.6，呈现"二、三、一"特征。1978 年工业在三产中占绝对主导地位，三产比例为 11.7：63.3：25.0。在此期间，武汉先后建设了 13 个工业区，分别为青山工业区、中北路（答王庙）工业区、石牌岭（钵盂山）工业区、白沙洲工业区、堤角工业区、易家墩工业区、庙山工业区、关山工业区、余家头工业区、七里庙工业区、唐家墩工业区、鹦鹉洲工业区和葛店化工区。

工业区的布局逐步从顺江走向离岸，从沿江、沿河靠近港口布局发展到沿湖、沿路多方向伸展。"一五"期间，为了充分利用沿江便利的水上运输条件，武汉工业用地主要分布在长江和汉江两岸，工业区大多选址于城市郊区，受地质条件和防洪内涝等因素影响，工业区呈点状分布，既没有连成片，也没有向老城纵深区域发展。同时，沿汉口老铁路线也有零星分布的工业点。

"二五"期间，工业区布局趋于集中，逐步从沿江伸展为主发展到沿湖、沿路多方向伸展。武昌主要沿着关山路、中北路、和平大道、冶金大道方向伸展；汉阳主要沿着汉阳大道、鹦鹉大道方向伸展；汉口主要沿着解放大道西段向西伸展。兴建并扩大的工业区如青山工业区、关东工业区、堤角工业区多集中在城市的边缘地带，这些工业区的建立使工业布局趋向合理，并促进工业与居住的逐步分离。

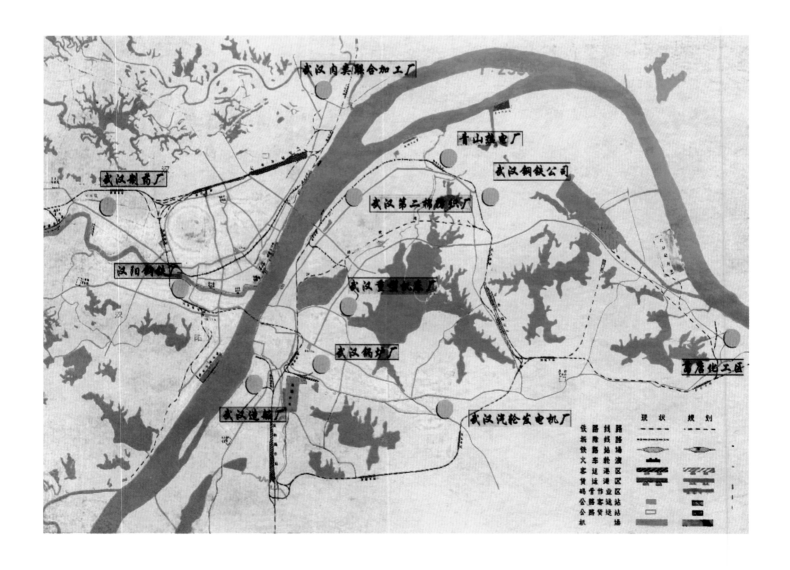

1

"一五""二五"时期重要工业项目建设示意图（编者自绘，底图为1960年武汉市对外交通规划图）

Important industrial projects in the "First and Second Five-Year Plan" period (drawn by the author; the base drawing is Wuhan External Traffic Plan of 1960)

2-3

1954年和1979年武汉市工业区分布图（编者自绘，底图为1954年和1979年武汉市城市总体规划图）

Distribution of Industrial Zones in Wuhan of 1954 and 1979 (drawn by the author; the base drawing is Wuhan City Master Plan of 1954 and 1979)

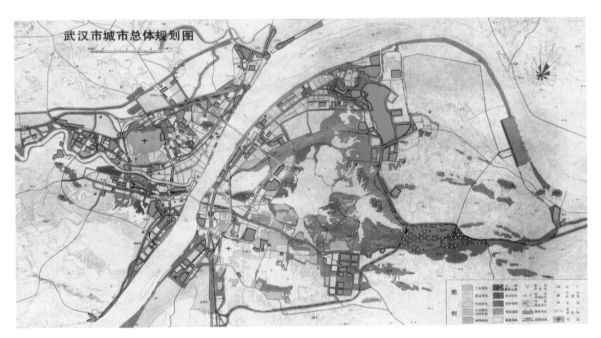

In the early days of People's Republic of China, China's economy was still dominated by agriculture. In 1952, the proportion of agricultural output value was 57.7%, while the industrial output value was only 19.5%. Under the guidance of the Soviet model of "catch-up policy giving priority to heavy industry", China quickly established a relatively complete industrial system and realized the transformation from an agricultural country to an industrial country.

Historically, Wuhan was a traditional commercial city, so the proportion of tertiary industry in the early days of People's Republic of China was very high, exceeding 50%, and the tertiary industry was ahead of the primary and secondary industries. With the implementation of the strategy of giving priority to heavy industry and curbing consumption, industry surpassed service industry for the first time in 1956, and the ratio of three industries were 13.9 : 45.5: 40.6, with the secondary industry ahead of the primary and tertiary industries. In 1978, industry was absolutely dominant in the three industries, with a ratio of 11.7 : 63.3 : 25.0. During this period, Wuhan successively built 13 industrial zones, namely, Qingshan Industrial Zone, Zhongbei Road (Dawangmiao) Industrial Zone, Shipailing (Boyushan) Industrial Zone, Baishazhou Industrial Zone, Dijiao Industrial Zone, Yijiadun Industrial Zone, Miaoshan Industrial Zone, Guanshan Industrial Zone, Yujiatou Industrial Zone, Qilimiao Industrial Zone, Tangjiadun Industrial Zone, Yingwuzhou Industrial Zone and Gedian Chemical Industrial Zone.

The industrial zones gradually moved from the riverside to the offshore area, from places along rivers and near ports to places along lakes and roads. During the "First Five-Year Plan" period, to make full use of the convenient water transportation conditions along rivers, Wuhan's industrial land was mainly distributed on both sides of the Yangtze River and the Hanjiang River. Most industrial zones were located in the suburbs. Affected by geological conditions, flood control, water logging and other factors, they were distributed in dots, neither in patches nor in deep areas of the old city. At the same time, there were scattered industrial sites along old railway lines in Hankou.

During the "Second Five-Year Plan" period, the layout of industrial zones tended to be centralized, gradually expanding from the riverside area to places along lakes and roads. Wuchang mainly extended along Guanshan Road, Zhongbei Road, Heping Avenue and Yejin Avenue. Hanyang mainly extended along Hanyang Avenue and Yingwu Avenue. Hankou mainly extended westward along the west section of Jiefang Avenue. Industrial zones built and expanded were mostly in the edge of the city, such as Qingshan Industrial Zone, Guandong Industrial Zone and Dijiao Industrial Zone. The establishment of these industrial zones has made the industrial layout more reasonable and promoted the gradual separation of industry and residence.

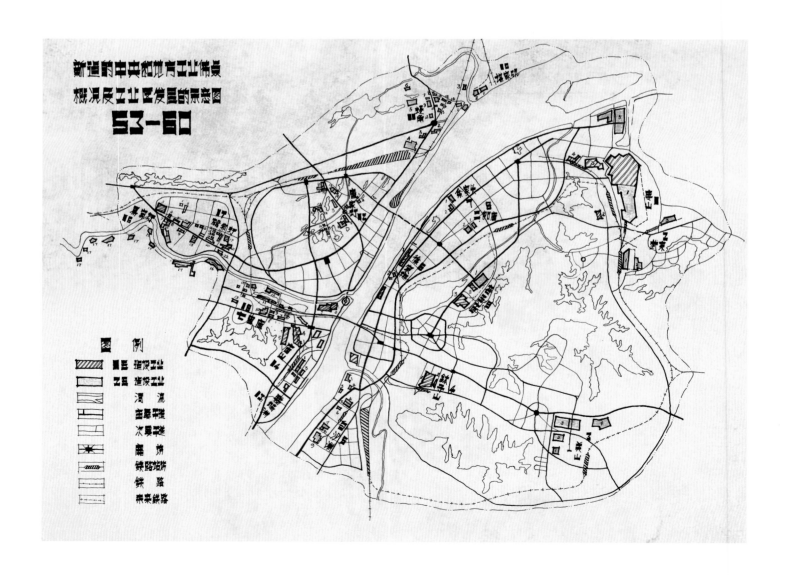

1

新建中央和地方工业布点
概况及工业区发展示意图
（1953年）

Overview of new central and
local industrial distribution
and development of industrial
zones (1953)

2

武汉钢铁公司厂区生活平面
图（1983年）

Factory and Living Quarter
Plan of Wuhan Iron and Steel
Corporation (1983)

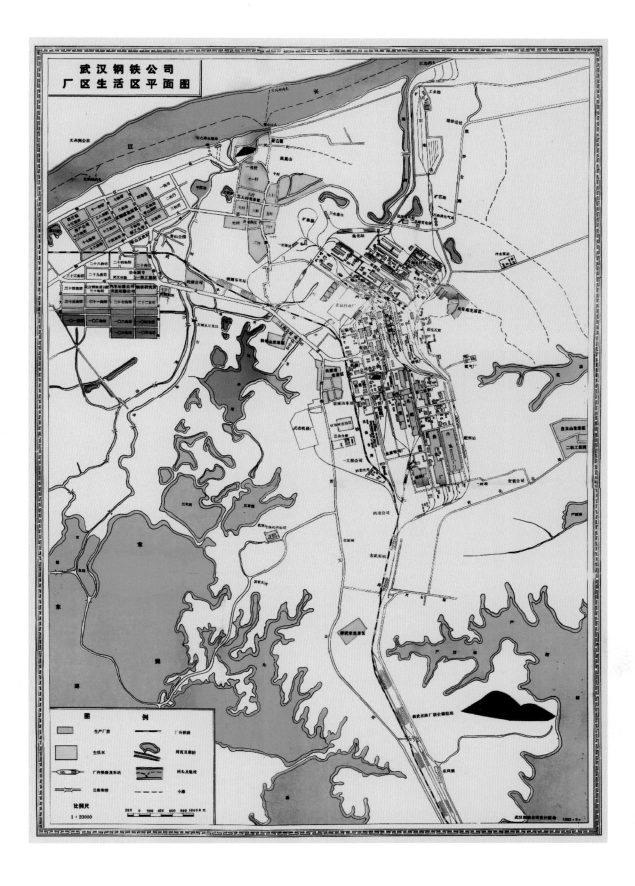

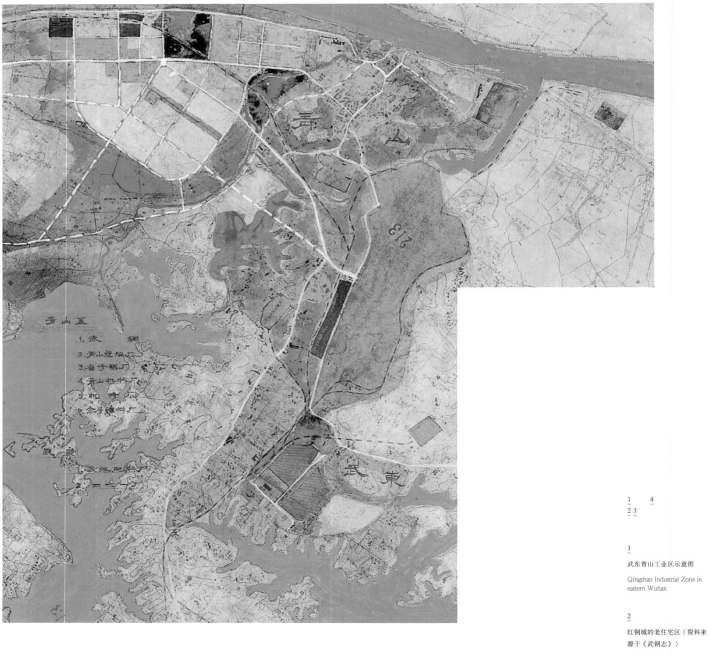

1	4
2	3

1

武东青山工业区示意图

Qingshan Industrial Zone in eastern Wuhan

2

红钢城的老住宅区（资料来源于《武钢志》）

Old residential area in Honggangcheng (data source: *The History of Wuhan Iron and Steel*)

3

20世纪70年代新建的住宅区（资料来源于《武钢志》）

New residential area in the 1970s (data source: *The History of Wuhan Iron and Steel*)

4

堤角工业区示意图

Dijiao Industrial Zone Plan

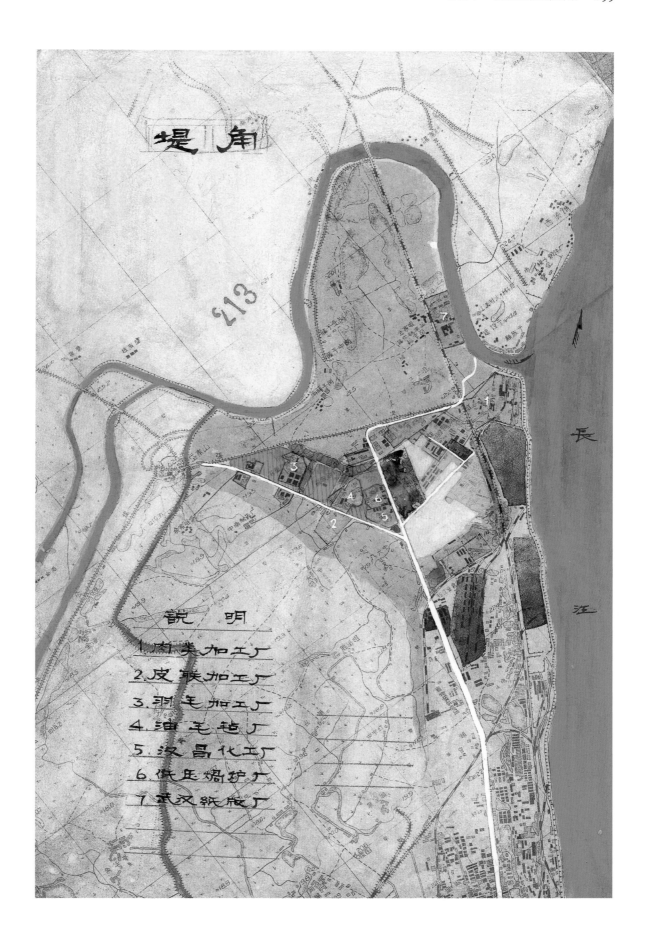

CHAPTER THREE • Cohesion of the Main City

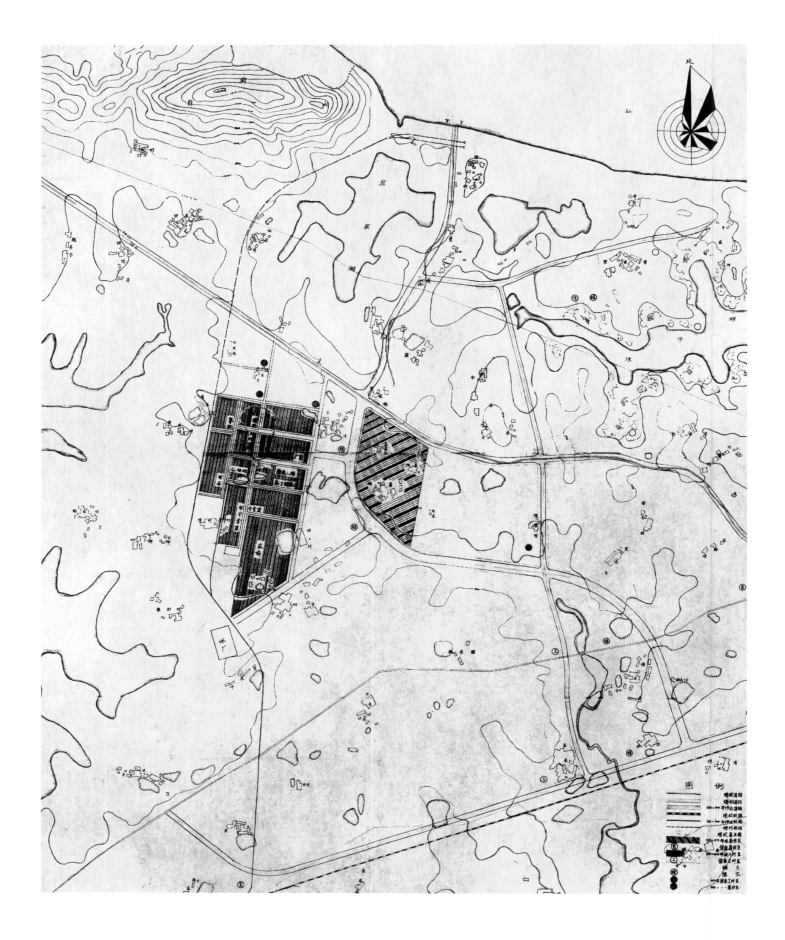

1

葛店工业区规划平面图
（1958年）

Gedian Industrial Zone Plan
(1958)

2

十里铺工业区示意图

Shilipu Industrial Zone Plan

3

中北路工业区现状图

Current status of Zhongbei
Road Industrial Zone Plan

4

关山工业区示意图

Guanshan Industrial Zone
Plan

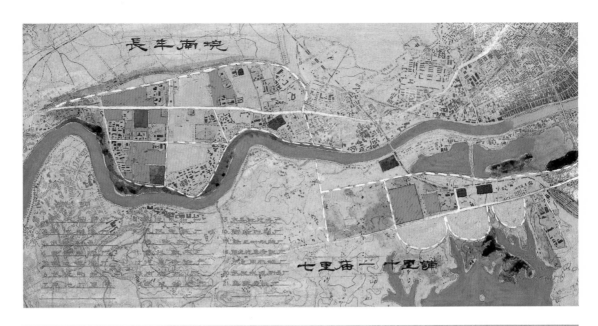

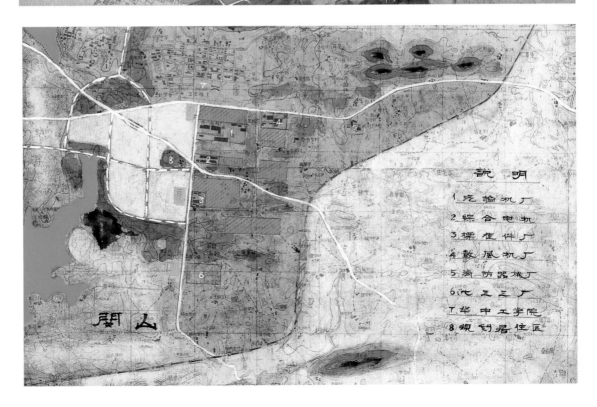

3 - 重大市政工程建设
Construction of Major Municipal Projects

一 武汉长江大桥建设（1957年）

修建第一条跨越长江的公铁两用大桥是中国人民近百年来的夙愿和梦想。1906年，京汉铁路全线通车，粤汉铁路也在修建中。万里长江横贯东西，京广、粤汉铁路纵穿南北，在中国的经济版图上，两条大动脉的黄金交汇处就是武汉。湖广总督张之洞首次提出，在武汉建一座长江大桥，跨越长江、汉水，连接京汉、粤汉两路的构思并为各方所关注。

1913年，詹天佑曾带领北京大学工科德国籍教授乔治·米勒及13名土木系毕业生来汉对长江大桥桥址展开初步勘测和设计实习；1919年孙中山在《实业计划》中写道："在京汉铁路线于长江边第一转弯处，应穿一隧道过江底，以连接两岸。"1929年，时任武汉特别市市长刘文岛曾邀美国工程师华德尔博士研商建桥之事；1934年，茅以升主持对长江大桥桥址进行测量钻探，并请当时驻华莫利纳德森工程顾问团拟定新的公铁两用桥建设计划，公路中间还设计了电车轨道；1946年，武汉区域规划委员会提出铁路和公路合并的建桥意见；新中国成立后，京汉铁路和粤汉铁路之间的运输全部由武昌和汉口的驳船、轮渡接转，随着货运量剧增，同时轮渡模式由于受天气影响较大，已无法满足经济发展的迫切需要。

1950年初，政务院决定修建武汉长江大桥，责成铁道部进行武汉长江大桥的勘测和设计。在1953年制定的《武汉市城市规划草图》方案中，政务院再次提出了长江大桥桥址计划，规划桥址三处，即龟蛇山之间、汉口分金炉对徐家棚、青山至谌家矶。

当时市政府决定修建武汉长江大桥之后，铁道部开展了桥址线的勘测和研究工作，先后提出了八个桥址线方案，采用左岸龟山及右岸蛇山，即由一个斜层的褶皱过渡到另一个斜层的褶皱的第五方案，并于1954~1955年完成了长江大桥的技术设计，于1955年7月经政府审核批准。

1953年11月，汉水铁路桥动工兴建，两岸铁路联络线及跨线桥工程也开始进行。1955年7月18日，国务院批准了武汉长江大桥的技术设计，正桥长1156米，八墩九孔，每孔128米，武昌岸引桥211米，汉阳岸引桥303米，桥宽为22.5米（车行道18米，两边人行道各2.25米），同年9月1日，大桥作为"一五"计划重点工程开始施工。1957年9月30日建成，10月15日正式通车，从此实现了"一桥飞架南北，天堑变通途"。

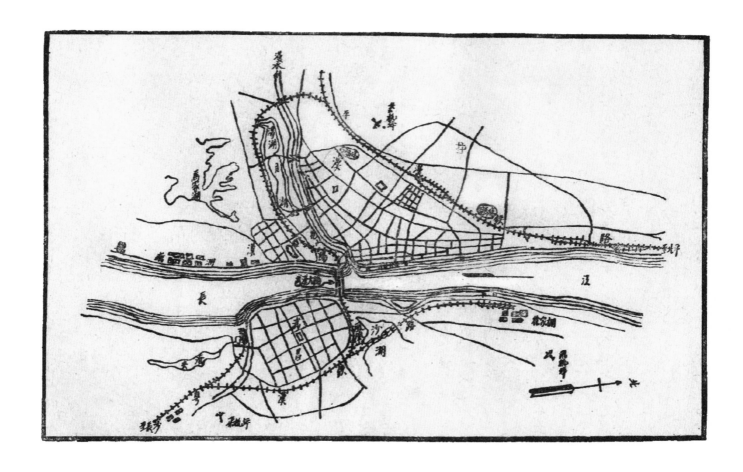

1	2	3
4		

1 武汉长江大桥施工情景（资料来源于武汉市档案信息网）

Wuhan Yangtze River Bridge under construction (datasource: Wuhan Archives Information Network)

2 武汉长江大桥桥头仰视图（模型，现存于武汉大桥局）

An upward view of the head of Wuhan Yangtze River Bridge (model, now stored in China Railway Major Bridge Engineering Group, Wuhan)

3 武汉大桥选址示意图（1947年，现存于武汉市档案馆）

Wuhan Bridge Site Selection Plan (1947, now stored in Wuhan Archives)

4 武汉长江大桥施工全景（1956年10月，现存于武汉大桥局）

A panoramic view of Wuhan Yangtze River Bridge under construction (October 1956, now stored in China Railway Major Bridge Engineering Group, Wuhan)

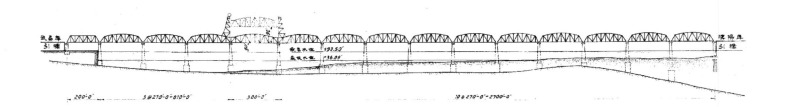

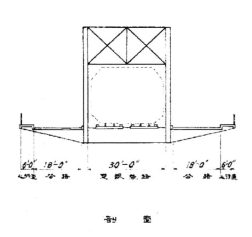

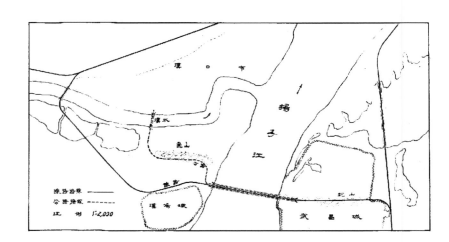

1	2
3	
4	

1 华德尔博士建桥计划图（1934年，现存于武汉市档案馆）
Dr. Waddel's bridge construction plan(1934, now stored in Wuhan Archives)

2 武汉长江大桥桥址总平面图（现存于武汉大桥局）
General Plan of Wuhan Yangtze River Bridge Site (now stored in China Railway Major Bridge Engineering Group, Wuhan)

3 武汉长江大桥桥址选线比较方案（现存于武汉大桥局）
Comparison of site and route selection of Wuhan Yangtze River Bridge (now stored in China Railway Major Bridge Engineering Group, Wuhan)

4 武汉长江大桥全桥建筑总图（现存于武汉大桥局）
General construction drawing of the Wuhan Yangtze River Bridge (now stored in China Railway Major Bridge Engineering Group, Wuhan)

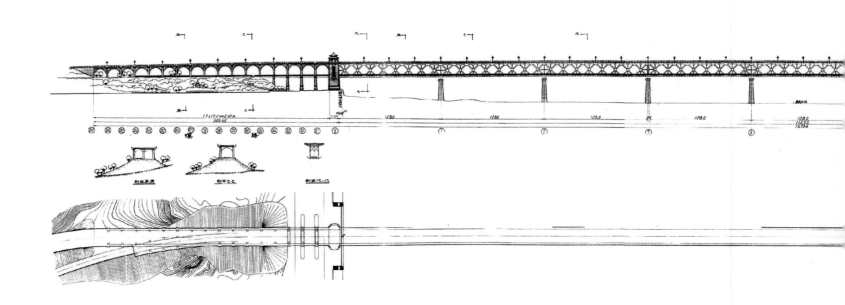

第叁章 · 主城内聚发展时期　105

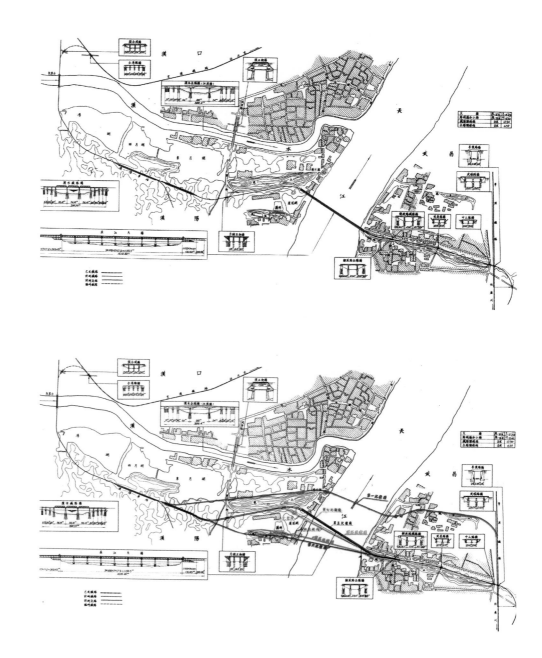

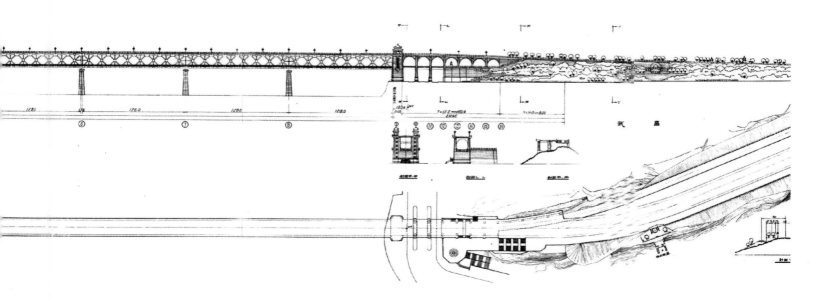

— Construction of Wuhan Yangtze River Bridge (1957)

The construction of the first road-rail bridge across the Yangtze River has been the long-cherished wish and dream of the Chinese people for nearly a century. In 1906, the Beijing-Hankou Railway was opened to traffic, and the Guangzhou-Hankou Railway was under construction. The Yangtze River runs from east to west, and the Beijing-Guangzhou Railway and Guangzhou-Hankou Railway run from north to south. On China's economic map, the golden junction of the two railways is Wuhan. Governor-General of Hu-Guang, Zhang Zhidong, proposed for the first time the idea of building a bridge over the Yangtze River in Wuhan to cross the Yangtze River and the Hanjiang River to connect the Beijing-Hankou Railway and Guangzhou-Hankou Railway, which attracted attention from all sides.

In 1913, Zhan Tianyou led German professor of engineering from Peking University, George Miller, and 13 graduates from the School of Civil Engineering to Wuhan for preliminary survey and design practice of the site of the Yangtze River Bridge. In 1919, Sun Yat-sen wrote in Material Reconstruction, "At the first turning of the Beijing-Hankou Railway by the Yangtze River, a tunnel should be built to connect the two sides." In 1929, Liu Wendao, mayor of Wuhan Special City, invited Dr. Waddel, an American engineer, to discuss bridge construction. In 1934, Mao Yisheng presided over the measurement and drilling of the Yangtze River Bridge site, and asked Molinadson Engineering Advisory Group in China to draft a new road and railway dual-purpose bridge construction plan. In 1946, the Wuhan Regional Planning Commission put forward the idea of merging railways and roads in bridge construction. After the founding of People's Republic of China, the transportation between the Beijing-Hankou Railway and the Guangzhou-Hankou Railway was connected by barges and ferries in Wuchang and Hankou. With the sharp increase in freight volume and the great influence of weather on ferries, the ferry mode could no longer meet the urgent needs of economic development.

1	2	5
	3	6
4		7

1
建设中的武汉长江大桥
（资料来源于武汉市档案信息网）

Wuhan Yangtze River Bridge under construction (data source: Wuhan Archives Information Network)

2
武汉长江大桥通车典礼（1956年）（资料来源于武汉市档案信息网）

Opening ceremony for the Wuhan Yangtze River Bridge (1956, data source: Wuhan Archives Information Network)

3
武汉长江大桥全桥鸟瞰，现存于武汉大桥局

A bird's-eye view of Wuhan Yangtze River Bridge (now stored in China Railway Major Bridge Engineering Group, Wuhan)

4
武汉大桥及联络路线鸟瞰图（1934年，现存于武汉市档案馆）

A bird's-eye view of Wuhan Bridge and connecting line (1934, now stored in Wuhan Archives)

5
桥台内部透视图，现存于武汉大桥局

A perspective view of the abutment (now stored in China Railway Major Bridge Engineering Group, Wuhan)

6
武汉江底隧道计划图（1934年，现存于武汉市档案馆）

Wuhan River Tunnel Plan (1934, now stored in Wuhan Archives)

7
桥址及联络路线平面图（1934年，现存于武汉市档案馆）

Plan of Bridge Site and Connecting Line (1934, now stored in Wuhan Archives)

In early 1950, the Government Administration Council decided to build the Wuhan Yangtze River Bridge and instructed the Ministry of Railways to carry out the survey and design of the Wuhan Yangtze River Bridge. Wuhan City Plan Draft drawn up in 1953 proposed the site selection plan of the Yangtze River Bridge and gave three bridge sites, i.e., the place between Turtle Mountain and Snake Mountain, Xujiapeng opposite Fenjinlu in Hankou, and the area from Qingshan to Shenjiaji.

After the municipal government decided to build the Wuhan Yangtze River Bridge, the Ministry of Railways carried out survey and research on the bridge site, and successively put forward eight site schemes. The fifth scheme was adopted, i.e. the transition from one fold of inclined layers to another fold of inclined layers with Turtle Mountain on the left bank and Snake Mountain on the right bank. The technical design of the Yangtze River Bridge was completed in 1954—1955 and approved by the government in July 1955.

In November 1953, the construction of the Hanjiang Railway Bridge was started, and the cross-river railway connecting line and the over-railway bridge project also began. On July 18, 1955, the State Council approved the technical design of Wuhan Yangtze River Bridge. The main bridge is 1156 meters long, with 8 piers and 9 openings; each opening is 128 meters long; the approach bridge to Wuchang is 211 meters long; the approach bridge to Hanyang is 303 meters long and 22.5 meters wide (18 meters for the roadway and 2.25 meters for sidewalks on both sides). On September 1 of the same year, the construction of the bridge began as a key project of the "First Five-Year Plan". It was completed on September 30, 1957 and officially opened to traffic on October 15, 1957. From then on, "the bridge connected the north and the south, and a natural barrier became a thoroughfare."

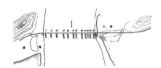

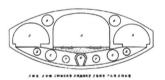

— 汉丹铁路建设（1958年）

随着武汉长江大桥的贯通和我国铁路网络格局的加密，武汉的铁路和水运优势逐渐削弱，枢纽中转的垄断地位下降为节点地位。汉丹铁路的建设开启了我国铁路西拓的步伐，强化了武汉向西的经济联系。汉丹铁路由武汉的汉西至丹江口，南接京广、武大铁路；在襄樊与焦枝、襄渝二铁路交会，是湖北省中部与西北部联系的交通干线。汉丹铁路全长412公里，1958年10月开工，1966年通车。该铁路的建成使武汉初步形成了"十"字形铁路网络格局。

— 东西湖围堤工程（1957年）

为了提高东西湖防洪能力、保障武汉市副食品供应、拓展武汉西向农业发展空间，"一五"计划时期提出围垦东西湖农场的设想。1955年12月，长江水利委员会编制了《东西湖蓄洪垦殖工程》；1957年8月，经国务院批准，东西湖围垦工程列入国家"二五"计划，于1957年11月开工，1958年4月竣工。东西湖大堤堤线从张公堤三金潭为起点，经张家墩、大李家墩、石头埠、北泾嘴及东山对岸，沿沧河经辛安渡至新沟，接汉江干堤。堤顶设计高程29.85～30.98米，堤顶宽5米。

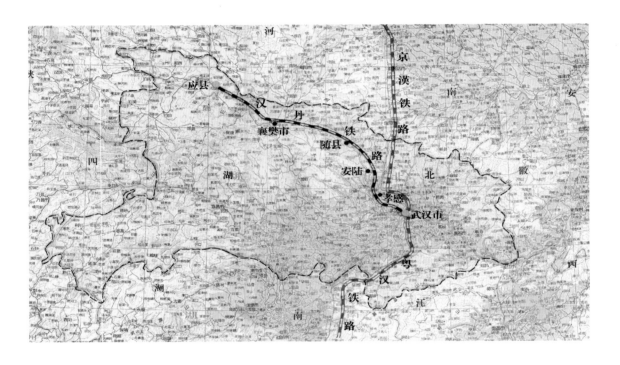

1

汉丹铁路建设（1958年）
（编者自绘，底图为1958年湖北省全图）

Construction of the Hankou-Danjiangkou Railway (1958), (drawn by the author; the base drawing is the general map of Hubei Province of 1958)

2

东西湖围堤示意图（编者自绘，底图为《武汉城市规划志》插图）

Dongxihu reclamation plan (drawn by the author; the base drawing is an illustration of *Wuhan Urban Planning Records*)

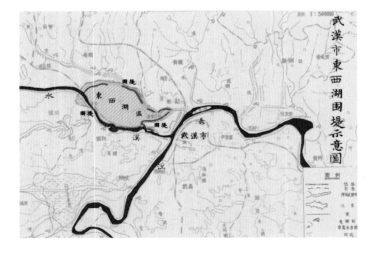

— Hankou-Danjiangkou Railway (1958)

With the completion of Wuhan Yangtze River Bridge and the densification of China's railway network, Wuhan's advantages in railway and water transportation have gradually weakened, and it was reduced to a node from a monopoly hub. The construction of the Hankou-Danjiangkou Railway started China's railway expansion to the west and strengthened Wuhan's economic ties to the west. The Hankou-Danjiangkou Railway runs from Hanxi in Wuhan to Danjiangkou, connects the Beijing-Guangzhou Railway and the Wuchang-Daye Railway in the south, and intersects the Jiaozuo-Zhicheng Railway and the Xiangyang-Chongqing Railway in Xiangfan; it is the main traffic line connecting central and northwest Hubei. The Hankou-Danjiangkou Railway has a total length of 412 kilometers and its construction began in October 1958; it was opened to traffic in 1966. With the completion of the railway, a "cross-shaped" railway network was initially formed in Wuhan.

— Dongxihu reclamation project (1957)

To improve the flood prevention capacity of Dongxihu, ensure the supply of non-staple food in Wuhan and expand the agricultural development space in western Wuhan, the Dongxihu reclamation plan was put forward during the "First Five-Year Plan" period. In December 1955, the Changjiang Water Resources Commission compiled the Dongxihu Flood Storage and Reclamation Project. In August 1957, with the approval of the State Council, the Dongxihu reclamation project was included in the national "Second Five-Year Plan". It was started in November 1957 and completed in April 1958. The dike line of Dongxihu starts from Sanjintan in Zhanggongdi, reaches the other side of Dongshan via Zhangjiadun, Dalijiadun, Shitoubu and Beijingzui, leads to Xingou via Xin'an along the Lunhe River, and connects with the Hanjiang River main dike. The design elevation of the dike top is 29.85～30.98 meters, and the dike top is 5 meters wide.

4 – 洪山广场规划建设
Planning and Construction of Hongshan Square

1954年编制的《武汉市城市总体规划》中规划了洪山广场，洪山广场先定位为中南区行政中心广场，后定位为湖北省的行政中心广场，广场四周布置了省级党政机构和大会堂。中心广场用地面积约9公顷，可承担全市性大型集会活动。广场四周的路网结构采用"环形加放射"形式，围绕广场有环形道路，车流可绕中心而过。

Hongshan Square was planned in the Wuhan City Master Plan compiled in 1954. It was first positioned as the administrative center square of South Central, and then as the administrative center square of Hubei Province. The square was surrounded by provincial Party and government organizations and the grand hall. The square covers an area of about 9 hectares and can accommodate large-scale city-wide gatherings. The road network structure around the square is in the form of "ring plus radiation". There are ring roads around the square to allow the traffic flow to bypass the square.

1　洪山中心区公共建筑分布图（1960年）
Distribution map of public buildings in Hongshan Central District (1960)

2　洪山中心区近期修建街坊规划方案图（1960年）
Recent construction plan of Hongshan Central District (1960)

5 - 苏式公共建筑布局
Layout of Soviet-Style Public Buildings

1959年版《武汉市城市建设规划（修正草案）》对大型公共建筑提出了分布原则：大型公建分布要江南、江北兼顾，以武昌为重点；要有便利的交通；要适当集中并有利于形成良好的城市面貌。因此，大型政治、经济、文化科技、体育等建筑设在洪山中心地区。该规划设想在长江大桥、武珞路、中南路、徐东路、规划的长江公路桥、解放大道、武胜路和汉阳桥头形成的环线区域，分布一些建筑艺术水平较高的建筑群，以突出城市良好的建筑面貌和轮廓。此外，还设想在沿江河湖泊的岸边也布局大型公共建筑，主要有：武汉展览馆、武汉体育馆、洪山宾馆、武汉剧院、武汉商场、武汉饭店、新华电影院、新华路体育馆等。

In 1959, the distribution principles were put forward for large-scale public buildings in Wuhan City Construction Planning (Amended Draft): large-scale public buildings shall be laid out in both regions south of the Yangtze River and regions north of the Yangtze River with Wuchang as the key point; there shall be convenient transport; it is required to properly concentrate and be in favor of forming a good urban appearance. Therefore, large-scale political, economic, cultural and science and technology, and sports buildings and other large-scale buildings were set up in Hongshan Central Region. It was imagined to distribute some building groups with higher building art level in the ring road region formed by Yangtze River Bridge, Wuluo Road, Zhongnan Road, Xudong Road, planning Yangtze River Road Bridge, Jiefang Avenue, Wusheng Road and Hanyang Bridgehead to stand out good urban building appearance and outline. Besides, it was also imagined to lay out large-scale public buildings on the shore along rivers and lakes, mainly including: Wuhan Exhibition Hall, Wuhan Gymnasium, Hongshan Hotel, Wuhan Theater, Wuhan Shopping Mall, Wuhan Restaurant, Xinhua Cinema, Xinhua Road Gymnasium and so on.

主要公共建筑建设示意图（1950年代）（编者自绘，底图为1963年武汉市城区现状图）

Schematic Diagram for Construction of Main Public Buildings (in 1950s, drawn by the author, the base drawing is Urban Status Map of Wuhan City in 1963)

第叁章 • 主城内聚发展时期

1	2	3
4	5	6
7	8	

1 1986年武汉剧院（资料来源于武汉市档案信息网）
Wuhan Theater (1986, data source: Wuhan Archives Information Network)

2 今洪山宾馆（资料来源于武汉市档案信息网）
Hongshan Hotel nowadays (data source: Wuhan Archives Information Network)

3 今新华电影院（资料来源于武汉市档案信息网）
Xinhua Cinema (nowadays) (data source: Wuhan Archives Information Network)

4 今新华路体育馆（资料来源于武汉市档案信息网）
Xinhua Road Gymnasium (nowadays) (data source: Wuhan Archives Information Network)

5 武汉体育馆（资料来源于武汉市档案信息网）
Wuhan Gymnasium (data source: Wuhan Archives Information Network)

6 武汉饭店（资料来源于武汉市档案信息网）
Wuhan Restaurant (data source: Wuhan Archives Information Network)

7 武汉商场（资料来源于武汉市档案信息网）
Wuhan Shopping Mall (data source: Wuhan Archives Information Network)

8 1950年代武汉展览馆（资料来源于武汉市档案信息网）
Wuhan Exhibition Hall in 1950s (data source: Wuhan Archives Information Network)

6 – 配套生活区建设
Construction of Supporting Living Quarters

随着大型工业基地的建成，从"有利生产、方便生活"的角度出发，武汉市开展了大规模的配套生活区建设。为快速改善产业工人的居住条件，生活区建设采用"拿来主义"，仿照苏联的建设模式和施工技术进行建设。

"一五"计划时期为配合武钢重工业基地的建设，武钢、一冶等大型企业在青山区工厂西侧集中修建了大型生活居住区，在短时间内解决了十几万工人的居住问题，满足了当时人们的住房要求。

青山"红房子"就是这种大型生活居住区的典型代表，它始建于1950年代，是武钢最大规模的住宅区，是我国"一五"计划时期的工业文化遗产。经过近20年的建设，逐渐形成了红砖外墙、红色坡屋顶的"红房子"居住建筑群，集中分布于红钢城片和红卫路片，共16个街坊，建筑规模20万平方米，使用年限已超过50年。

"红房子"街坊按"囍"字形布局，平面上中轴对称，采取合院式院落空间，若干个街坊组合成整体，区域划分采取横平竖直的棋盘式布局，房屋排列整齐划一。居民楼沿街道走向布置，公共建筑在居住区中心，住宅在四周，周边配套学校、医院、电影院等。

"红房子"建筑为2~4层砖木结构，景观广场位于围合空间内部，每个围合空间有6个出入口，其中2个为主入口，4个为次入口。按照标准化范式排列起来的高度类型化居住小区均衡分配给职工，建筑户型基本一致，从生活上维持着人与人之间的公平。

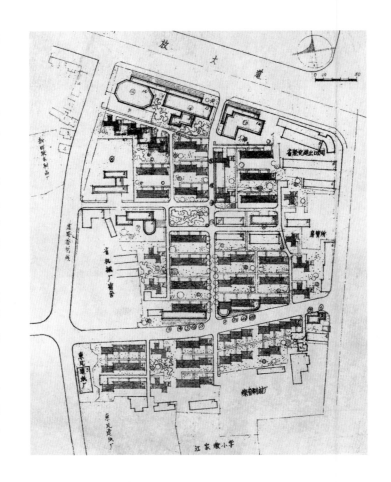

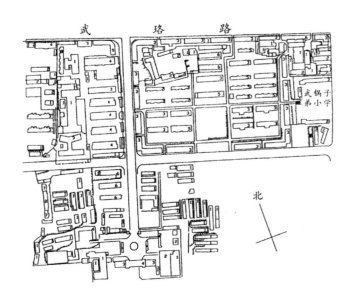

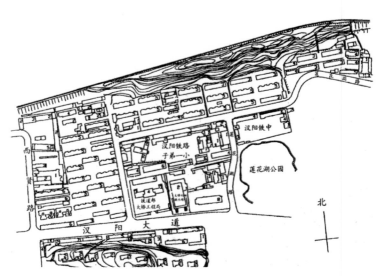

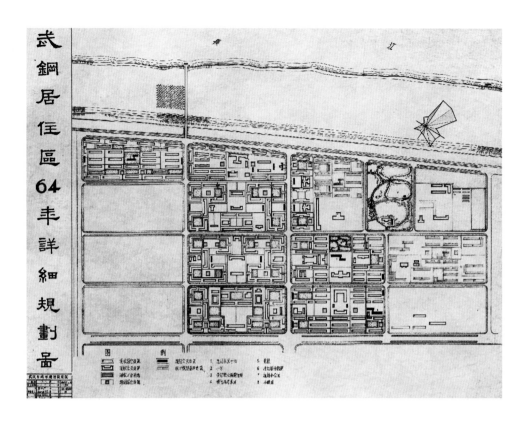

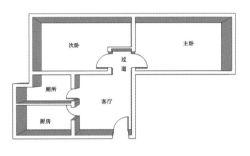
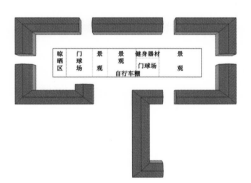

1	4
2 3	5 6

1

简易宿舍小区规划总平面图（1982年）

General Layout of District Planning in Simple Dormitaries (1982)

2

武汉锅炉厂居住区规划图（1956年）（资料来源于《武汉市城市规划志》）

Residential District Planning Map of Wuhan Boiler Factory (1956, (data source: *Urban Planning of Wuhan City*)

3

20世纪50年代中期建桥新村居住区规划图（资料来源于《武汉市城市规划志》）

Residential District Planning Map of Jianqiao New Village in the middle 1950s (data source: *Urban Planning of Wuhan City*)

4

武钢居住区详细规划图（1964年）

Detailed Planning Map of Wuhan Iron and Steel's Living Quarter (1964)

5–6

青山"红钢城"户型及平面布局图（资料来源于《艺术与设计（理论）》）

House types and plane layout of "Red Steel City" in Qingshan (Data source from *Art and Design* (*Theory*))

Upon the completion of a large-scale industrial base, to start from the angle of "being advantageous for production and convenient for life", Wuhan City has carried out large-scale construction of supporting living quarters. To rapidly improve industrial workers' living conditions, "copinism" is adopted in the construction of living quarters to follow the former Soviet Union's construction mode and construction technique in construction.

During the First Five-Year Plan Period, to coordinate with construction of Wuhan Iron and Steel Heavy Industry Base, Wuhan Iron and Steel, CFMCC and other large-scale enterprises built large-scale living quarters in a concentrated way in the west of factories in Qingshan District, which solved the housing problems of more than one hundred thousand workers and satisfied people's housing requirements then.

As a typical representative of the large-scale living quarter, commenced in 1950s, "Red House" in Qingshan was a living quarter with the largest scale in Wuhan Iron and Steel and the industrial culture heritage during the First Five-Year Plan Period in our country. Through construction for nearly 20 years, "Red House" living building group was gradually formed with red bricks, outer walls and red slope roofs and with a concentrated distribution in Red Steel City and Hongwei Road. There are a total of 16 neighbors with an architectural scale of 200,000 m² and a service life period of over 50 years.

"Red House" neighbors are laid out in a character shape of "囍" with axial symmetry on the plan. A courtyard space is adopted. Several neighbors are combined as a whole. The regional division is a checkerboard-type layout to be flat horizontally and straight vertically and houses are laid out in an orderly manner. Residential buildings are arranged along street direction, public buildings are in the center of dwelling district, residences are surrounding and schools, hospitals, cinemas and so on are matched in the surroundings.

A "Red House" building is a masonry-timber structure with 2-4 stories. Landscape square is inside the enclosed space. Each enclosed space has 6 exits and entrances including 2 primary entrances and 4 secondary entrances. Highly typed residential communities arranged according to normative paradigm are assigned uniformly to workers with basically consistent building types, maintaining justice among people in life.

二、改革时期的规划
Planning in Age of Reform

1978-1999

1978年3月，国务院在北京召开第三次全国城市工作会议，制定了关于加强城市建设工作的意见，强调了城市在国民经济发展中的重要地位和作用，提出了"控制大城市规模，合理发展中等城市，积极发展小城市"的方针，预示着中国正式进入城市化的发力阶段，开启了中国城市化高速发展的"上半场"进程。1982年，国家第一批历史文化名城名单公布，预示着以工业为中心的建城模式悄然扭转。

改革开放以来的20年，城市发展思路经历了"从小城镇到大城市"的转变。改革开放初期，乡村是重点，城市是试点，沿海地区乡村工业化的快速发展带动了一批小城镇的迅猛发展，城市化水平从17.92%增加到30.40%。这一时期，"小城镇论"占主导地位，在这种形势下，1982年版《武汉市城市总体规划（1982~2000年）》、1988年版《武汉市城市总体规划修订方案》均提出了控制三镇地区用地规模和发展空间，加速发展葛店化工城、纸坊、蔡甸等小城镇的布局设想。受此影响，1996年版《武汉市城市总体规划（1996~2020年）》提出"主城+7个重点镇"的城镇发展体系，在有控制地发展主城基础上，距离主城15~25公里左右布置了7个卫星城镇，接纳主城疏散人口和农村城市化人口。

1990年代末，随着城市土地市场改革和房地产市场的兴起，市场化和全球化的全面发展，在"开发区、新区"模式的驱动和基础设施投资的拉动下，早期乡村工业化逐渐失去了发展动力，"小城镇论"一统天下的局面开始逐渐被打破。地方政府对城市的改造实践和对城市化的推进并未按照"十五"计划的设想展开，中国城市化进程开始大大加速，城市化水平从30.40%提高到46.59%。

In March, 1978, the State Council convened the 3rd National Urban Working Conference in Beijing, in which the opinions about strengthening urban construction work were formulated, the important positions and roles of cities in national economic development were stressed and the policy "to control the scale of big cities, to reasonably develop middle cities and to positively develop small cities" was put forward, which indicated that China was formally entering into a rapid development stage of urbanization and opened the first half course of rapid development in China's urbanization. In 1982, the list of famous historic and cultural cities in the first batch was published, which indicated that the urban construction pattern centered with industry was turned over silently.

For 20 years since the reform and opening-up, the urban development thought has experienced the transformation "from small towns to large cities". At the initial stages for reform and opening-up, the village was the key point and the city was the pilot. The rapid industrialized development of villages in coastal regions drove the swift and violent development of a batch of small towns with the urbanization level increased from 17.92% to 30.40%. During this time period, the "theory of small cities and towns" played a dominant role. Under this situation, Wuhan Overall Plans of 1982 and revised edition of 1988 both proposed a layout idea of controlling scale of land use and development space of the three towns and quickening to develop Gedian Chemical Industry City, Paper Workshop, Caidian and other small cities and towns. Affected by this, Overall Planning of 1996 proposed the urban development system of "major city+7 key towns" to lay out 7 satellite towns about 15～25 kilometers away from the major city to accept population evacuated from the major city and population urbanized from rural areas based on developing the major city with control.

At the end of 1990s, rural industrialization at early stages gradually lost the motive force of development and the situation that the "theory of small cities and towns" played a dominant role started to be broken through along with the emerging urban land market reform and real estate market, the comprehensive development of marketization and globalization, the driving of patterns "development district and new district" and pulling of infrastructure investment. The local government's urban transformation practice and urbanization promotion were not carried out according to what was imagined in the "Tenth Five-Year" Plan, Chinese urbanization process started to greatly accelerate and urbanization level increased from 30.40% to 46.59%.

1 – "控制大城市、多搞小城市"引导下的三版城市总体规划

Overall Urban Planning in Three Versions Under the Guidance of "Controlling Big Cities and Developing More Small Cities"

一 恢复国民经济发展需要的武汉市城市总体规划（1982年）

1980年代，为改变城市建设完全围绕重点工业项目的固定模式，十一届三中全会提出"调整、改革、整顿、提高"的方针，大力发展"有计划的商品经济"，强调了城市在经济建设中的龙头地位，首次将满足人民生活需要放在了政府发展的核心地位上。在此背景下，当时武汉市城市规划管理局在1973～1985年总体规划方案的基础上，对过去单一发展生产城市的观点进行了纠偏，于1979～1981年编制完成了《武汉市城市总体规划（1982～2020年）》，1982年获得国务院正式批准，这也是新中国成立后武汉市首次获国家正式批准的总体规划。该规划确定武汉的城市性质是："武汉是湖北省政治、经济、科学、文化中心，是全国重要的水陆交通枢纽之一，是以冶金、机械工业为主，轻、化、纺、电子工业都具有一定规模的综合性的大城市。"

改革开放以后，大批知识青年和下放干部回城，迁移增长人口占净增人口的48.76%，形成武汉市第二次人口增长高峰。城市人口的迅速增加，使得城市生产、生活不堪重负。因此，严格控制人口、发展小城镇成为基本解决途径。到2000年，规划城区人口规模控制在280万人左右，建成区面积控制在200平方公里以内。

《武汉市城市总体规划（1982～2000年）》提出总体布局设想：一是调整三镇布局，对武昌地区留有发展余地，汉口地区要严加控制，汉阳地区尚可安排一些规模不大的工业、仓库以及生活居住设施；加强江南地区运输设施和商业服务网点的建设，使长江南北地区相对独立，各项设施分别自行配套。二是加速发展小城镇，近期建设葛店化工城、纸坊、蔡甸等工业卫星城镇，远期发展金口、新沟以及黄陂、孝感、咸宁等区外工业城镇。同时，该规划进一步明确了近郊已有的易家墩、唐家墩、堤角、白沙洲、石牌岭、中北路、余家头、青山、关山、鹦鹉洲、庙山、七里庙等12个工业区的性质和建设原则。

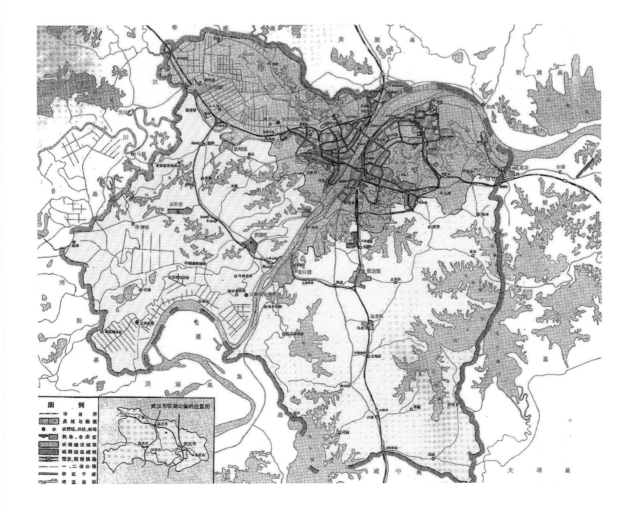

武汉市工业城镇分布规划图（1982年）

Distributed Planning Map of Industrial Cities and Towns in Wuhan City (1982)

116 CHAPTER THREE • Cohesion of the Main City

武汉市城市总体规划图
（1982年）

Overall Urban Planning Map in Wuhan City (1982)

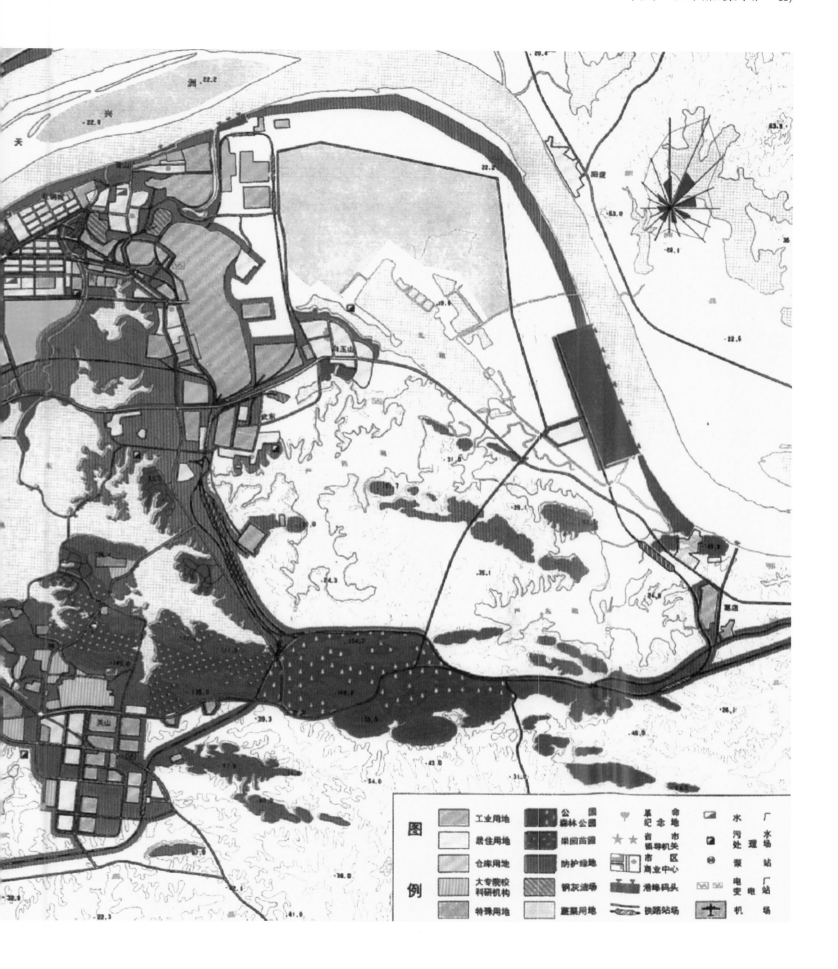

— **Wuhan Urban Master Planning required to recover national economic development (1982)**

In 1980s, to change the fixed pattern that urban construction completely surrounded key industrial projects, the 3rd Plenary Session of the 11th Central Committee of CPC put forward the policy of "adjustment, reform, reorganization and enhancement" and vigorously developed "planned commodity economy", which emphasized the leading position of cities in economic construction and put meeting people's living needs in the core position of government development for the first time. Under this background, based on the overall planning schemes during 1973 to 1985, Wuhan Urban Planning Administration Bureau corrected errors for viewpoints to develop production cities in a single way in the past and compiled and completed Wuhan Overall Urban Planning during 1979 to 1981 and gained an official approval of the State Council in 1982, which was also an overall plan of Wuhan City to be approved officially for the first time by the country after the founding of P.R.China. In the plan, the designated function of Wuhan City is confirmed: "Wuhan is the political, economic, science and cultural center of Hubei Province, one of important water and land transportation junctions nationwide, a comprehensive metropolis primary with metallurgy and machinery industry and light industry, chemical industry, textile industry, and electronic industry in a certain scale."

After the reform and opening-up, a large number of educated youths and cadres transferred to do manual labor in the countryside returned to the city and the population with growth from migration occupied 48.76% of net increase of population, which formed the second peak of population development in Wuhan City. The rapid increase of urban population enabled urban production and life to be overwhelmed. Therefore, strictly controlling population and developing small cities and towns became a basic solution. In 2000, the population size of planned urban areas was controlled to be about 2.8 million and the construction land area was controlled within 200 square kilometers.

The imagined entire distribution was proposed in the plan: firstly, it is required to adjust the layout of three towns, to leave room for development in Wuchang Region and to strictly control Hankou Region, to arrange some industries, warehouses and life housing facilities with small scale in Hanyang Region. It is necessary to strengthen the construction of transport facilities and business service branches in the south region of Yangtze River and make the north region and the south region of Yangtze River relatively independent. All the facilities were respectively matched by themselves. Secondly, it is required to quicken to develop small cities and towns, to build Gedian Chemical City, Paper Workshop, Caidian and other industrial cities and towns in near future. Meanwhile, it was further made clear the properties and construction principles of 12 industrial districts including Yijiadun, Tangjiadun, Dijiao, Baishazhou, Shipailing, Zhongbeilu, Yujiatou, Qingshan, Guanshan, Yingwuzhou, Miaoshan and Qilimiao in suburbs in the plan.

1 | 2
  | 3

1
武汉市城市现状图
（1982年）
Urban State Map of Wuhan City (1982)

2
武汉市园林绿化规划图
（1982年）
Landscape Planning Map of Wuhan City (1982)

3
武汉市对外交通规划图
（1982年）
External Transportation Planning Map of Wuhan City (1982)

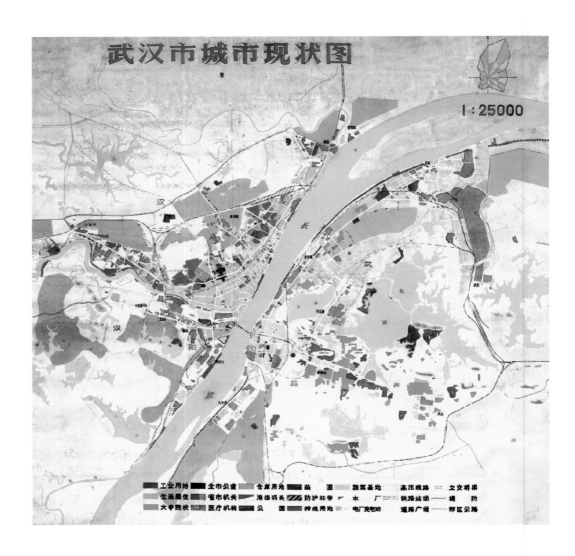

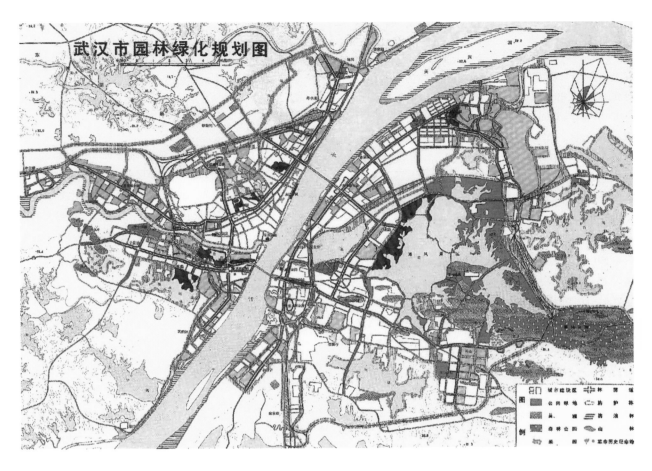

一 "两通突破"战略引导的武汉市城市总体规划修订方案（1988年）

与1982年版《武汉市城市总体规划（1982~2000年）》仍然具有较强的计划体制特点相比，1988年版《武汉市城市总体规划修订方案》是一次在市场经济体制下的重大突破。1980年代，随着黄陂县和新洲县划归武汉市行政辖区，武汉市被确定为国家综合改革试点城市后，为实现"对外开放，对内搞活"的改革方针，市委、市政府提出了武汉市经济体制改革应以"交通"和"流通"为突破口，把武汉建成"内联华中、外通海洋的经济中心"。1988年，根据国务院转批的《武汉市经济体制改革试点实施方案》中"要抓紧修订武汉市城市总体规划"的要求，武汉市编制了《武汉市城市总体规划修订方案》。根据该修订方案，确定武汉的城市性质为："武汉市是湖北省会，是座富有革命传统的历史文化名城，是全国重要的水陆空交通枢纽、通信中心和对外通商港口；是我国重要的钢铁、机械、轻纺、化工、电子等传统工业生产基地，并将逐步发展为我国光纤、微电子、激光、生物工程、新材料等新兴产业基地之一，在改革开放中将形成华中地区和长江中游的商业、贸易、金融、科技、文教、信息中心。

该修订方案还预测，到2000年全市人口规模控制在740万以内，中心城人口规模控制在350万人左右，建成区面积控制在245平方公里以内，城市规划区面积约为650平方公里。

在市域规划布局中，该修订方案提出中心城、卫星城、县城关镇、县辖镇、集镇五个层次的城镇网络。整个城镇空间布局将长江和京广铁路呈十字轴线展开，以中心城为轴心，汉丹、武九铁路、国道、汉江为放射线，在轴线和放射线上分布不同功能、不同层次和规模的城镇群。同时，对市域绿化、风景旅游区和蔬菜与副食品生产基地进行规划。

汉口布局商业贸易、金融信息和对外交通，汉阳布局旅游和涉外设施，武昌布局教育科研和新型产业，青山以钢铁工业为主，并提出建设武汉经济技术开发区和阳逻工业港。

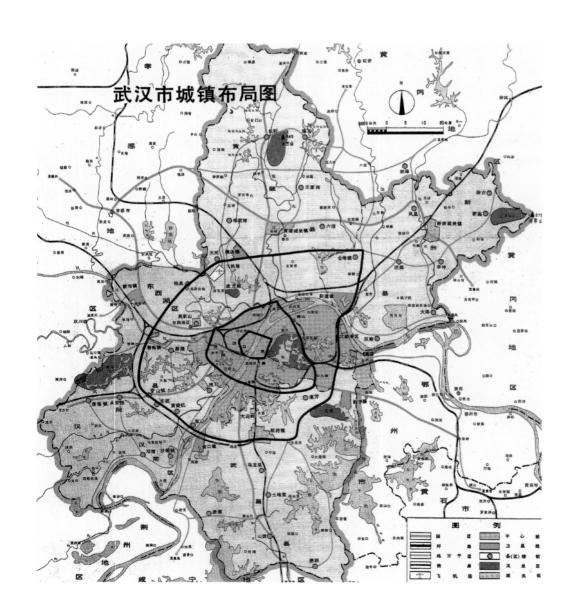

1

武汉市城镇布局图
（1988年）

Layout of Cities and Towns in Wuhan City (1988)

2

武汉市城区规划结构示意图（1988年）

Structure Diagram in Urban Planning of Wuhan City (1988)

— Revised Proposal of Wuhan Urban Overall Planning under the Guidance of the Strategy "Circulation and Traffic Breakthrough" (1988)

The Overall City Planning in 1982 still had strong planning system features while the overall city planning was a major breakthrough under market economic system. In 1980s, after Huangpi County and Xinzhou County was put under administration of Wuhan City and Wuhan City was confirmed to be a national pilot city for national comprehensive reform, it was proposed that the reform of the economic system in Wuhan City should set "traffic" and "circulation" as breakthroughs and it was necessary to build Wuhan as "an economic center to connect China with seas and oceans" to realize the reform policy of "opening to the outside world and enlivening the domestic economy". In 1988, Wuhan City compiled Revised Proposal of Wuhan Urban Overall Planning according to the requirement that "it is required to seize the time to revise the Wuhan Urban Overall Planning" in the Pilot Implementation Plan of Economic System Reform in Wuhan City transmitted and approved by the State Council. According to the revised proposal, the designated function of Wuhan City was confirmed to be: "as the provincial capital of Hubei Province, Wuhan City is a famous historic and cultural city full of revolutionary tradition and an important transportation junction, communication center and external commercial intercourse port by sea, land and air; it is an important production base in steel and iron, machinery, light texture, chemical engineering, electronics and other traditional industries and will gradually develop into one of emerging industry bases in optical fiber, micro-electronics, laser, bioengineering, new materials and other industries. It will form a commercial, trade, finance, science and technology, culture and education and information center in Central China and Middle Reaches of Yangtze River in the reform and opening up.

It is estimated that by 2000, the population size of the whole city will be controlled within 7.4 million, the population size of the central city will be controlled about 3.5 million, the construction land area will be controlled within 245 square kilometers and the area of city planning areas will be controlled about 650 square kilometers."

In the urban planning layout, the urban network was proposed for five levels including central city, satellite city, county town, town under administration of county and market town. The whole spatial layout of cities and towns spreads the Yangtze River and the Beijing-Guangzhou Railway in a cross-shaped axis centering with the central city and in radioactive rays with Handan, Wuchang-Jiujiang Railway, National Road and Hanjiang. The town clusters with different functions, levels and scales are distributed along axles and radioactive rays. Meanwhile, plans are made for urban greening, resort district and production bases of vegetables and non-staple food.

It is required to lay out commercial trade, finance information and external traffic in Hankou, tourism and facilities about foreign affairs in Hanyang, education and scientific research and emerging industries in Wuchang and primarily steel and iron industry in Qingshan. Wuhan Economic Technological Development Zone and Yangluo Industrial Port is put forward to be built in Qingshan.

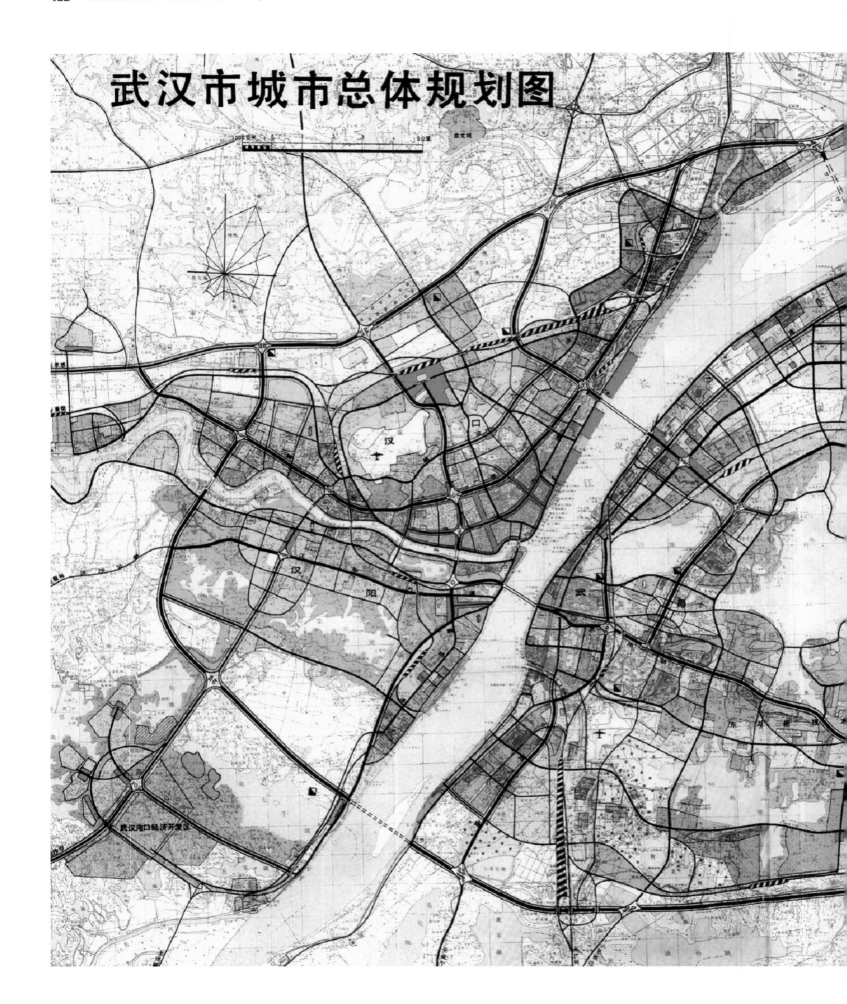

第叁章 · 主城内聚发展时期

武汉市城市总体规划图
（1988 年）

Overall Urban Planning Map
of Wuhan City (1988)

1
武汉市城区园林绿化规划图（1988年）
Wuhan Urban Landscape Planning Map (1988)

2
武汉市城区道路系统规划图（1988年）
Wuhan Urban Road System Planning Map (1988)

3
武汉市城市总体规划·主城园林绿地系统规划图（1996年）
Wuhan Urban Master Planning: Major City Landscape and Green System Planning Map (1996)

一 "四城雄踞、三区崛起"战略影响的武汉市城市总体规划（1996年）

1990年代，武汉相继被批准为对外开放城市、开放港口，标志着武汉进入了新的重要发展阶段，随即把建设四城雄踞（钢铁城、商业城、科技城、汽车城）、三区崛起（东湖新技术开发区、武汉经济技术开发区、阳逻经济开发区）、两通发达（交通、流通）的现代化国际性城市作为目标。

随着"小城镇论"一统天下的局面开始逐渐被打破，内聚式城市发展满足不了武汉远郊区的发展需求。在城市经济实力得到极大提升，可以开展大规模的基础设施建设的情形下，外围新城的规模化建设得到重视和关注。为推进市域一体化发展，1996年编制完成的《武汉市城市总体规划（1996~2020年）》，于1999年获国务院批复，构建了"圈层+轴向"的空间结构和"主城+7个卫星镇"的市域城镇发展体系，成为市域范围第一次全覆盖的空间规划。在国务院的批复中，提出了建设"生态环境良好、社会高度文明，并具有滨江、滨湖城市特色的现代化城市"的发展目标，这也为1998年建设山水园林城市提供了目标指引。

该规划不再争论"是建工业城市，还是消费城市，而是强调提升城市综合竞争力"，因此将武汉的城市性质确定为湖北省省会，我国中部重要的中心城市，全国重要的工业基地和交通、通信枢纽。规划到2020年，武汉主城常住人口为505万，建设用地面积427平方公里。

同时，该规划强调从区域范围内重新组织城市空间结构，采纳"花园城市"理论，在城市地区进行集约发展，形成有机统一的城市总体功能格局，构建了"主城+7个卫星镇"的城镇体系。在主城区构建了"圈层+轴向"的空间结构，采取了"环形+放射"的交通模式和"多中心、组团式"的布局结构，布局了武昌、汉口2个核心区、10个中心区片和10个综合组团，突出了"两江交汇、三镇鼎立"的地理格局。

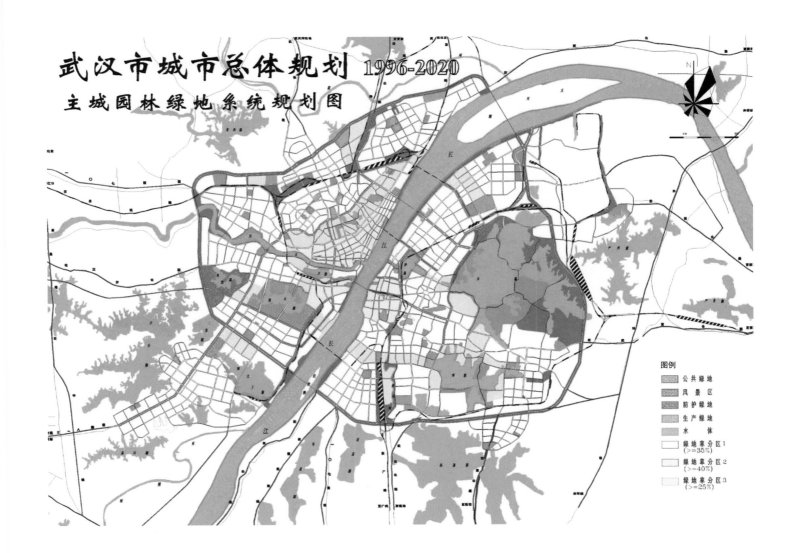

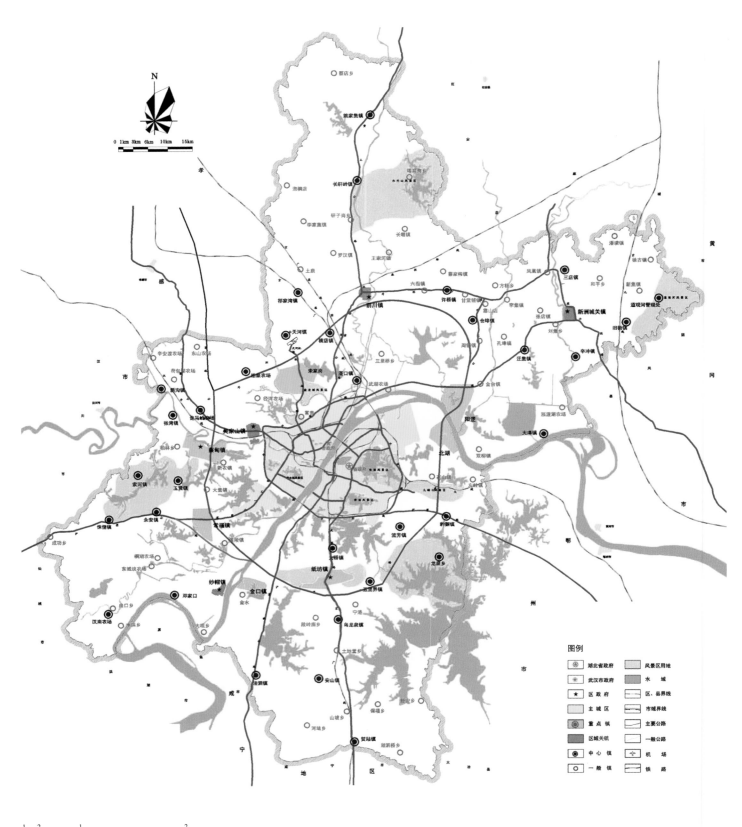

<u>1</u> <u>2</u>

<u>1</u>
武汉市城市总体规划·城镇体系规划图（1996年）

Wuhan Urban Master Planning: Urban System Planning Map (1996)

<u>2</u>
武汉市城市总体规划·主城历史文化名城保护规划图（1996年）

Wuhan Urban Master Planning: Major City Historic Cultural City Conservation Planning Map (1996)

— Wuhan Overall Urban Planning affected by strategy of "Four Cities Being Prominently Located and Three Districts Rising Sharply" (1996)

In 1990s, Wuhan was successively approved as a city opening to the outside world and an open port, which marked Wuhan to enter into a new important development stage. Then, it set building a modernized international city with four aims (steel and iron city, commercial city, science and technology city and motor city) being prominently located, three districts (Donghu New Technology Development District, Wuhan Economic and Technological Development District, and Yangluo Economic Development District) rising sharply, and traffic and circulation well-developed as a goal. After the situation that the "theory of small cities and towns" played a dominant role was broken, cohesive urban development could not satisfy the development demand of outer suburban district in Wuhan. Under the circumstance that the urban economic strength is greatly promoted, infrastructure construction in a large scale can be carried out and the large-scale construction of new cities in the surroundings draws close attention. To promote urban integrative development, Wuhan Overall Urban Planning (1996–2020) compiled in 1996 was approved by the State Council in 1999, which built spatial structure "circle layer + axis direction" and urban town development system of "major city + 7 satellite towns" and became a spatial planning for the first time with overall coverage in the urban scope. In the written approval of the State Council, a development goal to build "a modernized city with good ecological environment, highly civilized society and rivershore and lakeshore city features" was put forward, which also provided a target guideline for building a landscape garden city in 1998.

The plan no longer argued "to build an industrial city or a consumption city but emphasized to promote city comprehensive competitiveness", so the designated function of Wuhan City was confirmed to be the provincial capital of Hubei Province, an important central city in Central China and an important industrial base and an important traffic and communication hub nationwide. It is planned that by 2020, the permanent resident population of the major city in Wuhan will be 5.05 million and the construction land area will be 427 square kilometers.

The plan emphasizes reorganizing urban spatial structure within the regional scope, adopting the theory of "garden city", carrying out intensive development in urban areas, forming an organic and unified urban overall functional pattern, and constructing the town system of "major city + 7 satellite towns". In the main urban area, the spatial structure of "circle layer + axis direction" is constructed, the traffic mode of "ring + radial" and the layout structure of "multi-center and group" are adopted and two core areas, and 10 central areas and 10 comprehensive groups are arranged in Wuchang and Hankou, highlighting the geographical pattern of "two rivers crossing and three towns standing side by side".

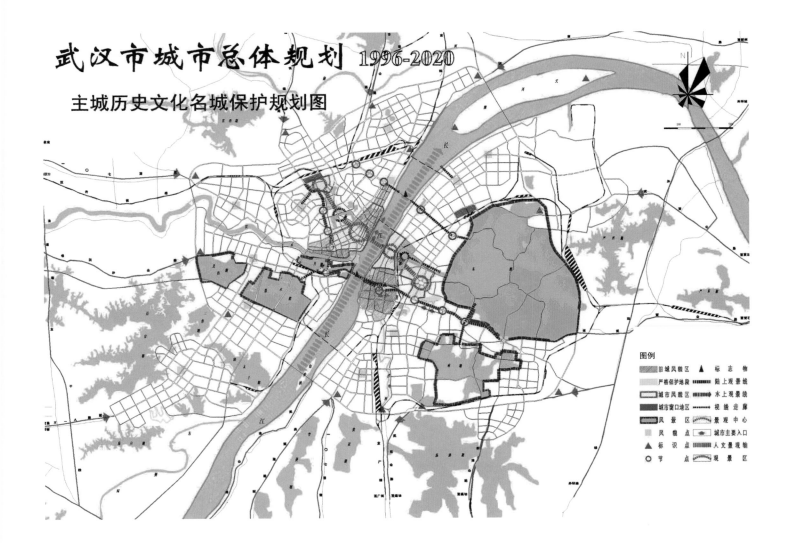

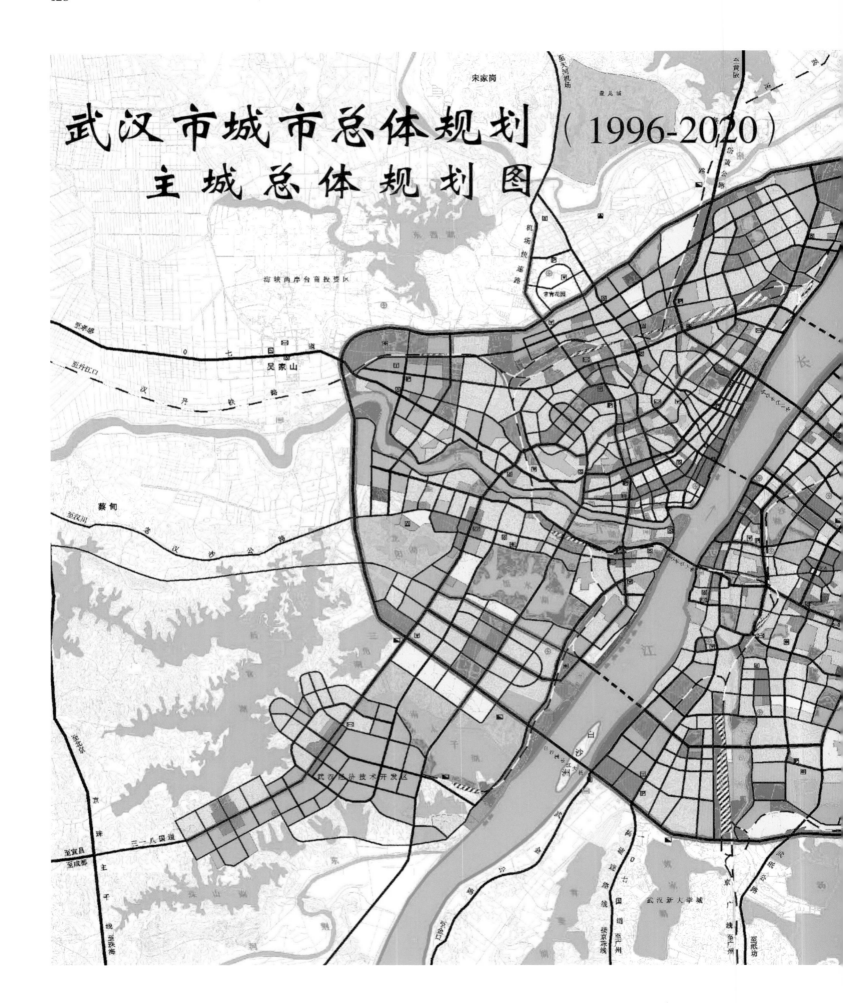

3
武汉市城市总体规划·主城总体规划图（1996年）

Wuhan Urban Master Planning: Major City Overall Planning Map (1996)

2 – 重大交通设施建设
Construction of Major Means of Tansportation

"九省通衢"是"水运时代"武汉地理优势的最好概括。随着武汉长江大桥的贯通和我国铁路网络格局的加密（郑州、济南、长沙等铁路枢纽的兴起），横贯武汉东西方向的沿江铁路迟迟没有建成，导致武汉作为内陆最大的水陆交通枢纽地位逐步下降。在进入"航空时代"后，受制于新机场建设滞后、航空公司运营管理体制的变化，武汉的航空优势也未能充分发挥，武汉在20世纪上半叶树立的"九省通衢"的交通区位优势已逐渐减弱。

— "两通起飞"交通战略构想

1980年代，市委、市政府结合武汉地理位置适中、商业基础较好、工业物质技术基础雄厚、高等学府科研机构较集中、水电能源丰富等特点和优势，制订了"两通起飞（交通和流通）"的城市发展战略。以"交通"和"流通"为突破口，把武汉市建成内联华中、外通海洋的经济中心、商业中心和交通运输中心。

在对外交通方面，武汉以铁路、水运为主转向铁路、水运、公路、航空综合体系发展。港口方面，规划新建武汉港和汉口新客运站。铁路方面，规划完成汉口旧城区京汉铁路线外移规划，为城市外拓提供新空间，随着建设大道、发展大道、青年路相继建成，使城市得以大规模沿路向纵深腹地发展。同时，建立以武汉铁路枢纽为中心，以京广、京九、汉丹和沿江铁路为骨干的铁路交通系统，强化武汉沟通南北、承东启西的作用。机场方面，规划新建天河国际机场，使武汉地区的民用航空机场设施得到根本性改善。公路方面，规划形成以国、省道干线为主骨架，"环状+放射"的干支相连、四通八达的网络格局。城区道路交通方面，主城规划路网以江河为天然轴线，一方面强化平行和垂直江河向外拓展布置，另一方面也同步提出了构建城市"环线"的设想。

— 南湖机场、天河机场

十一届三中全会以后，城市基础设施有了较快发展。当时武汉有两个机场：其一为王家墩机场，是军用机场；其二为武昌南湖机场，供民航使用，但因长度和宽度不足，不能起降大型客机。为实现华中地区经济中心、商业中心和交通运输中心的目标，响应"两通起飞"的战略部署，武汉急需建设一处大型机场。但因场址未能统一意见，机场建设被搁置。

1984年，机场选址于华容、江夏流芳岭、东西湖矛庙集、三店等地理位置对比后，民航总局提出将场址定于黄陂丰荷山，却因该方案跑道与王家墩机场跑道距离太近、空域相互干扰未被采纳。后将跑道向西推移5公里至天河镇。1985年，国务院、中央军委批复同意场址定于黄陂天河镇。1988年版《武汉市城市总体规划修订方案》明确提出建设天河机场。1990年机场动工，1995年初机场一期完工，同年4月中旬投入运行，武汉民用航空机场设施得到根本性改善。

— 汉口京汉铁路外迁（含1992年轻轨规划）

京汉铁路的建设促成了武汉半个多世纪的繁荣。随着城市的进一步发展，京汉铁路对城市的分隔、沿线居民的生活以及解放大道和中山大道间的交通联系等造成了较大干扰，1954年版《武汉市总体规划方案》、1959年版《武汉市城市建设规划》和1982年版《武汉市城市总体规划（1982~2000年）》均提出汉口旧有铁路的外迁问题。1985年，铁道部和武汉市政府以[1985]铁计字784号、武政文[1985]144号文《关于汉口客站外迁工程设计任务书的报告》报请国家计委审批，报告要求"七五"期间实现汉口客运站外迁工程，得到了国家肯定。

根据《汉口旧城区京汉铁路外移规划》（1985年），将京汉铁路丹水池至太平洋段外迁，新线沿规划的发展大道以北敷设，整体绕过原王家墩机场后向南接铁路汉江桥，在金家墩附近设新汉口火车站，同步配套建设汉口车辆整备所和江岸西编组站。铁路隔离解除后，形成了宽约200米、长近10公里的狭长地带，总用地面积约4.8平方公里。为此，武汉市组织编制了《汉口旧铁路沿线地区用地（调整）规划》（1992年），形成"一线串五组团"的总体结构。"一线"即利用现有旧铁路线改造为城市轻轨交通线，"五组团"分别为永清居住及办公组团、老车站商贸组团、展览馆博览展销组团、体育馆娱乐体育组团和太平洋工业仓储组团。该规划指导了原京汉铁路地区的城市更新，在后续城市轻轨建设及沿线地块开发建设中得到了落实，对于汉口核心区交通、综合环境、居住条件的改善，起到了重要的促进作用。

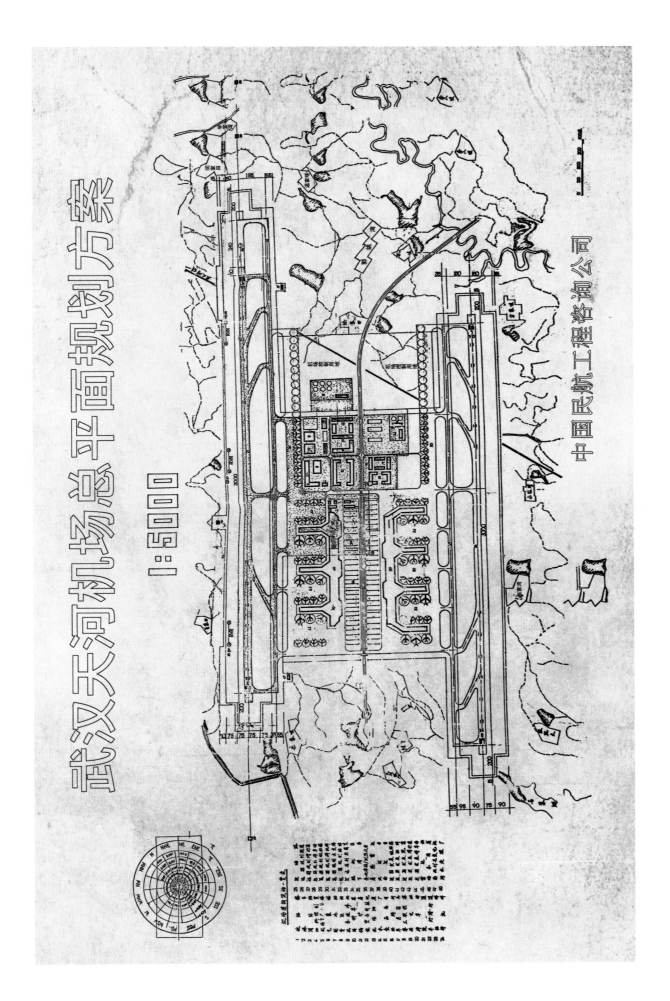

武汉天河机场总体规划平面图

Wuhan Tianhe Airport Master Plan

"Wuhan has the convenient transportation to reach the outside" is the best summary of Wuhan's geographical advantages in the "Era of Waterway Transportation". With the completion of the Wuhan Yangtze Great Bridge and the densification of China's railway network pattern(the rise of railway hubs in Zhengzhou, Jinan and Changsha and others), the delayed completion of the east-west Yangtze River railway across Wuhan has led to the gradual decline of Wuhan's status as the largest inland land and water transportation hub. After entering the "Era of Aviation", Wuhan's advantage in aviation has not been fully brought into play due to the lag of new airport construction and the changes in airline operation and management system. Wuhan's advantage in traffic location of "Wuhan has the convenient transportation to reach the outside" established in the first half of the 20th century has gradually diminished.

— Traffic Strategic Conception of "Take-off in Transportation and Circulation"

In the 1980s, the Municipal Party Committee and the Municipal Government formulated an urban development strategy of "Take-off in (Transportation and Circulation)" based on Wuhan's characteristics and advantages such as moderate geographical location, better commercial basis, strong industrial, material and technological foundation, concentrated scientific research institutions in institutions of higher learning, and abundant hydroelectric energy, taking the transportation and circulation as the breakthrough and forging Wuhan into a center of economy, commerce and transportation that connects not only China but also foreign countries.

In terms of external transportation, railway and water transportation are changed into a comprehensive system of railway, water transportation, highway and aviation. For ports, new Wuhan Port and Hankou Passenger Terminal are built as planned. In view of railway, the planned relocation of the Beijing–Hankou Railway in Hankou's old urban area will provide new space for the expansion of the city. With the completion of the Jianshe Avenue, Development Avenue and Qingnian Road, the city will be able to develop in depth and breadth along the way on a large scale. Meanwhile, a railway transportation system with Wuhan Railway Terminal as the center and Beijing-Guangzhou Railway, Beijing-Kowloon Railway, Hankou-Danjiang Railway and railways along the Yangtze River as the backbone will be established to strengthen Wuhan's role in connecting the North and South and linking the East to the West. As for airports, the Tianhe International Airport is planned for construction so as to fundamentally improve the facilities of civil aviation airports in Wuhan. As regards highways, it is planned to form a network pattern with national and provincial trunk lines as the main framework and "Ring Plus Radiation" trunk and branch lines connected and extending in all directions. In regard to urban road traffic, the planned road network of the main city takes rivers as its natural axis. On one hand, it strengthens the outward expansion and arrangement of parallel and vertical rivers, and on the other hand, it simultaneously puts forward the idea of construction of an urban "loop".

— Nanhu Airport and Tianhe Airport

After the Third Plenary Session of the 11th Central Committee of the Chinese Communist Party, the urban infrastructure has been developed rapidly. At that time, there were two airports in Wuhan: Wangjiadun Airport was a military airport and Nanhu Airport in Wuchang, used by civil aviation, which was not available for takeoff and landing of large passenger planes due to its insufficient length and width. In order to realize the goal of center of economy, commerce and transportation in central China and to respond to the strategic deployment of "Take-off in Transportation and Circulation", Wuhan urgently needs to build a large airport. However, the construction of the airport was put on hold because the consensus wasn't reached on the site.

In 1984, after comparing the site selection of the airport with the geographic positions of Huarong County, Liufangling Mountain in Jiangxia District, Maomiaoji in Dongxihu District and Sandian Street, the Civil Aviation Administration of China proposed to locate the site at Fenghe Mountain in Huangpi District, but it was not adopted because the runway of the plan was too close to that of Wangjiadun Airport and the airspace interference between each other. Later, the runway was moved to the west by 5 kilometers to Tianhe Town. In 1985, the State Council and the Central Military Commission approved the site to be located in Tianhe Town in Huangpi District. It was clearly proposed in the 1988 edition of the *Wuhan Urban Overall Planning* (Revised Version) that the construction of Tianhe Airport. The airport was started in 1990, the first phase of the airport was completed in early 1995, and it was put into operation in mid-April of the same year. The facilities in civil aviation airports in Wuhan areas were fundamentally improved.

— External Relocation of Beijing-Hankou Railway in Hankou (1992 Light Rail Plan included)

The construction of Beijing-Hankou Railway has contributed to the prosperity of Wuhan for more than half a century. With the further development of the city, the Beijing-Hankou Railway has caused great interference to the separation of the city, the lives of the residents along the railway, and the traffic links between Jiefang Avenue and Zhongshan Avenue. the 1954, 1959 and 1982 versions of Wuhan Urban Overall Planning all raised the issue of the relocation of the old railways in Hankou. In 1985, the Ministry of Railways and the Wuhan Municipal Government submitted to the State Planning Commission for approval the "Report on the Design Specification of Hankou Passenger Station Relocation Project" with documents [1985] Tie Ji Zi No.784 and [1985] Wu Zheng Wen No.144. It was required in the report that Hankou Passenger Station Relocation Project be implemented during the "Seventh Five-Year Plan", which was approved by China.

The Danshuichi-Taipingyang section of the Beijing-Hankou Railway will be relocated according to the Plan for the External Relocation of the Beijing-Hankou Railway in Hankou's Old Urban Area in Hankou. The new line is laid along the planned Development Avenue to the north, bypassing the former Wangjiadun Airport and connecting the Hanjiang Railway Bridge to the south. A new Hankou Railway Station will be set up near Jinjiadun, and Hankou Vehicle Service Station and Jiangan West Marshalling Station will be constructed simultaneously. After the railway isolation was lifted, a narrow strip with a width of about 200 meters and a length of nearly 10 kilometers was formed and total area of used land was about 4.8 square kilometers. To this end, Wuhan organized and compiled *Land Use (Adjustment) Plan for Areas along Old Railways in Hankou* to form the overall structure of "One Line Through Five Clusters". The "One Line" is to transform the existing old railway line into an urban light rail transit line. The "Five Clusters" are for Yongqing residence and office, old station trade, exhibition center, gymnasium entertainment and sports and Taipingyang industrial storage respectively. The plan guided the urban renewal of the former Beijing-Hankou Railway area and was implemented in the later urban light rail construction and the development and construction of plots along the line, which played a significant role in promoting the improvement of traffic, comprehensive environment and living conditions in core areas of Hankou.

第叁章·主城内聚发展时期　133

1/2

1
武汉铁路枢纽总布置图
(1985年，资料来源于《武汉市城市规划志》)

Wuhan Railway Hub
General Arrangement (1985)
(data source: *Urban Planning of Wuhan City*)

2
汉口车站用地规划图
(1985年)

Hankou Railway Station
Land Use Plan (1985)

一　长江二桥

兴建武汉长江二桥的意向由来已久，早在《武汉市城市建设规划（1959年）》中便有建设武汉长江二桥的宏伟蓝图。1990年代以来，市内机动车过江需求增大，过境货车大量涌入，长江大桥的运量和运输能力不能担负运输任务的增长，难以满足跨江交通发展的需要，"三镇交通一线牵"的交通格局已成为制约武汉经济社会发展的瓶颈。为此，武汉市政府果断决策，通过自筹资金和引进外资的方式，建设武汉长江二桥，1982年国务院批准《武汉市总体规划（1982~2000年）》，同意武汉长江二桥建设规划意见。1986年，武汉长江二桥的建设列入国家"七五"计划基本建设重点项目前期工作计划。1988年国家计委在关于武汉长江二桥项目建议书复函中"同意建设武汉长江二桥"。1991年，武汉长江二桥列入国家"八五"期间重点建设项目，并提出"实用、经济、美观、先进"的建桥原则。

武汉长江二桥位于武汉长江大桥下游约7公里处，西北接汉口黄浦大街、跨越长江后向东南连武昌徐东大街，是城区规划的7条过江通道之一和内环线的重要组成部分，全长约4公里，其中正桥长1876米、宽29.7米，最大跨径为400米。武汉长江二桥是当时国内已建成跨度最大的预应力钢筋混凝土斜拉桥，是武汉第一座自主设计和自主施工的桥梁。1995年6月建成通车以来，打通了武汉市内环线"主动脉"，结束了武汉长江两岸"一线牵"的历史，使数十年来形成的以"航大线"为中心的城市发展格局发生了根本性转变，城市中心逐渐向东北转移。

1　2
3

1
武汉长江二桥鸟瞰图（1985年，现存于武汉市城建档案馆）

Aerial view of the Second Wuhan Yangtze River Bridge (1985, now stored in Wuhan Urban Construction Archives)

2
武汉长江二桥总平面绿化布置图（1985年，现存于武汉市城建档案馆）

General layout of greening of the Second Wuhan Yangtze River Bridge (1985, now stored in Wuhan Urban Construction Archives)

3
武汉长江二桥立面图（1985年，现存于武汉市城建档案馆）

Elevation of the Second Wuhan Yangtze River Bridge (1985, now stored in Wuhan Urban Construction Archives)

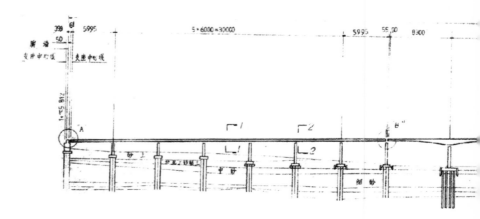

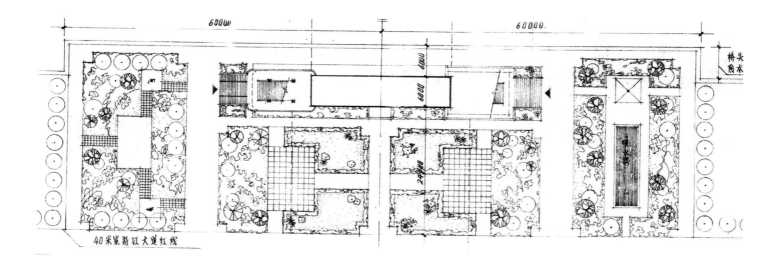

— The Second Wuhan Yangtze River Bridge

There has been a long-standing intention of construction of the Second Wuhan Yangtze River Bridge and grand blueprint for it in the Wuhan Municipal Construction Planning (1959). Since the 1990s, the urban motor vehicles' demand for crossing the river has increased, and a large number of cross-border trucks have poured in. The traffic volume and transportation capacity of the Wuhan Yangtze Great Bridge cannot meet the increase in transportation tasks and cannot meet the needs of the development of cross-river transportation. The traffic pattern of "one traffic line for three towns" has become a bottleneck restricting economic and social development of Wuhan. To this end, the Wuhan Municipal Government was decisive to build the Second Wuhan Yangtze River Bridge through self-raised funds and the introduction of foreign capital. In 1982, the State Council approved the overall planning of Wuhan City and agreed upon advice for construction and planning for the Second Wuhan Yangtze River Bridge. In 1986, the construction of the Second Wuhan Yangtze River Bridge was included in the preliminary work plan of the national key infrastructure projects of the "Seventh Five-Year Plan". In 1988, the State Planning Commission "agreed to build the Second Wuhan Yangtze River Bridge" in its reply to the proposal for the Second Wuhan Yangtze River Bridge Project. In 1991, the Second Wuhan Yangtze River Bridge was included in the national key construction projects during the Eighth Five-Year Plan and the principle of "practical, economical, beautiful and advanced" bridge construction was put forward.

The Second Wuhan Yangtze River Bridge is located about 7 kilometers downstream of the Wuhan Yangtze Great Bridge. It is connected to Huangpu Street in Hankou in the northwest and Xudong Street in Wuchang in the southeast after crossing the Yangtze River. It is one of the seven pathways crossing rivers planned in the city and an important part of the inner ring road. Its total length is about 4 kilometers, of which the main bridge is 1876 meters long, 29.7 meters wide and 400 meters in maximum span. The Second Wuhan Yangtze River Bridge was the prestressed reinforced concrete cable-stayed bridge with the longest span in China at that time, and was the first bridge designed and constructed independently by Wuhan. Since it was completed and opened to traffic in June 1995, it has opened up the "main artery" of Wuhan's inner ring road, ending the history of "one line only" on both sides of Yangtze River in Wuhan. It has made a fundamental change in the development pattern of the city centered on the "Hangda Line" formed over the past decades, and the urban center has gradually shifted to the northeast.

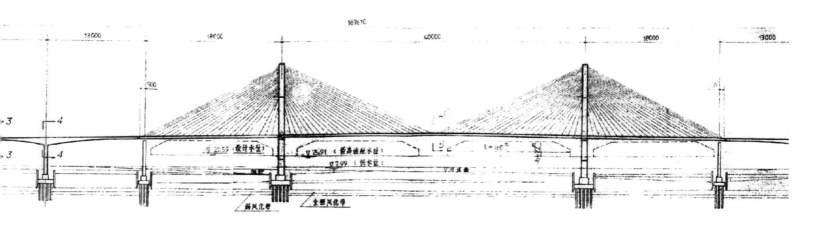

武汉市新区开发旧城改造规划分布图 1:25000

3 – 住宅建设"三部曲"

Three Stages of Residential Construction

改革开放之初的武汉城市建设是从解决欠账过多的市民住房困难开始的。按照"新区开发与旧城改造并举，商品房与保障性住房同进"的思路，全市组团式推进旧城改造和新区开发，住宅小区建设朝着成片化、多类别、综合配套、环境舒适的方向发展。这段时期，武汉市住房发展大体可以总结为"以旧城改建起步、以安居工程改善、以土地市场驱动"等三个阶段。

一 "零星式"旧城改建

1990年代以前，武汉市按照"加强维修、合理利用、适当利用、适当调整、提高水平"的方针，以改造破、危、板壁房为主，以排危、原地、原面积、见缝插针的"零星式"改造拉开旧城改造序幕。1990年共改造57处，危破率达55.9%。

1990年代以后，房地产逐渐起步，旧城改造在"零星式"基础上扩大改造范围，对棚户区、城中村和"三旧"（旧城区、旧厂房、旧村庄）采取点、线、面相结合的"成片改造"模式。通过成片改造换取了城市中心区的发展空间，拉动了房地产的发展，盘活了部分企业的存量资产，完善了旧城道路和基础设施。通过实施旧城改造，逐渐消除了城镇危房，使绝大多数住房困难的居民远离了"瓦屋竹楼千万户"的时代。

1

武汉市新区开发旧城改造规划分布图（1985年）

Map of planning and distribution of old city reconstruction and new district development in Wuhan City (1985)

2

江岸区九一、九二危改片分布图

Map of distribution of dangerous buildings necessary for reconstruction in Jiang'an District in 1991 and 1992

At the beginning of the reform and opening up, the urban construction in Wuhan started from solving the housing difficulties of citizens with excessive debts. According to the idea of "simultaneous development of new district and reconstruction of old city, and synchronism of commercial residential building and indemnificatory house". The whole city has organized clusters to promote the old city reconstruction and new district development. The residential district construction is developing in the direction of comprehensiveness, multiple categories, comprehensive matching and comfortable environment. During this period, the housing development in Wuhan can be summarized into three stages: starting with the reconstruction of the old city, improving the affordable housing project, and being driven by the land market.

— "Fragmentary" Reconstruction of Old City

Before the 1990s, Wuhan City, in accordance with the policy of "strengthened maintenance, rational utilization, proper utilization, proper adjustment, and improved standards", started the reconstruction of the old city by mainly transforming dilapidated, dangerous, and wood paneled houses, and by "fragmentary" transformation of the old city in order to eliminate the danger, in situ, in the original area, and at every opportunity. In 1990, 57 houses with a dangerous failure rate of 55.9% were renovated.

Since the 1990s, the real estate has gradually started. The reconstruction of the old city has expanded its transformation form on the basis of "fragment" and adopted a "comprehensive reconstruction" model combining dots, lines and areas for shantytowns, villages in the city and the "old towns, factories, villages". The comprehensive reconstruction has brought the development space in the urban center, promoted the development of real estate, revitalized the stock assets of some enterprises and improved the roads and infrastructure in the old city. The implementation of reconstruction of old city has gradually eliminated dangerous buildings in cities and towns so that most residents with housing difficulties can stay away from the era of "ten million households living in the tile-roofed houses and bamboo buildings".

1

江汉区九一、九二危改片
分布图

Map of distribution of
dangerous buildings necessary
for reconstruction in Jianghan
District in 1991 and 1992

2

青山区九一、九二危改片
分布图

Map of distribution of
dangerous buildings necessary
for reconstruction in
Qingshan District in 1991
and 1992

3

洪山区九一、九二危改片
分布图

Map of distribution of
dangerous buildings necessary
for reconstruction in
Hongshan District in 1991
and 1992

1

硚口区九一、九二危改片分布图

Map of distribution of dangerous buildings necessary for reconstruction in Qiaokou District in 1991 and 1992

2

汉阳区九一、九二危改片分布图

Map of distribution of dangerous buildings necessary for reconstruction in Hanyang District in 1991 and 1992

3

常青花园小区总平面规划图

The master planning map of the Evergreen Garden Residential Quarter

— "片区式"安居新区

1990 年代，为了改善居住条件和城市面貌、应对住房短缺的问题，武汉市集中规划了一批"安居工程"项目，先后编制了常青花园居住区、汉口后湖地区、汉口新火车站地区、楚雄花园等"安居工程"项目。常青花园是市政府"居者有其屋"的重要项目，是由 15 个小区组成的居住新区，规划总用地面积 266.6 公顷，规划总建筑面积 360 万平方米。2000 年 9 月竣工，迄今为全国规模最大的试点小区之一。

— "平方公里式"地产开发

2000 年以来，随着土地收购储备供应制度的建立，经营城市土地资产成为政府工作重点之一，土地一、二级市场建立并逐步完善，市场规模和成交金额逐年呈倍数增长，旧城改造规模按"平方公里"组织和推进的意识增强，开发模式从"临街剥皮"式转变为纵深发展式，从经济实用转变为营造环境、时尚舒适，实现了从量变到质变的重大转变。这一时期，建设了御江苑、金桥世家、锦绣长江、润园等一批精品地产项目。

— "Fragmentarily" New Comfortable Housing Area

In the 1990s, in order to improve the living conditions and the urban appearance and to cope with the housing shortage, Wuhan City centrally planned a number of "housing projects" and successively compiled Evergreen Garden Residential Quarter, Hankou Houhu Area, Hankou New Railway Station Area, Chuxiong Garden and others. The Evergreen Garden is an important project of the "Home Ownership Scheme" of the Municipal Government with a total planning land area of 266.6 hectares, a new residential area consisting of 15 residential quarters and a total planning construction area of 3.6 million square meters. Completed in September 2000, it is by far the largest pilot residential quarter in China.

— Development of Real Estate in Square Meters

Since 2000, with the establishment of the land acquisition reserve supply system, the management of urban land assets has become the focus of the governmental work. The primary and secondary markets for land have been established and gradually improved. The market scale and trading volume have increased by multiples year by year. The awareness of "organization and promotion" of the reconstruction of the old city according to the "square kilometers" has been enhanced. The development mode has changed from the street peeling type to the in-depth development type and the economical and practical type to the creation of environment, fashion and comfort, realizing a major change from quantitative change to qualitative one. During this period, a number of exquisite real estate projects such as Yujiang Garden, Jinqiao Aristocratic Family, Splendid Yangtze River and Runyuan were built. During this period, a number of exquisite real estate projects such as Yujiang Garden, Jinqiao Shijia, Splendid Yangtze River and Runyuan and others were built.

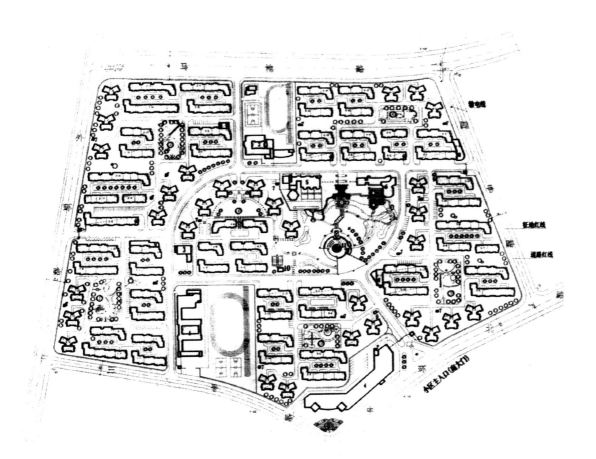

4 – 历史风貌的恢复
Restoration of Historical Features

1982年，随着国家第一批历史文化名城名单公布和《中华人民共和国文物保护法》的颁布，预示着以工业为中心的建城模式悄然扭转。按照第三次全国城市工作会议的精神，每个城市要根据地方特点，扬长避短，凸显城市特色。在这种背景下，武汉从1980年代初期开启了大规模的历史文化建筑重建和修葺工作，其中尤以黄鹤楼、晴川阁和古琴台为代表。

一 黄鹤楼重建规划（1980年）

黄鹤楼是我国重要的历史文化遗产。黄鹤楼与龟山、蛇山、长江构成了武汉城市意象中心，成为武汉市重要的城市标志，其景观形象早已渗透到民间传说、文学、诗歌、绘画、工艺美术等艺术形式之中，具有强大的生命力和深远的影响力，是武汉市历史文化名城景观与文化的精华。

黄鹤楼始建于三国吴黄武二年（公元223年），历代屡有兴废，宋代的黄鹤楼是多栋两层的楼群，明、清两代的黄鹤楼都是单栋三层的高楼。清光绪十年（1884年）黄鹤楼毁于大火，1907年在遗址上修建了奥略楼。1954年长江大桥选线于蛇山，工程线占用了黄鹤楼旧址，因此奥略楼等历史古迹也全部被拆除。1985年1月，为顺应广大群众愿望和要求，政府决定在蛇山中部高地，参照宋、清两代式样重建黄鹤楼。黄鹤楼公园建筑群分布在黄鹄山（蛇山）的三层平台上，重建的黄鹤楼建筑群由主楼、配亭轩廊、牌坊等组成，主楼耸立于第三平台中央，共五层，高51.4米，平台采用正方折角，四面对称，即"四座如一"。

尽管黄鹤楼重建形式并不是最佳的历史建筑保护方式，但此时期大批古迹修复不仅及时抢救了一批濒临消失的历史遗迹，还探索了文物保护的分级分类保护方式，改变了建设工业城市和"以农为纲"时代盛行的"劈山填湖""向湖泊要粮"等错误概念，山水资源和历史文化首次成为武汉城市特点纳入建设重点。

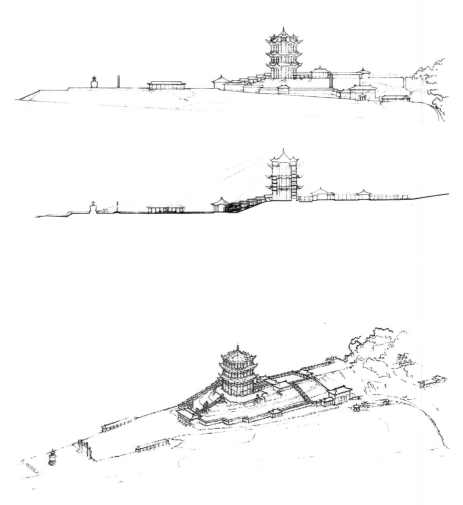

1	
2	
3	5 6 7
4	8

1
古黄鹤楼风景
Scenery of ancient Yellow Crane Tower

2–3
黄鹤楼立面图
Elevation of the Yellow Crane Tower

4
古黄鹤楼透视图（现存于武汉市城建档案馆）
The perspective view of ancient Yellow Crane Tower (now stored in Wuhan Urban Construction Archives)

5
黄鹤楼旧影
The old image of the Yellow Crane Tower

6
古黄鹤楼一角（现存于武汉市城建档案馆）
A Corner of ancient Yellow Crane Tower (now stored in Wuhan Urban Construction Archives)

7
古黄鹤楼立面图（现存于武汉市城建档案馆）
Elevation of ancient Yellow Crane Tower (now stored in Wuhan Urban Construction Archives)

8
黄鹤楼总平面布置图（现存于武汉市城建档案馆）
General layout of the Yellow Crane Tower (now stored in Wuhan Urban Construction Archives)

In 1982, with the publication of the list of the first batch of famous historical and cultural cities and the promulgation of the Law of the People's Republic of China on the Protection of Cultural Relics in China, the model of urban construction centered on industry was foreshadowed to be quietly reversed. In accordance with the spirit of the Third National Conference on Urban Work, each city should, according to its local characteristics, foster strengths and circumvent weaknesses to highlight its features. Under this background, Wuhan has started a large-scale reconstruction and repair of historical and cultural buildings since the early 1980s, especially the Yellow Crane Tower, Qingchuan Pavilion and Ancient Lute Platform.

— Reconstruction Plan of the Yellow Crane Tower (1980)

The Yellow Crane Tower is an important historical and cultural heritage in China. The Yellow Crane Tower, Turtle Hill, Snake Hill and Yangtze River form the urban image center and become the important urban symbol of Wuhan. Its landscape image has already been expressed in folklore, literature, poetry, painting, arts and crafts and other artistic forms. It has strong vitality and far-reaching influence and is the essence of landscape and culture of the famous historic and cultural city, Wuhan.

The Yellow Crane Tower was built in the second year of Wu and Huang Wu's reign in the Three Kingdoms (223). It has been frequently used and abandoned in past dynasties. The Yellow Crane Tower in the Song Dynasty is a multiple two-storied complex, and in the Ming and Qing Dynasties it was a single three-storied building. In the 10th year of the reign of Emperor Guangxu of the Qing Dynasty (1884) it was destroyed by fire. In 1907, the Aolue Building was built on the site of the Yellow Crane Tower. In 1954, the Wuhan Yangtze Great Bridge was located in the Snake Hill. The boundary line of the project occupied the former site of the Yellow Crane Tower, so the Aolue Tower and other historical sites were all demolished. In January 1985, in order to meet the demands and wishes of the public, the government decided to rebuild the Yellow Crane Tower in the central highland of the Snake Hill by following the pattern of the Song and Qing dynasties. The Yellow Crane Tower Park Complex is distributed on the three-storey platform of the Yellow Pigeon Mountain (Snake Hill). The reconstructed Yellow Crane Tower Complex consists of the main building, the supporting pavilion, porch and corridor, the memorial archway and so on. The main building stands in the center of the third platform with a total of five floors and a height of 51.4 meters. The platform is adopted with square corners and is called "four in one".

Although the reconstruction of the Yellow Crane Tower is not the best way to protect historical buildings, a large number of historic sites were restored at this time, not only to rescue a number of historical sites that are on the verge of extinction, but also to explore the methods for classification and protection of cultural relics, changing the erroneous ideas of "splitting mountains and filling lakes" and "feeding on lake" prevalent in the era of building an industrial city and "only growing the food with economic crop excluded". For the first time, landscape resources and historical cultures have become the focus of Wuhan's urban construction with its characteristics included.

 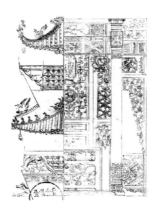 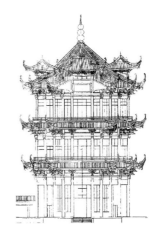

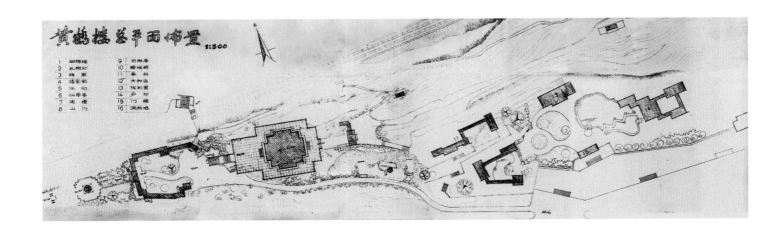

— 晴川阁重建规划（1983 年）

晴川阁位于龟山东麓功矶上，始建于明嘉靖年间（公元 1547～1549 年），以唐代崔颢题黄鹤楼"晴川历历汉阳树"的诗意命名。清咸丰年间毁于兵乱，清同治三年（公元 1864 年）汉阳郡守钟谦钧重建，为古晴川阁的最后形制。1935 年农历八月，晴川阁被大风刮倒。1983 年 12 月，武汉市政府组织在原址按原形式重建，于 1986 年 6 月落成。全部景区占地 0.8 公顷（包括禹王庙），其中，晴川阁建筑面积 600 平方米。

— 黄鹤楼视线及开敞空间保护规划（1990 年）

1980 年代末期，随着经济建设蓬勃兴起，忽略了对蛇山山体的保护，大量建筑杂乱无章地向山体发展。为控制黄鹤楼周边地区的建筑高度，保护武汉市历史文化名城中的名楼名山和旧城风貌特色，分别于 1990 年、1993 年、1996 年、2004 年、2015 年编制了五轮"黄鹤楼视线及开敞空间保护规划"。

在黄鹤楼视线保护相关规划中，1990 年版规划沿江地段仅可建 2～3 层建筑；控制范围北部看到汉口江汉关，南部看到汉阳老关，武昌地区控制范围面积约 138 公顷。1993 年版规划提高了视点和望江视线，使沿江地段可建 5～6 层建筑，司门口以北中华路地区可建 12～15 层建筑，控制范围北部可看到汉口黄浦路，南部可看到汉阳钟家村，前景地区控制范围面积约 145 公顷，此外增加了从武昌桥头堡看黄鹤楼的背景视线控制。

2004 年版规划选取的视点为黄鹤楼第三层（黄海高程 77.7 米），前景视线控制标准为看长江江心常年水位（黄海高程 17.3 米）。视线控制范围的前景为武昌以黄鹤楼为中心，临江大道、大成路口至大堤口约 100° 扇面；背景为从武昌桥头堡望黄鹤楼半径为 0～3 公里、角度为 54° 的扇面。

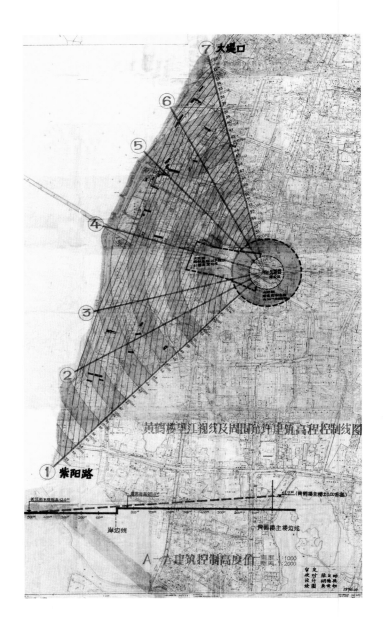

| 1 2 | 4 5 |
| 3 | |

1
民国前的晴川阁（资料来源于《武汉旧影》）
Qingchuan Pavilion before the Republic of China (data source: *Old Pictures of Wuhan*)

2
晴川阁（今）
Qingchuan Pavilion (now)

3
黄鹤楼望江视线及周围允许建筑高程控制线图（1990 年）
Map of river view of the Yellow Crane Tower and allowable building height control line nearby (1990)

4
黄鹤楼视线及开敞空间保护规划（2004 年）
Plan for Protection of Sight and Open Space of the Yellow Crane Tower (2004)

5
视廊及开敞空间分析图（2004 年）
Map of analysis on viewing corridor and open space (2004)

— Qingchuan Pavilion Reconstruction Plan (1983)

Located in the Merits Stone at the eastern foot of the Turtle Mountain, the Qingchuan Pavilion was built in the reign of Jiajing of the Ming Dynasty (1547—1549) and named after the poetry of "The clear river reflects each Hanyang tree" of The Yellow Crane Tower by Cui Hao in the Tang Dynasty. During the reign of Xianfeng in the Qing Dynasty, it was destroyed by a riot of soldiers. In the third year of reign of Tongzhi (1864), Zhong Qianjun, the County Chief of Hanyang, rebuilt it into the final form of the ancient Qingchuan Pavilion. In August of 1935, Qingchuan Pavilion was blown down by the wind. In December 1983, the Wuhan Municipal People's Government organized the reconstruction at the original site in its original form, which was completed in June 1986. All scenic spots cover an area of 0.8 hectares (including King Yu's Temple), of which Qingchuan Pavilion has a building area of 600 square meters.

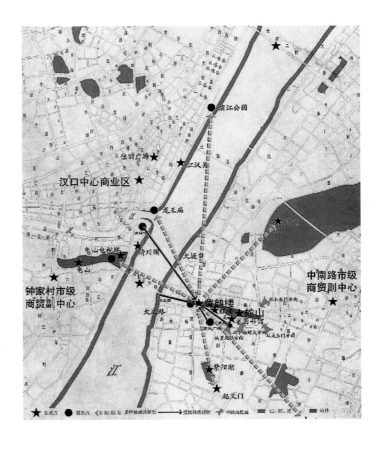

— Plan for Protection of Sight and Open Space of the Yellow Crane Tower (1990)

In the late 1980s, with the vigorous development of economic construction, the protection of Snake Mountain was neglected and a large number of buildings developed toward the mountain in a disorderly way. In order to control the building heights in the areas around the Yellow Crane Tower and protect the famous towers and mountains and features and characteristics of old cities in the famous historic and cultural city of Wuhan, five editions of "Plan for Protection of Sight and Open Space of the Yellow Crane Tower" were compiled in 1990, 1993, 1996, 2004 and 2015 respectively.

In 1990, it was planned that only 2 to 3 floors of buildings could be built along the river. The controlled areas include Jianghanguan Building in the north and Laoguan Village of Hanyang District in the south. The controlling range of Wuchang District covers an area of 138 hectares. In 1993, the Plan raised the view point and the river view line, making it possible to build 5 to 6 floors along the river and 12 to 15 floors in the area of Zhonghua Road in the north of the Simenkou. Huangpu Road in Hankou can be seen in the north and Zhongjia Village in Hanyang in the south of the controlled area. The area of the controlling range of the front view is about 145 hectares. In addition, the background sight control of the Yellow Crane Tower from the bridgehead of Wuchang was increased. The point of sight selected in the plan in 2004 was the third floor of the Yellow Crane Tower (with the Huanghai Vertical Datum of 77.7 meters), and the control standard of front view sight is the observation of the perennial water level in the center of Yangtze river (with the Huanghai Vertical Datum of 17.3 meters). The controlling range of sight: the front views are Wuchang centering the Yellow Crane Tower, the Linjiang Avenue, the Dacheng Road and the Dadikou, forming a sector of about 100°; the background is the sector with an angle of 54° and a radius of zero to three kilometers while seeing from the bridgehead of Wuchang to the Yellow Crane Tower.

5 – 重大市政工程建设
Major Municipal Engineering Construction

十一届三中全会以后，国家经济建设的加快发展对自来水事业提出了更高的要求，水厂新、改、扩建步入快车道，特别是1980年代实施的"湖改江"工程，进一步提升了城市饮用水的品质。与此同时，水环境污染和治理的压力逐步加大，渍水成灾的情况也多次发生。在这样的情况下，武汉实施了以黄孝河综合治理和东湖截污为代表的水体治污工程规划。

一　城市给水工程建设规划

1978年以前，国民经济处在恢复阶段，城市供水工程建设主要是对原有水厂生产进行整顿改造，增设供水管道，这段时期先后新、扩建了国棉一厂水厂、青山地下水水厂、老团山水厂、堤角水厂、新团山水厂、宗关水厂、白沙洲水厂、琴断口水厂等8处水厂，但总体上仍难以满足用水量增长的需求。

1980~1990年间，武汉自来水事业进入大发展时期，城市供水水源、净水厂、供水管网的建设和改造取得较快发展，日供水量由118.6万吨提升至242.7万吨，使此前的"位于长江边上无水吃"的状况得到解决，基本保证了城市供水需求。特别是1986年以来将饮用湖水改为水质较好的长江水的"湖改江"工程得以完成，改善了东湖沿线14所院校和周边居民的饮用水品质。

一　黄孝河综合治理工程

武汉地势低洼，河湖港汊交错密布，城市拓展的方向避开山水，选择地势较高的平地，因此布局较为分散，填湖造地成为就近扩大城市可建区的一条重要途径。但由于汛期江河水位高于地面，大部分用地都存在排渍防洪的压力，城市扩张的过程也是不断治水的过程。其中黄孝河综合治理工程是1980年代城市大建设时期的一项重大工程，它不仅是排渍排污的治理工程，同时也清除了汉口城区向纵深方向扩展的障碍。

黄孝河是在明清时期黄陂和孝感间的天然水道基础上随城市发展逐步改建形成的。张公堤的兴建，使黄孝河切断了清水源头，从水运渠道变成

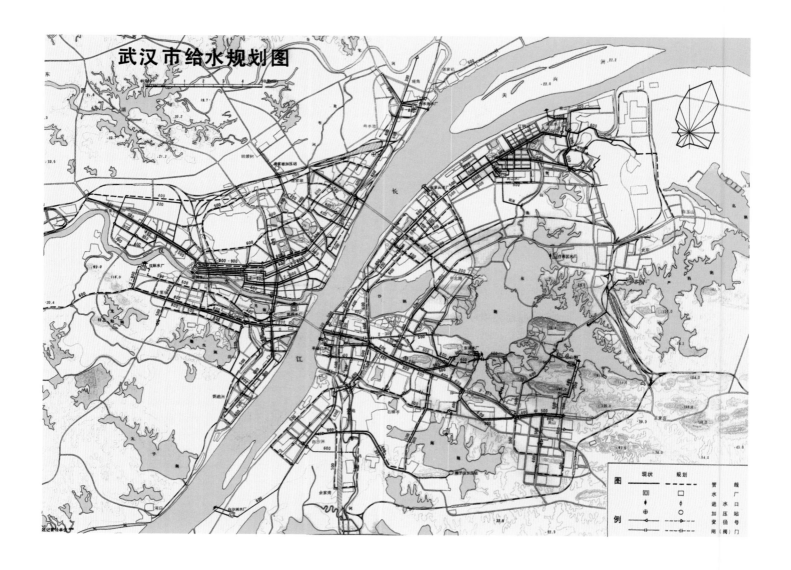

武汉市给水规划图

1 武汉市给水规划图
Wuhan water supply plan

2 武汉市湖塘水体保护用地图（1981年）
Lake Pond Water Protection in Wuhan（1981）

当时汉口市的排水渠道。随着城市建成区不断扩大，黄孝河污染环境日益严重，并多次溃水成灾。1983年10月，武汉市委市政府以"全面规划、综合治理、分段建设、逐年实施"为总方针，作出全面治理黄孝河的决定。黄孝河综合治理采取了排水、道路、居住、上水、煤气、电力、电信等工程同时进行，其中排水时采取"分区排水、分散出口"的原则，将黄孝河流域分为三大排水系统：沿江排水系统、机场河系统和新华路系统。工程自1985年10月开始，历时8年，共建成大型地下排水箱涵长达1.6万米，明渠万余米，规划还开辟了建设大道，成为贯穿汉口城区的主干道。同时还建成黄孝河路，可直通岱家山，并与黄陂县（今黄陂区）连通，成为城市的重要对外道路之一。

— 东湖截污工程建设规划

1990年白沙洲水厂扩建前，东湖一直作为饮用水源地，为武钢、青山电厂等单位提供生产用水和沿湖居民生活用水，但由于东湖流域内城市建设迅速发展，排水配套设施与城市发展不匹配，导致了东湖水质日益下降。

武汉市分别于1973年、1980年编制了《东湖水源保护及沙湖排渍规划》和《武汉市东湖水源保护规划》，按照分段、分期的原则规划敷设沿湖截污干管24公里，建设2座污水泵站、3座污水处理厂，总处理能力达13万吨/日。1984年启动了环湖截污工程，随着多年持续不断的建设，东湖流域的水质逐渐好转。

After the Third Plenary Session of the Eleventh Central Committee, the accelerated development of national economic construction has put forward higher requirements for the water industry. The government has accelerated the new-built, renovation and expansion of waterworks. The "changing lake to river" project implemented in the 1980s has greatly improved the quality of urban drinking water. At the same time, the water environmental pollution was deteriorating. The pressure of pollution control became more and more difficult. Damages caused by waterlogging also increased. Under such circumstances, Wuhan has implemented the water pollution control project planning represented by comprehensive Control of Huangxiao River and sewage interception of Donghu Lake.

— Construction Planning of Urban Water Supply Project

Before 1978, the national economy was going under recovery. Urban water supply mainly focused on the rectification and transformation of the original waterworks, building more water supply pipelines. During this period, eight water plants, including Guomian No. 1 Water Plant, Qingshan Groundwater Plant, Old Tuanshan Water Plant, Dijiao Water Plant, New Tuanshan Water Plant, Zongguan Water Plant, Baishazhou Water Plant and Qinduankou Water Plant, were successively built and expanded. But it was still difficult to meet the increasing demand of water consumption on the whole.

Between 1980 and 1990, Wuhan's water supply industry has entered a period of expansion. The construction and transformation of urban water supply sources, water purification plants and water supply pipelines have made rapid progress. The daily water supply has increased from 1.186 million tons to 2.427 million tons, which solved the situation of "having not enough water even living by the Yangtze River". The urban water supply need was basically met. The "changing lake to river" project that uses water from the Yangtze River with better quality as drinking water instead of from lakes has been finished since 1986, which improved the drinking water quality of 14 colleges and universities along the East Lake and the neighbor residents.

— Comprehensive Control Project of Huangxiao River

Wuhan has a low-lying terrain, interlaced rivers, lakes and harbors. The direction of urban development avoids mountains and rivers and prefers to locate on high-lying flat land. Therefore, the layout is rather dispersed. The major method to expand the urban areas is filling the nearby lakes to get more land. However, the higher the water level of rivers during the flood season becomes a threat to drainage and flood control in most areas. The comprehensive control project of Huangxiao River is a major project during the period of urban construction in the 1980s. It not only controlled drainage and sewage disposal, but also removed the obstacles of the urban expansion in the north and south of Hankou.

Huangxiao River was excavated on the foundation of the natural waterway between Huangpi and Xiaogan in Ming and Qing dynasties. The construction of Zhanggongdi cut off the head stream of the Qingshui River and changed Huangxiao River into the drainage channel of Hankou City. With the continuous expansion of urban built-up areas, the pollution of Huangxiao River became more and more serious. Waterlogging disasters occurred many times. In October 1983, the Municipal Party Committee and Municipal Government made a decision to comprehensively control the Huangxiao River's pollution under the general policy of "overall planning, comprehensive management, construction by stages and implementation year by year". The comprehensive control of Huangxiao River was done by the rectification and transformation of drainage, road, residential, water supply, gas, electricity and telecommunications at the same time. The drainage rectification adopted the principle of "zonal drainage and decentralized outlets", dividing the drainage system of Huangxiao River Basin into three parts: drainage system along the river, airport river system and Xinhua Road system. Since October 1985, the project has lasted for eight years. It has built a large underground drainage box-type culvert with a length of 16,000 meters and an open channel of more than 10,000 meters. The planning has also constructed Jianshe Avenue, which became the main road through Hankou urban area. At the same time, it has built Huangxiaohe Road, which directly connected with Daijiashan and Huangpi County, and became the city's exit road.

— Construction Planning of East Lake Sewage Interception Project

Before the expansion of Baishazhou Water Work in 1990, Donghu Lake had been a drinking water source for factories like Wuhan Iron and Steel Company and Qingshan Power Plant and the residents along the lake. However, the rapid development of urban construction in the Donghu Basin and the mismatch between drainage facilities and urban development led to the deterioration of the water quality of Donghu Lake.

In 1973 and 1980, Wuhan City compiled the Water Conservation and Drainage Planning of Sand Lake and Water Conservation Planning of Donghu Lake in Wuhan City. In accordance with the principle of stage by stage and period by period, 24 kilometers of sewage interception pipe along the lake were set up, and 2 sewage pumping stations and 3 sewage treatment plants were built. The total capacity of dealing with polluted water has reached 130,000 tons per day. The pollution interception project around the lake was launched in 1984. With the continuous construction for years, pollution interception in the East Lake Basin has been gradually realized, benefiting a lot of areas and residents.

6 – 山水园林城市的创建
Construction of Garden City with Mountains and Rivers

改革开放后，沿海地区通过环境美化改造引来了大量外来投资，增加了城市就业并实现了产业升级，取得了较好的建设成效。因此，一段时期以来，各地掀起了一场城市环境美化运动。

1990年，武汉曾提出"三年消灭荒山荒滩，五年绿化武汉""十年建成山水园林城市"的基本构想，其后在当时国家建设部评选"园林城市"和钱学森倡议建设"山水城市"的推动下，1997武汉正式提出"初步建成山水园林城市"的总体目标，同年底，武汉提出"大力推动环境创新、构筑城市发展的新优势"，建设具有滨江、滨湖特色的现代化城市。

— 武汉市创建山水园林城市综合规划（1998年）

1998年，武汉市第九次党代会和十届人大一次会议正式提出了"经过五年的努力，初步建成山水园林城市"的目标。在此背景下开展了规划编制工作，力争通过5年的奋斗，在城市核心区建成汉口滨江市政广场和沿长江滨水绿化风光带，形成两大现代化景观环线，整体保护山、水两条景观轴线，完善东湖、南湖、龙阳湖和汉口中部五湖4片绿心，整治10条城市主要出入道路，构成以"一点一带、两轴两环、四心十线"为主体的山水园林城市基本框架，使武汉市成为绿化系统纵横通透、城市形象特色鲜明、环境质量舒适优越、城市文化气息浓郁的现代化山水园林城市。

— 武汉市主城山体湖泊保护界定规划（1999年）

1998年长江流域发生特大洪水后，全社会逐渐意识到湖泊调蓄功能的重要性，也认识到维护河湖水系自然连通的重要性，在此情况下湖泊保护的观念空前提高，因此武汉市开展了山体湖泊保护界定规划工作。《武汉市主城山体湖泊保护界定规划》（1999年）将山体湖泊进行分级排类后，明确了保护措施和方法，如将27个湖泊和31座山体进行界桩保护，提出了"蓝线""绿线"的保护概念，实施了不同的建设限制要求。这为以后山体湖泊的系统保护打下了良好基础。对"紫线"建筑、城市广场、公园也都进行了粗略布局规划，并根据参差不齐的现状采取了修整恢复、扩大完善、新建等不同建设措施。

— 街景整治和公园开放系列规划

1999年开始，在"创建山水园林城市"的目标指导下，武汉市开展了70余项街景整治和公园开放规划，如东湖环湖景观建设，东湖风景区听涛景区开放，中山大道环境改造，江汉路步行街商业环境景观改造，中山公园开放，解放公园改造等。城市景观规划将城市空间按照点、线、面、路径、标志等要素进行整合设计，引入了商业步行街、主题广场和开放式公园等城市公共场所的概念，对打造整洁、舒适、优美的城市环境发挥了重要作用。

1
黄孝河流域排渍工程示意图（1975年）
Schematic drawing of Waterlogging Project in Huangxiao River Basin (1975)

2
武汉市创建山水园林城市综合规划·总体建设规划图（1998年）
Comprehensive Planning for the Construction of Garden City with Mountains and Rivers — Overall Construction Plan (1998)

1

武汉市创建山水园林城市综合规划·主城绿地系统规划图（1998年）

Comprehensive Planning for the Construction of Garden City with Mountains and Rivers — City Parkland System Planning Map of Downtown (1998)

2

武汉市创建山水园林城市综合规划·大气环境治理规划图（1998年）

Comprehensive Planning for the Construction of Garden City with Mountains and Rivers — Atmospheric Environment Control Plan (1998)

3

武汉市主城湖泊及山体保护界定规划·龟山山体界定图（1999年）

Definition Planning for Mountains and Lakes Protection in the Downtown of Wuhan — Turttle Mountain Definition Map (1999)

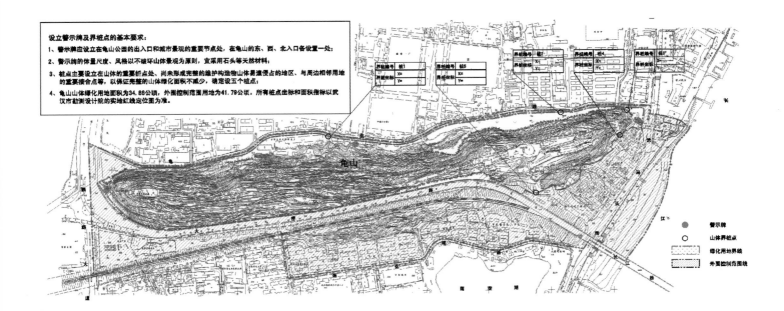

After the reform and opening up, coastal regions have attracted a lot of foreign investment through transforming and improving the environment. The urban employment has increased and the industrial upgrading has accelerated, making great achievement in development. Therefore, a movement of urban environmental beautification has been launched all over the country.

In 1990, Wuhan put forward the basic idea of "spending three years to eliminate barren hills and wastelands and five years to plant trees in Wuhan" and "build a landscape city within ten years". Then with the push of selection of landscape city by the Ministry of Construction and the proposal of "landscape city" by Qian Xuesen, Wuhan formally put forward "preliminary construction of garden city with mountains and rivers" in 1997. At the end of the same year, Wuhan proposed to "vigorously promote environmental innovation and create new advantages for urban development", aiming at building a modern city with riverside and lakeside characteristics.

— Comprehensive Planning for the Construction of Garden City with Mountains and Rivers in Wuhan (1998)

In 1998, the Ninth Party Congress and the First Session of the Tenth People's Congress of Wuhan formally put forward the goal of "building preliminarily a garden city with mountains and rivers within five years ". In this context, the work of planning has been carried out. The plan proposed to build the Hankou Riverside Municipal Plaza and the Yangtze River Riverside Greening Landscape Belt in the core area of the city in five years, forming two major modern landscape loops. It planned to protect the two landscape axes, the mountain and river from the overall picture, improving the four green core areas of Donghu, Nanhu, Longyang Lake and the five lakes in the middle of Hankou. Ten main access and exit roads of cities were to be renovated, forming the basic framework of "one point, one area, two axes, two rings, four core areas and ten lines" as the main body of a garden city. The plan aimed at making Wuhan a modern garden city with thorough green system, distinctive city feature, comfortable living environment and rich urban culture.

— Definition Planning for Mountains and Lakes Protection in the Main City of Wuhan (1999)

After the catastrophic floods in the Yangtze River Basin in 1998, the whole nation gradually realized the importance of the flood storage function of lakes as well as the importance of maintaining the natural connectivity of river and lake systems. The society has laid unprecedented attention on lake protection. Thus, the definition planning of mountain lake protection has been carried out. After classifying the mountain lakes in *Definition Planning for Mountains and Lakes Protection in the Downtown of Wuhan*, the protection measures and methods were defined. For example, 27 lakes and 31 mountain bodies were protected by boundary piles. The protection concepts of "Blue Line" and "Green Line" were put forward, and different construction restrictions were implemented. These measures have laid a good foundation for the systematic protection of mountains and lakes in the future. The layout of the "Purple Line" buildings, city squares and parks was roughly planned, and different construction measures such as renovation and restoration, expansion and improvement, and new construction were adopted based on the uneven situation.

— Streetscapes Renovation and Park Opening Series Planning

Since 1999, under the guidance of "creating a garden city with mountains and rivers", Wuhan has carried out more than 70 streetscapes renovation and park opening plans, such as the construction planning of Donghu Lake Rim, the opening of Tingtao Scenic Area of Donghu Scenic Area, the environmental renovation of Zhongshan Avenue, the environment renovation of Jianghan Road Commercial Pedestrian Street, and the opening of Zhongshan Park. Fang, renovation of Jiefang Park, etc. Urban landscape planning integrates urban space in accordance with points, lines, planes, paths and signs, introducing the concepts of public place like commercial pedestrian streets, theme squares and open parks. It plays an important role in creating a clean, comfortable and beautiful urban environment.

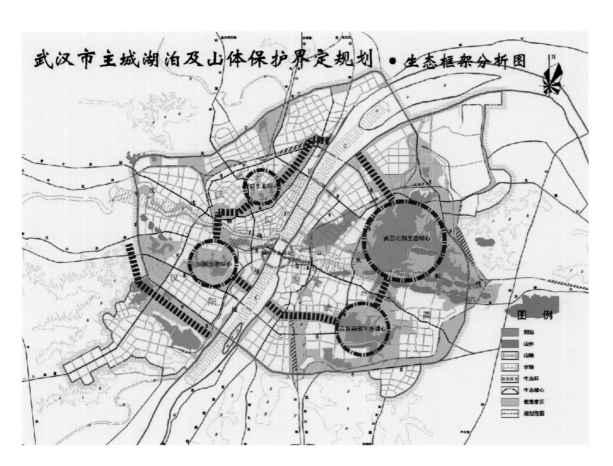

1

武汉市主城湖泊及山体保护界定规划分布图（1999年）

Definition Planning for Mountains and Lakes Protection in the Downtown of Wuhan — Distribution Map (1999)

2

武汉市主城湖泊及山体保护界定规划生态框架分析图（1999年）

Definition Planning for Mountains and Lakes Protection in the Downtown of Wuhan — Ecological Framework Analysis Map (1999)

3

江汉路步行街环境综合整治规划结构图（1999年）

Comprehensive Environmental Renovation Planning of Jianghan Road Pedestrian Street — Planning Structural Map (1999)

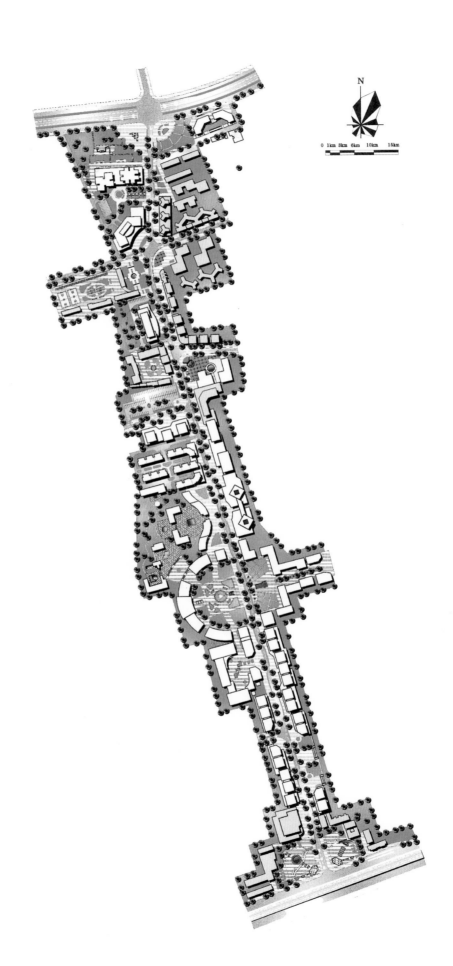

7 - 工业布局模式的变迁
The Change of Industrial Distribution Model

武汉作为传统的老工业基地，在改革开放的前二十年里，经济发展一直在"调整中发展、发展中调整"，其工业空间布局的重点经历了"十三个工业区——四城三区多园"等阶段，每一阶段与当时的经济发展战略、产业结构调整、支柱产业选择等密切相关。

一 工业区延续：十三个工业区

1980年代初期，选择一个合适的突破口来启动经济体制综合改革，是武汉市亟待解决的重大问题。为此开展了多轮经济社会发展战略研讨会：如于光远的《关于改革与战略问题》，童大林的"建立长江大流域经济新模式，武汉如何成为长江流域的特级市场"，宦乡在《武汉要发挥全国内陆最大中心城市的作用》提到的武汉在"东靠西移，南北对流"中的战略地位等，这些大讨论最终引出了把"两通"作为经济发展的战略重点，发展国内大市场，让武汉在华中地区和长江流域发挥出中心城市的应有作用。"两通起飞"战略的实施，对活跃市场起到了积极的推动作用，汉正街一跃成为国内商贸市场的排头兵。但与之相对的是，工业发展力度不够，过于强调"门类齐全、综合发展"的发展模式，规模效益无法实现，农业、轻工业、重工业结构调整受到一定影响，重工业比重仍然较高，钢铁、机械和纺织仍然是传统的支柱产业。

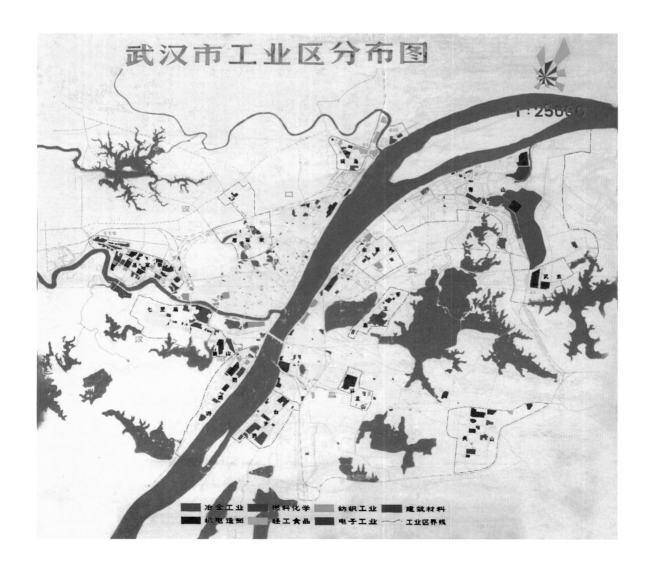

1
武汉市工业区分布图
（1978年）
Distribution of Industrial Zones in Wuhan (1978)

2
东湖新技术开发区总体规划图（1990年）
Master Plan of Donghu New Technology Development Park (1990)

3
武汉经济技术开发区总体规划图（1991年）
Master Plan of Wuhan Economic and Technological Development Park (1991)

这一期间工业布局沿袭计划经济时期的"十三个工业区"格局,主要表现以下特征:一是重心高度集聚主城,70%左右的企业数量集中在江汉、江岸等7个中心城区;二是南北展开较均衡,工业在长江南北两岸比较均衡地展开布局,北岸企业户数、人数略大于南岸;三是主城外围环状分布,在老城区外围形成了一条宽窄不等的圆环状工业密集带。

一 开发区兴起:四城三区多园

进入1990年代,武汉确定为沿江开放城市,引进外资速度加快,通过实施改革企业产权制度和推进"壮大放小"战略,工业经济持续快速增长,工业运行环境逐步优化。在此期间,武汉提出了四城雄踞(钢铁城、商业城、科技城、汽车城),三区崛起(东湖新技术开发区、武汉经济技术开发区、阳逻经济开发区)的发展战略,四城中的三城以及三区都与工业布局有关,城市发展重点又重回到工业这一经济的基石上来。产业结构发生了一些本质性的变化,呈"重快轻慢"的格局,重工业增长速度比轻工业高出16%。虽然原有的一些品牌产品已优势不再,但形成了以"钢、车、机、新"为支柱的工业发展新格局。进入21世纪,武汉又形成了光电子信息产业基地、现代制造业产业基地、钢材及新材料生产基地、生物工程和新医药产业基地、环保产业基地。

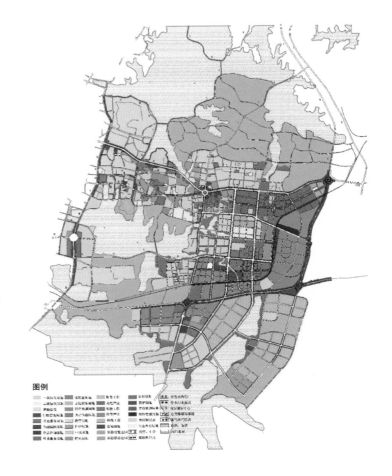

As a traditional old industrial base, in the first two decades of reform and opening-up, Wuhan's economic development has been "developing when adjusting". Its industrial spatial layout has undergone stages as "13 industrial zones-four cities, three districts and many parks". Each stage is closely related to the economic development strategy, industrial restructuring and the choice of pillar industries at that time.

— Industrial Parks Continuation: Thirteen Industrial Parks

In the early 1980s, choosing a suitable breakthrough to start the comprehensive reform of the economic system is a major problem to be solved urgently in Wuhan. Several rounds of seminars on economic and social development strategies have been held to find a proper approach, such as Yu Guangyuan's On Reform and Strategy, Tong Dalin's Establishing a New Economic Model of the Yangtze River Basin, the Way Wuhan becomes a Significant City of the Yangtze River Basin, and the strategic position of Wuhan in "connecting the eastern region to the west, linking the northern region with the south" in Huan Xiang's Wuhan Should Play the Role of the Largest Central City in the Inland of China. These discussions eventually led to make the "connection and link" as the strategic focus of economic development. Developing large domestic markets allows Wuhan to play its due role as a central city in central China and the Yangtze River Basin. The implementation of the "connection and link" strategy has played a positive role in promoting the market, and Hanzheng Street has become the leading force in the domestic business market. In contrast to the prosperous markets, the industrial development was rather insufficient. The development mode of "complete and comprehensive development" was overemphasized, resulting in the scale merit

1
制造业空间布局体系图
Manufacturing space layout system diagram

2
武钢集团汉阳钢厂用地规划图
WISCO Group Hanyang Steel Plant Land Plan

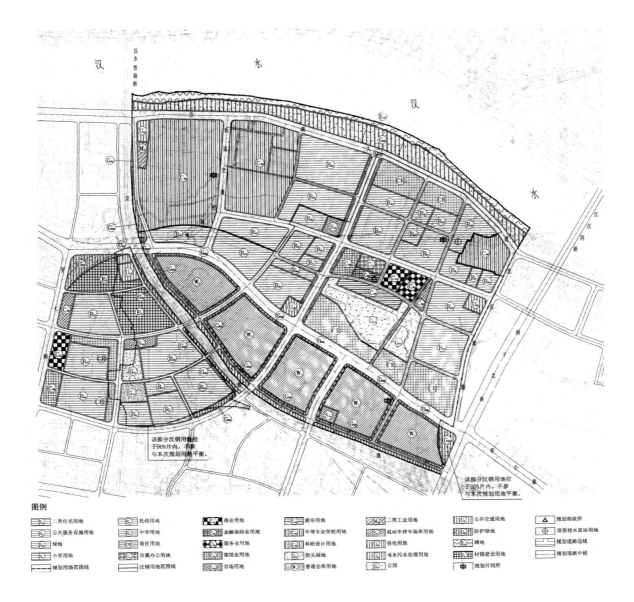

almost unable to realize. The reconstructing of agriculture, light industry and heavy industry were also affected. The heavy industry still held high proportion. Steel, machinery and textile were still the traditional pillar industries.

During this period, the industrial layout followed the plan of "thirteen industrial parks" in the planned economy period, which mainly showed the following characteristics: firstly, the focus was in the downtown. About 70% of the enterprises were in seven central cities, such as Jianghan and Jiang'an. Secondly, the expansion in the north and south was more balanced. The industrial layout was roughly equivalent on both sides of the Yangtze River. And the number of enterprises in the North Bank is slightly larger than that in the South Bank. Thirdly, there was a circular distribution of industrial parks around the downtown, forming a circular industrial belt around the old city.

— The Rise of Development Parks: Four Cities, Three Districts and Multiple Parks

In the 1990s, Wuhan was designated as an open city along the Yangtze River. The speed of introducing foreign capital was accelerated. Through the implementation of the reform of enterprise property rights system and the promotion of the strategy of "providing political supports to the larger enterprises and giving more right to minor enterprises", the industrial economy continued to grow rapidly and the industrial operating environment became more favorable. During this period, Wuhan put forward the development strategy of continually developing four functions of the city (iron and steel city, commercial city, science and technology city, automobile city) and supporting three districts (Donghu New Technology Development Park, Wuhan Economic and Technological Development Park and Yangluo Economic Development Park). Three of the four functions and the three districts were all related to the industrial layout. The focus of urban development has returned to industry, the cornerstone of the economy. Some essential changes have taken place in the industrial structure, showing a pattern of "rapid development in heavy industries and relative slow development in light industries". The growth rate of heavy industry was 16 percentage points higher than that of light industry. Although some of the original brand products no longer had advantages, Wuhan has formed a new industrial development pattern with "steel, cars, machinery, new industry" as the pillar. In the 21st century, Wuhan has developed the base of optoelectronic information industry, modern manufacturing industry, steel and new materials production base, bioengineering and new medicine industry, and environmental protection industry.

CHAPTER
第 肆 章
FOUR
2000-PRESENT

区域一体化时期的规划
Planning During Regional Integration

时代浩荡而行。改革开放40年来，中国经历了史无前例的大规模、快速城镇化进程。中国城镇化水平从1978年的17%快速提升至2018年的59.58%，特别是2011年超过了50%，宣告中国从"农村社会"进入"城市社会"，也宣告了中国城镇化进程从长期滞后于世界发达国家到迅速迎头赶上的转变。

这一时期，"小城镇、大战略"一统天下的局面开始逐渐被打破，随着城市土地市场和房地产市场的兴起、市场化和全球化的全面发展、"开发区、新区"模式的驱动、基础设施投资的拉动，早期乡村工业化逐渐失去发展动力，大城市迅速成为中国城镇化的主导空间。应该来说，中国城镇化的"上半场"以"高速"带来了"繁荣"，是典型的增长主义发展模式。由于城镇化历史进程较短，不可避免地带来"大城市病"、环境污染加大、乡村凋敝等矛盾。

与此同时，在全球化浪潮席卷的近十年，国内特大城市掀起了一场"圈、群运动"，城市之间的竞争愈发取决于城市群和大都市地区之间的竞争。而近年来全球范围出现了出口贸易骤降、工业低成本优势丧失、土地财政环境收缩、老龄化社会加速到来等内外环境的剧烈变化，"高速增长"的动力基础持续疲软，发展环境全面转入"换挡减速"。中国城镇化的"下半场"以"高速增长主义模式的终结"开启，这是武汉等后发城市面临的巨大挑战，既要解决发展体量不够大、能级辐射不够强等做强发展问题，还要解决"大城市病"、资源环境压力等转型发展难题。

在"速度城镇化"向"深度城镇化"转变过程中，在"增量规划"向"存量规划"的转型过程中，武汉城镇化的"下半场"必然转向以人为本的城镇化。如果说"上半场"是"灰色"的城镇化，那么"下半场"一定是"绿色"的城镇化。在这一过程中，以城市空间承载生活美好、以城市远见奠定世纪风貌，是武汉规划人矢志不渝的理想追求。

The tide of the times is surging ahead. Over the past four decades of reform and opening-up, China has experienced an unprecedented process of large-scale and rapid urbanization. The level of urbanization in China rose rapidly from 17% in 1978 to 59.58% in 2018, and exceeded 50% in 2011, marking that China has moved from a "rural society" to an "urban society" and was catching up fast after a long period of lagging behind the world's urbanization process.

During this period, the dominance of "small towns and grand strategies" began to break down. With the rise of urban land and real estate markets, the all-round development of marketization and globalization, the impetus of "development zones and new districts" mode and the promotion of infrastructure investment, the early rural industrialization gradually lost its momentum, and the big cities quickly became the leading space of China's urbanization. It can be said that the "first half" of China's urbanization brought "prosperity" at a "high speed", which is a typical growth model. Due to the short history of urbanization, contradictions such as "urban diseases", serious environmental pollution and rural depression inevitably arose.

Meanwhile, in the recent ten years of globalization, there has been a "circle and cluster movement" in megacities in China. The competition between cities increasingly depends on the competition between city clusters and metropolitan areas. In recent years, there have been drastic changes in the internal and external environment worldwide, such as sharp drop in export trade, loss of low-cost advantage of industry, contraction of land finance environment, and accelerated arrival of an aging society. The dynamic basis for "rapid growth" remains weak and the development environment has shifted to a "slowdown". The second half of China's urbanization started with the "end of the rapid growth model". This is a huge challenge faced by latecomers like Wuhan. These cities need to solve not only the development problems such as insufficient development volume and insufficient influence, but also the transformation and development problems such as "urban diseases" and pressure on resources and environment.

During the transition from "rapid urbanization" to "deep urbanization" and from "incremental planning" to "stock planning", Wuhan's urbanization is bound to be people-oriented in the second half. If the first half of urbanization is "grey", the second half must be "green". In this process, it is Wuhan planners' unswerving ideal to bear the beauty of life in urban space and establish the style of the century with urban vision.

一、新世纪发展期的规划
Planning in the New Century

2000 - 2011

2000 年以后是中国城市规划行业发展大放异彩的时代。

这一时期，多中心大都市区、都市圈、全球城市区域等新的空间概念，以及战略规划、概念规划等新的规划手法被相继引入，最早在珠三角、长三角等地区的中心城市率先引入，吸引规划行业和政府决策人员的高度关注。广州的"南拓"和珠江新城的大手笔规划建设、上海"一城九镇"空间发展战略以及杭州"从西湖时代到钱塘江时代"发展的成功，使得多中心都市区已经成为一种先进规划理念和技术的图腾，并被广泛地讨论和效仿。

时光回放到 2003 年的武汉，胡锦涛总书记提出湖北要成为中部崛起重要的战略支点，国家建设"两型"社会试验区已经启动；蔡甸、江夏、黄陂、新洲等四县相继"撤县改区"，城区范围扩大，相应地要扩大规划管理的范围；国家在武汉地区布局建设了 4 条城际铁路、西气东输等一系列国家性、区域性的交通和基础设施，进一步强化了武汉的区域交通枢纽地位。在经济力量的推动下，武汉主城区与规划的重点镇地区形成连片发展，圈层式空间结构已转化为轴向发展趋势。为此，武汉在刚刚跨入 21 世纪之初即开始酝酿新中国成立后第六轮城市总体规划的修编工作，并采取了战略规划先导、总体规划后发的工作方式，明确了"中部地区中心城市"的定位，提出了"都市发展区"的规划模式，确立了"1+6"的空间发展格局，有力支撑了武汉经济社会发展和城市建设工作。

After the 2000s, China's urban planning industry has been developing brilliantly.

During this period, new spatial concepts such as multi-center metropolitan areas, metropolitan circles and global urban regions, as well as new planning methods such as strategic planning and conceptual planning, were successively introduced. They were first introduced in central cities in the Pearl River Delta and Yangtze River Delta regions, attracting high attention from the planning industry and government policymakers. The "southward expansion" and the grand planning and construction of Zhujiang New Town in Guangzhou, the spatial development strategy of "One City and Nine Towns" in Shanghai, and the success of "development from the West Lake to the Qiantang River" in Hangzhou have made the multi-center metropolitan area a totem of advanced planning concepts and technologies, and have been widely discussed and emulated.

Back to Wuhan in 2003, former General Secretary Hu Jintao proposed that Hubei should become an important strategic support for the rise of central China, and the construction of a "two-oriented" society experimental zone has started. Caidian, Jiangxia, Huangpi and Xinzhou have "converted from counties to districts". With the expansion of the urban area, the scope of planning and management should be expanded accordingly. The state has built four intercity railways and a series of national and regional transportation and infrastructure facilities in Wuhan, further strengthening Wuhan's position as a regional transportation hub. Driven by economic forces, the main urban area of Wuhan and the planned key town areas have developed as a continuum and the circle-layer spatial structure has been transformed into an axial development trend. For this reason, at the beginning of the new century, Wuhan started to prepare for the sixth round of revision of urban master plans after the founding of the People's Republic of China. It adopted the working method of master planning guided by strategic planning, defined its position as the "central city in central China", put forward the planning mode of "urban development area", and established the "1+6" spatial development pattern, strongly supporting Wuhan's economic and social development and urban construction.

1 – "大城市论、都市区论"引导下的总体规划

Master Planning guided by "Megacity and Metropolitan Area Theory"

— 基于不同空间场景下的城市战略规划

为博采众家之长、集思广益，高起点地编制城市总体规划，2004年，武汉市在总体规划前期组织了"武汉城市发展战略规划征集"工作。主要目的是利用"战略规划"良好的"纲领性""全局性""务虚性"和"动态性"，为武汉市下一步跨越式发展提出启发性思想、开拓性构架和突破性战略。征集工作邀请了北京大学、中国城市规划设计研究院、同济大学、武汉市规划研究院4家研究机构参与。

北京大学针对武汉宏观地位不断下降的趋势，提出了"做实都市区、借力于南北、发力于西部、目标在东方"的成"弓"战略，重点在于对接上海和长三角，并迈出国门、走向世界，加快打通以武汉为中点的沿江铁路、公路干线，建设武汉航空枢纽港，完善和周边城市交通联系，做强紧密腹地。实施"两轴三城"的空间优化战略，强化长江轴和汉水—武珞路轴两条产业发展和景观生态主轴，由功能"三镇"向独立"三城"转型。

武汉市规划研究院按照"中部地区首席城市""跨越中部、面向国际的区域中心城市"目标，提出"顺江离岸、三轴拓展"总体布局，顺江突进产业空间拓展，离岸推动居住空间拓展，安排"两快、两轨、一水"组成复合交通走廊。构筑"依山抱湖、三环六楔"的武汉生态网，实现滨江城市向滨江滨湖生态城市转型。

中国城市规划设计研究院提出，武汉要充分发挥区域整体优势和现有的传统产业优势，要极化武汉、跨越中部、连接四极，建设成为"中部崛起"的战略支点。提出市域"沿江拓展"战略，即沿长江向外形成产业新城拓展轴线；提出中心区"三镇双城"战略，即把汉口、汉阳融为一体，与武昌对江发展，形成两个特大城市，在都市区发展蔡甸、流芳两个新中心。

同济大学认为武汉应该重新承担起在整个国家城市体系中的责任与义务，目前最为重要的是"找回武汉，重塑江城"。规划提出"回到长江"，形成国际化的武汉"都市核"；以长江为城市空间拓展轴线，构建"顺江而展、一带多心"的带形城市空间，弱化环路，强化顺江轴向拓展，打破圈层式蔓延扩张的发展态势。

— Urban Strategic Planning in Different Spatial Scenarios

To draw on the strengths of others, pool collective wisdom and draw up the urban master plan from a high starting point, Wuhan organized the "solicitation of strategic plans for Wuhan's urban development" in the early stage of master planning in 2004. The main purpose was to put forward enlightening thoughts, pioneering framework and ground-breaking strategies for Wuhan's further leap-forward development with the "programmatic" "global" "discussing" and "dynamic" characteristics of "strategic planning". Peking University, China Academy of Urban Planning and Design, Tongji University and Wuhan Planning and Design Institute were invited to the solicitation.

In response to the declining macro position of Wuhan, Peking University has put forward a "bow" strategy of "building a metropolitan area, relying on the north and the south, making efforts in the west and targeting in the east". The key points are to link Shanghai and the Yangtze River Delta to the world, speed up the opening of main railway and highway lines along the river with Wuhan as the center, build an aviation hub in Wuhan, improve traffic links with surrounding cities, strengthen the city hinterland, implement the spatial optimization strategy of "two axes and three cities" and strengthen the two industrial development and landscape ecological axes of the Yangtze River axis and Hanjiang River-Wuluo Road axis, so as to transform the functional "three towns" into independent "three cities".

Toward the goals of building a "chief city in the central region" and "regional central city spanning the central region and facing the international community", the Wuhan Planning and Design Institute proposed the overall layout of "expanding along the river, in offshore areas and around three axes", i.e. promoting the expansion of industrial space along the river and residential space in offshore areas. It arranged "two expressways, two rails and one river route" to form a composite traffic corridor, built an ecological network of "mountains and lakes, three rings and six wedges" to realize the transformation from a riverside city to a riverside and lakeside ecological city.

The China Academy of Urban Planning and Design proposed that Wuhan should give full play to regional overall advantages and the advantages of existing traditional industries and Wuhan should be polarized, spanning the central region and connecting the four poles to build a strategic fulcrum for the "rise of central China". It put forward the strategy of "expanding along the river" for the city area, i.e. to form an expansion axis of the new industrial city along the Yangtze River, and the strategy of "three towns and two cities" for the central area, i.e. to integrate Hankou and Hanyang into a whole which develops with Wuchang across the river, to form two megacities, and to develop Caidian and Liufang into two new centers in the metropolitan area.

Tongji University believed that Wuhan should reassume its responsibilities and obligations in the urban system of China and the most important thing at present is to "find and reshape Wuhan". The plan proposed to "return to the Yangtze River", form an international "urban core" of Wuhan, construct a belt urban space of "expanding along the river with multiple centers in one belt" with the Yangtze River as the axis of urban space expansion, weaken the rings, expand axially along the river, and break the development trend of circle-layer expansion.

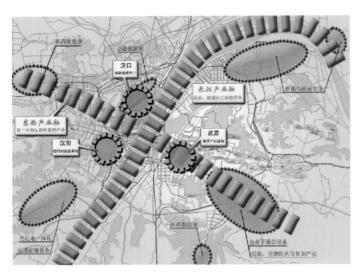
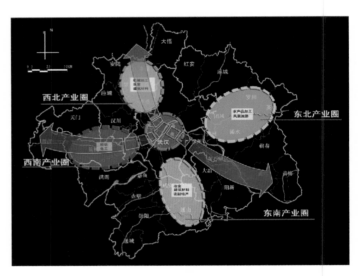

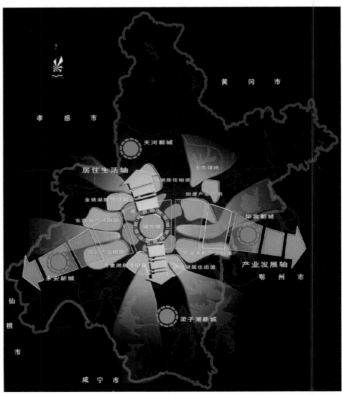

第肆章・区域一体化时期的规划 163

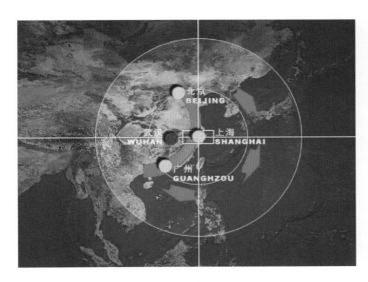

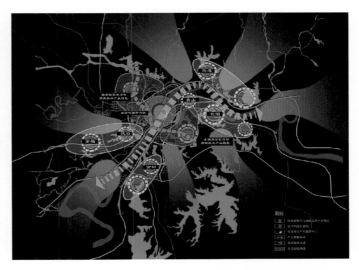

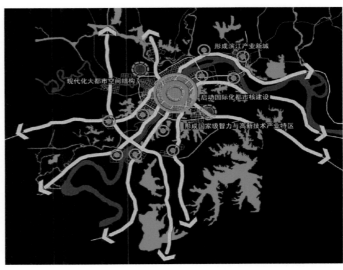

1	2
3	4

1
北京大学规划方案
The planning scheme by Peking University

2
武汉市规划研究院规划方案
The planning scheme by Wuhan Planning and Design Institute

3
中国城市规划设计研究院规划方案
The planning scheme by China Academy byUrban Planning and Design

4
同济大学规划方案
The planning scheme by Tongji University

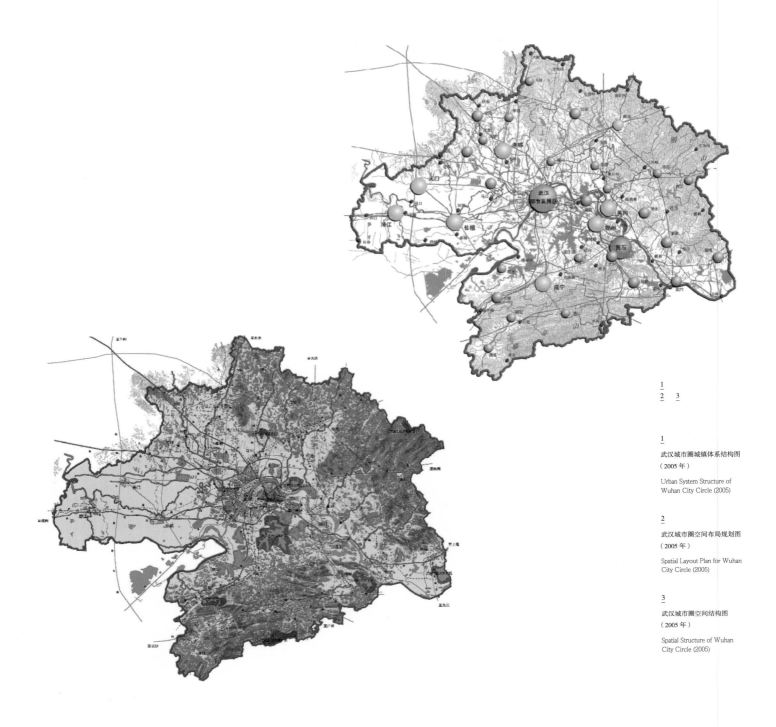

1
武汉城市圈城镇体系结构图
（2005年）
Urban System Structure of Wuhan City Circle (2005)

2
武汉城市圈空间布局规划图
（2005年）
Spatial Layout Plan for Wuhan City Circle (2005)

3
武汉城市圈空间结构图
（2005年）
Spatial Structure of Wuhan City Circle (2005)

一 基于区域视角的武汉城市圈空间布局规划

武汉城市圈规划研究起始于2003年，与《武汉市城市总体规划》修编同步编制《武汉城市圈城镇布局规划》，并作为《武汉城市总体规划（2006~2020年）》的专题之一。2007年12月，武汉城市圈被国家正式批准为全国"两型"社会建设综合配套改革试验区，为了推进城市圈一体化，省政府组织编制空间、交通、环保、社会事业等专项规划，其中以《武汉城市圈空间布局规划》（2005年）为骨干。

《武汉城市圈空间布局规划》确定武汉城市圈国土面积57822平方公里，规划将其建设成为我国重要的新型工业基地和科教创新基地、中部地区的现代服务业中心和特色农业区。

该规划预测到2020年，武汉城市圈将率先在湖北省和中部地区实现工业化、城镇化和基本现代化；人均GDP接近6万元，提前实现全面建设小康社会目标。相应，城市圈总人口将达3340万左右，城镇人口2335万左右，城镇化水平70%左右。按照"两型"社会建设的要求，规划制订了4大类、26项空间发展指标，以体现空间规划的引导与控制要求。

该规划构建"一核一带三区四轴"的区域发展框架和"一环两翼"的区域保护格局。按照主体功能区规划、城镇体系规划、土地利用规划"三规协调"的布局思路，在空间上协调落实城镇建设、产业园区、农林生产区、各类保护用地、交通与重大基础设施的规划布局，形成城市圈城乡空间的总体布局。

— Spatial Layout Planning of Wuhan City Circle from a Regional Perspective

The research on Wuhan city circle planning started in 2003, and the Town Layout Plan for Wuhan City Circle was compiled simultaneously with the revision of Wuhan City Master Plan, which was one of the special topics of Wuhan City Master Plan (2006—2020). In December 2007, the Wuhan city circle was officially approved by the state as the experimental zone for the comprehensive supplementary reform of national construction of a "two-oriented" society. To promote the integration of the city circle, the provincial government organized the preparation of special plans for space, transportation, environmental protection and social undertakings, with Spatial Layout Plan for Wuhan City Circle (2005) as the backbone.

Spatial Layout Plan for Wuhan City Circle determines that the Wuhan city circle covers an area of 57,822 square kilometers and plans to build it into an important new industrial base and scientific and educational innovation base in China, a modern service center and a characteristic agricultural area in central China.

The plan predicts that the Wuhan city circle will take the lead in realizing industrialization, urbanization and preliminary modernization in Hubei Province and central China by 2020, and its per capita GDP will be close to RMB 60,000, realizing the goal of building a well-off society in an all-round way ahead of schedule. Correspondingly, the total population of the city circle will be about 33.4 million and the urban population about 23.35 million, with an urbanization level of about 70%. According to the requirements of "two-oriented" society construction, the plan has formulated 26 spatial development indicators in four categories to reflect the guidance and control requirements of spatial planning.

It's planned to construct a regional development framework of "one core, one belt, three zones and four axes" and a regional protection pattern of "one ring and two wings". According to the layout idea of "coordination of three rules" in the main functional area planning, urban system planning and land use planning, the overall layout of urban and rural space in the urban circle will be formed by spatially coordinating the planning and layout of urban construction, industrial parks, agricultural and forestry production areas, various types of protected land, transportation and major infrastructure.

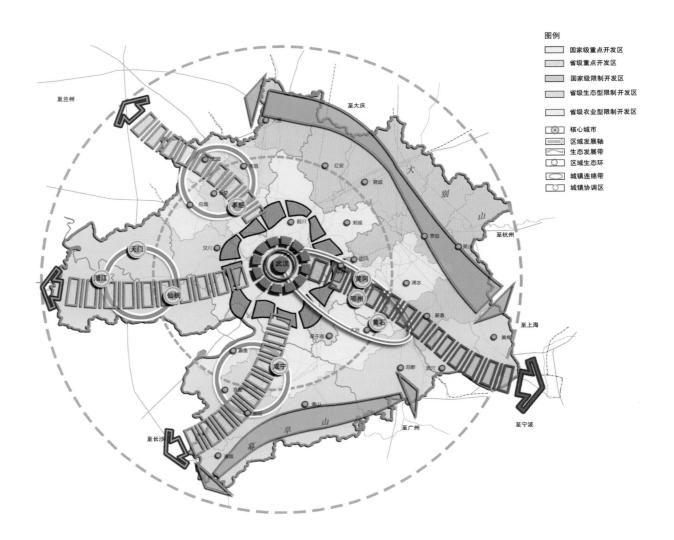

— "1+6"都市区主导下的 2010 版城市总体规划

为全面落实科学发展观，实施中部地区崛起战略，促进"两型"社会建设，引导经济又好又快发展，全面建设小康社会，武汉市于 2004 年开展了城市总体规划的编制工作，并编制了《武汉市城市总体规划（2009～2020年）》。该规划 2010 年获国务院正式批复，并获国际区域和城市规划师协会"杰出贡献奖"。

在该规划中，**城市发展定位由"重要"升华为"唯一"**。与 1999 年 2 月国务院对武汉市上轮总体规划批复的"我国中部重要的中心城市"相比，去掉了"重要"二字，升格为区域性综合中心城市，是武汉在区域和国家发展格局中地位、作用和使命的综合体现，反映了在"中部崛起"的过程中，国家对于武汉发挥中部地区龙头城市地位与作用的期望和肯定。未来十年，武汉不仅应朝着中部地区龙头城市的奋斗目标迈进，更应抓住机遇，着眼于用更长的时间恢复为一线城市或者说是国家中心城市。

空间发展理念从"城镇地区"走向"都市发展区"。规划在延续上轮总规人口和产业转移的思路基础上，考虑到外围组团已与主城连成一片的现实状况，因而将重点镇的概念拓展为由"新城+组团"合成的新城组群，并进一步提出了都市发展区（主城区+新城组群）的概念和市域、都市发展区、主城区三层次的规划范围，实现了 2256 平方公里城镇地区向 3261 平方公里都市发展区的转型。规划至 2020 年，市域常住人口为 1180 万人，城镇建设用地面积为 1030 平方公里。

空间形态结构从"圈层模式"调整为"轴带模式"。在 21 世纪之初，乘着城镇化快速发展的"东风"，贴近主城区的若干组团抢先一步发展起来，大有先轴线拓展、后圈层填充之势。在这种情势下，为了更好地引导城市空间的向外拓展，积极保护河湖密布的生态环境，提出了"以主城区为核、多轴多心"的组群式城市空间结构，以"双快一轨"构成的复合交通走廊为轴，沿常福、汉江、盘龙、阳逻、豹澥、纸坊方向布局 6 大新城组群，新城组群之间控制 6 大生态绿楔，形成有机生长的轴向组群结构。主城区规划结构调整为中央活动区、东湖风景区和 15 个城市综合组团。外迁主城区扰民工业企业，实施"退二进三"策略，调整优化工业用地布局。

生态框架结构由"环廊相间"扩展为"两环六楔"。规划延续利用南北水系、东西山系构建"十字型"山水生态轴的设想，扩大了生态环的保护范围，在城市三环线、外环线附近，构成两个环形"生态保护圈"，评价模拟武汉城市在各种气候条件下的环境效应，提出了贯穿多向城市风道，将 5 条生态走廊拓展为大东湖、汤逊湖、青菱湖、后官湖、府河和武湖等 6 大生态绿楔，形成楔入主城内部的 6 大通风廊道。

— The 2010 City Master Plan Guided by the "1+6" Metropolitan Area

To fully implement the scientific outlook on development, implement the rise of central China, promote the construction of a "two-oriented" society, guide sound and rapid economic development and build a well-off society in an all-round way, Wuhan launched the compilation of the city master plan and compiled Wuhan City Master Plan (2009—2020) in 2004. In 2010, the plan was officially approved by the State Council and awarded the "Outstanding Contribution Award" by the International Society of City and Regional Planners.

In this plan, **the orientation of urban development has been elevated from "important" to "only"**. Compared with the "important central city in central China" approved by the State Council in February 1999 for Wuhan's last master plan, the removal of the word "important" and its elevation to a regional comprehensive central city reflect Wuhan's position, role and mission in the regional and national development patterns and the state's expectation and affirmation of Wuhan's position and role as a leading city in the central region in the "rise of central China". In the next ten years, Wuhan should not only move towards the goal of being a leading city in the central region, but also seize the opportunity to restore itself to a first-tier city or a national central city in a longer period of time.

The concept of space development has changed from "urban area" to "urban development area". In the continuation of the population and industrial transfer in the previous master plan, in consideration of the fact that the peripheral clusters have been linked with the main city, the concept of key towns in this Plan has been expanded to a new town cluster composed of "new towns + clusters", and the concept of urban development area (main urban area + new town clusters) and the three-level planning scope of city area, urban development area and main urban area have been further proposed, realizing the transformation of the 2,256-square-kilometer urban area to the 3,261-square-kilometer urban development area. It is planned that by 2020, the permanent resident population in the city area will be 11.8 million and the area of urban construction land will be 1,030 square kilometers.

The spatial structure has been adjusted from "circle-layer mode" to "axis-belt mode". At the beginning of the 21st century, taking advantage of rapid urbanization, several clusters close to the main urban area developed first, with the trend of expanding along the axis first and filling the circle-layer later. Under such circumstances, to better guide the outward expansion of urban space and protect the ecological environment of rivers and lakes, a cluster-type urban space structure with "the main urban area as the core and multiple axes and multiple centers" was proposed. With the composite transportation corridor of "double expressways and one rail" as the axis, six new urban clusters were distributed along the Changfu,

第肆章 · 区域一体化时期的规划   167

都市发展区用地规划图
（2009年）

Land Use Plan for the Urban
Development Area (2009)

Hanjiang River, Panlong, Yangluo, Baoxie, and Zhifang. Six ecological green wedges were controlled among the new urban clusters to form an organic axial cluster structure. The planning structure of the main urban area has been adjusted to a central activity area, Donghu Lake Scenic Area and 15 comprehensive urban clusters. The industrial enterprises disturbing residents in the main urban areas were relocated, the strategy of "shitting from the second industry to the third industry" was implement and the layout of industrial land was adjusted and optimized.

The ecological framework structure has been expanded from "rings alternating with corridors" to "two rings and six wedges". The Plan continues the idea of building a "cross-shaped" landscape ecological axis using the river systems in the south and north the mountain systems in the east and west to expand the protection scope of the ecological ring. Two ring-shaped "ecological protection circles" were formed near the city's third ring road and outer ring road. The multi-directional urban air passages were connected by evaluating and simulating the environmental effects of Wuhan under various climatic conditions. The five ecological corridors were expanded into six green wedges, namely, Great Donghu Lake, Tangxun Lake, Qingling Lake, Houguan Lake, Fu River and Wuhu Lake, forming six ventilation corridors wedged into the main urban areas.

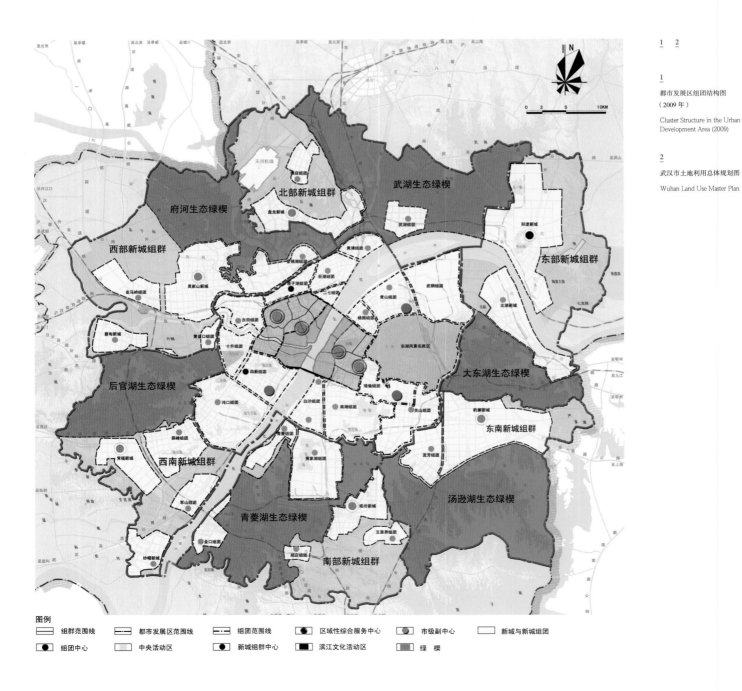

1

都市发展区组团结构图
（2009年）

Cluster Structure in the Urban Development Area (2009)

2

武汉市土地利用总体规划图

Wuhan Land Use Master Plan

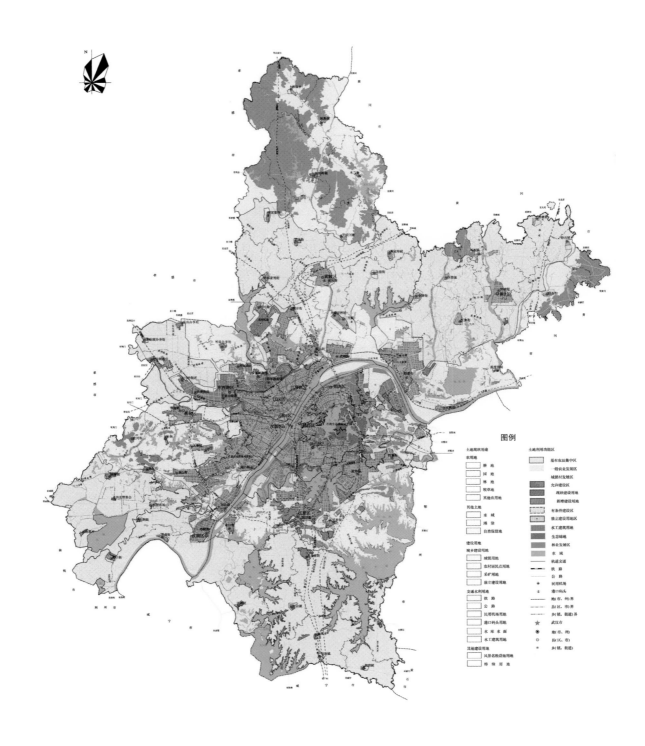

一 同步开展的土地利用总体规划

与城市总体规划编制工作同步，2003年武汉市国土资源和规划局启动了土地利用现状更新调查、专题研究、"四查清、四对照"等多项前期工作，2006年12月，国土资源部批复《武汉市土地利用总体规划修编大纲（2005～2020年）》，并以之为基础，深化完善，2009年编制完成了《武汉市土地利用总体规划（2006～2020年）》，2010年8月，国务院批准了该规划。

该规划立足于武汉市土地资源供需状况和规划期间武汉经济社会发展形势，提出了2006～2020年武汉市土地利用的方针及目标、土地利用结构调整、土地利用空间布局、耕地和基本农田保护、土地用途管制、土地整治重大工程、重点建设项目、土地生态环境保护和规划实施保障措施。该规划是指导武汉市土地资源保护利用和可持续发展的重要依据。

同时该规划明确提出，到2020年武汉市耕地保有量不低于338000公顷（507万亩），确保全市264500公顷（396.75万亩）基本农田数量不减少、用途不改变、质量有提高；到2020年，武汉市新增建设用地总量控制在58600公顷（87.9万亩）以内，新增建设占用耕地控制在36800公顷（55.2万亩）以内；具有重要生态功能的耕地、园地、林地、水域等占全市土地总面积的比例保持在75%以上。

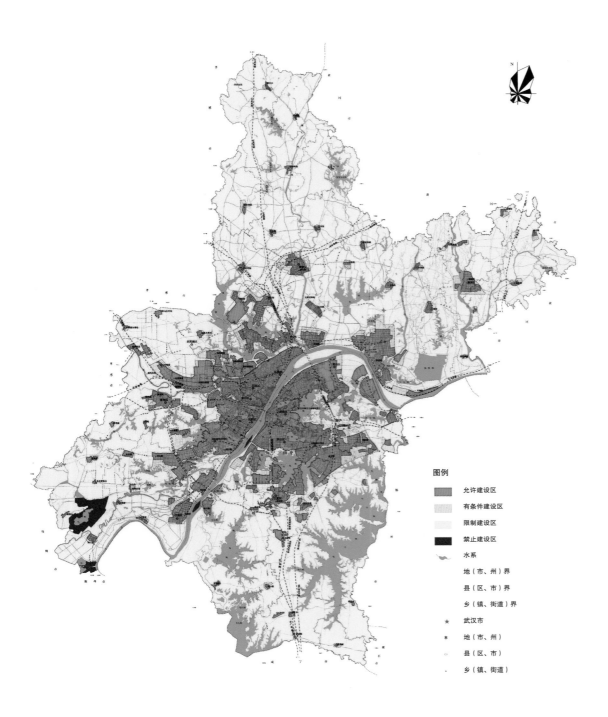

— Synchronous land use master planning

Synchronous with the preparation of the urban master plan, a series of preliminary work were launched in 2003, such as the update and investigation of land use status, special topic research, and "four investigations and four comparisons". In December 2006, the Ministry of Land and Resources approved the Revision Outline of Wuhan Land Use Master Plan (2005—2020) and deepened and perfected it on this basis. In 2009, Wuhan Land Use Master Plan (2006—2020) (hereinafter as the Plan) was prepared. In August 2010, the State Council approved the Plan.

Based on Wuhan's supply and demand of land resources and economic and social development during the planning period, the Plan proposes the policy and objectives of land use in Wuhan from 2006 to 2020, the adjustment of land use structure, the spatial distribution of land use, the protection of cultivated land and basic farmland, land use control, major land improvement projects, key construction projects, the protection of land ecological environment and the planned supporting measures. The Plan is an important basis for guiding the protection, utilization and sustainable development of land resources in Wuhan.

According to the Plan, by 2020, the amount of cultivated land in Wuhan won't be less than 338,000 hectares (5,070,000 mu); the basic farmland of 264,500 hectares (3,967,500 mu) should not be reduced; its use will not be changed and its quality will be improved. By 2020, the total amount of new construction land in Wuhan will be controlled within 58,600 hectares (879,000 mu), and the amount of farmland occupied by new buildings will be controlled within 36,800 hectares (552,000 mu); cultivated land, garden land, forest land and water areas with important ecological functions should account for more than 75% of the city's total land area.

— 主城区分区规划及控规导则

在总体规划阶段成果编制完成后，武汉市国土资源和规划局陆续开展了主城区分区规划及控规导则的编制工作。根据《武汉市城乡规划体系研究》（2008年）确定的"总规—分规—控规"三个层次的法定规划主干体系，为满足《中华人民共和国城乡规划法》实施后过渡期规划管理要求，主城区分区规划的任务是落实总体规划对19个组团（含中央活动区、东湖风景区等组团）的功能、人口、空间结构、主要公共设施、道路及市政基础设施的战略部署，深化统筹城市土地利用、人口分布和公共服务设施、基础设施的配置。划定控规编制单元，将分区规划管控体系传导至控规导则的编制。控规导则实现了主城区的全覆盖，从刚性和弹性控制相结合的角度出发，创新性提出了虚线、实线、点位和指标控制4种方式，并制定了用地建设强度、用地兼容性两个规定，为落实总体规划意图和规范用地规划管理奠定了坚实基础。

— "两规合一"的乡镇总规

为加强规划对城乡建设的引导，发挥武汉市国土和规划合一的优势，建立"两规合一、突出主线、多规支撑、城乡统筹"规划体系，武汉市抓住区、乡级土地利用总体规划修编契机，特在全市78个乡镇范围内开展乡镇总体规划编制工作。

乡镇规划主要是根据自身的资源条件和经济基础，结合区级规划分解下达各区的土地利用控制指标，确定乡镇功能定位和土地利用目标、镇村和重要的基础设施体系，划定镇域空间管制分区，落实基本农田保护目标，确定各类重点建设项目用地安排。

1　2

1
武汉市建设用地管制分区图
Zoning Plan for Construction Land Control in Wuhan

2
武汉市主城区分区编号图
（2008年）
Map with Zoning Number (2008)

— The Zoning Plan for the Main Urban Areas and Control Plan Guidelines

After the stage results of the master plan were compiled, the compilation of the zoning plan for the main urban areas and control guidelines was successively carried out. According to the three-level legal planning backbone system of "master plan—sub-plan—control plan" determined by the Wuhan Urban and Rural Planning System Research (2008), in order to meet the planning and management requirements in the transition period after the implementation of Urban and Rural Planning Law, the task of the zoning plan for the main urban areas is to implement the master plan's strategic deployment of the functions, population, spatial structure, main public facilities, roads and municipal infrastructure of the 19 clusters (including the central activity area, and Donghu Lake Scenic Area), and to deepen the overall planning of urban land use, population distribution and the allocation of public service facilities and infrastructure. The preparation units of the control plan have been defined to transmit the zoning planning control system to the preparation of control guidelines. The guidelines cover the whole main urban areas. By combining rigid and flexible control, the guidelines innovatively put forward four methods, namely, dotted line, solid line, point and index control, and formulated two regulations, namely, land use construction intensity and land use compatibility, which have laid a solid foundation for the implementation of the intent of the master plan and the standardization of land use planning and management.

— Village and Town Master Planning of "Integrating Two Plans"

To strengthen the guidance of the plan for urban and rural construction, give full play to Wuhan's advantages in integrating land and planning, and establish a planning system of "integrating two plans, highlighting the main line, relying on multiple plans, and coordinating urban and rural areas", Wuhan has seized the opportunity of revising the land use master plans at the district and township levels to compile the village and town master plan in 78 villages and towns in Wuhan.

Based on local resource conditions and economic basis and the district-level plan, the village and town plan has issued the land use control indicators of each district, determined the functional orientation of villages and towns, land use targets, villages and important infrastructure systems, delimited the spatial control zones of towns, implemented the basic farmland protection targets, and determined the land use arrangements for various key construction projects.

172 CHAPTER FOUR • Planning During Regional Integration

武汉市主城区分区规划·用地规划图（2008年）

1　2

1
武汉市主城区分区规划·用地
规划图（2008年）
Land Use Plan (2008)

2
黄陂区蔡家榨土地利用规划图
Land Use Plan for Caijiazha

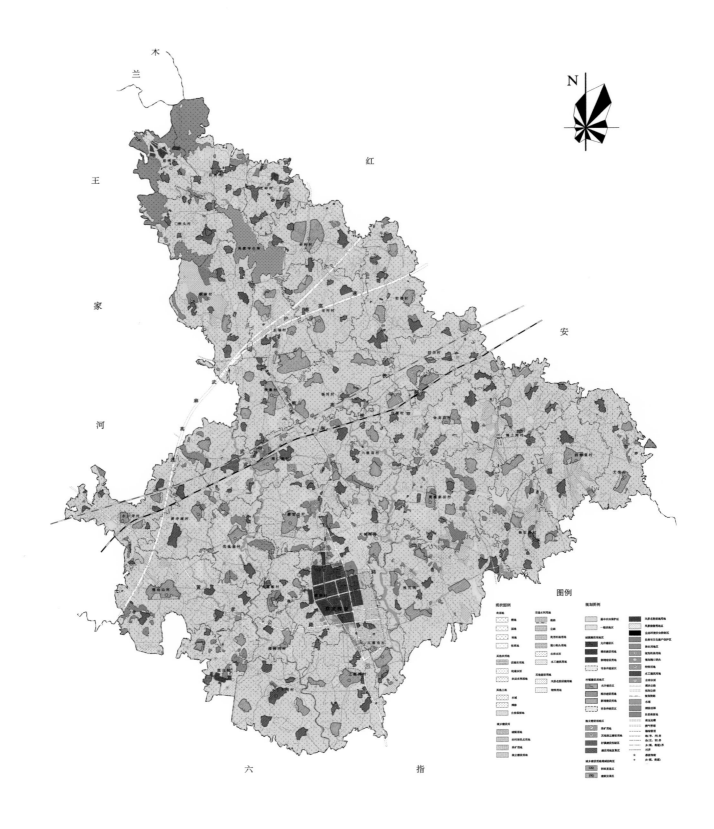

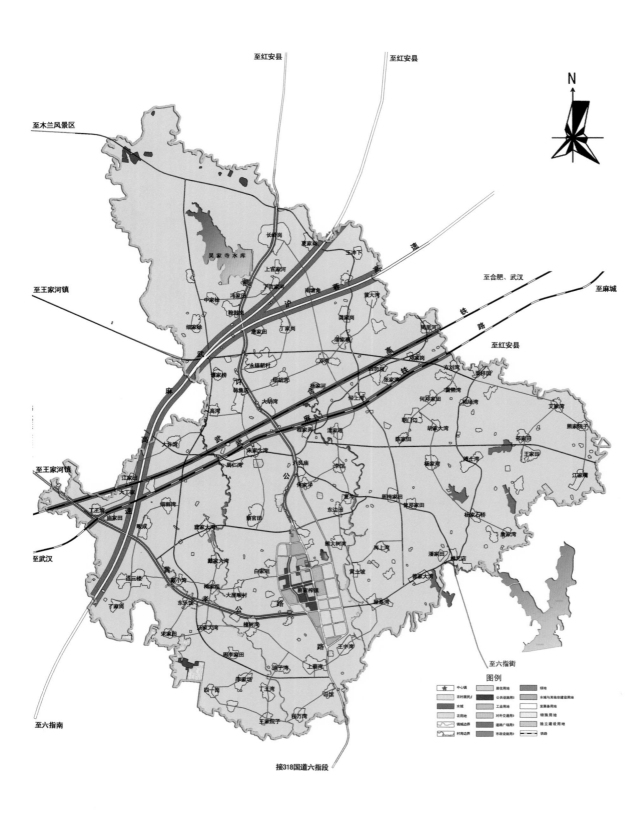

1

黄陂区蔡家榨街域总体规划图

Master Plan for Caijiazha Sub-district

2

江夏区郑店街土地利用总体规划图

Land Use Master Plan for Zhengdianjie Sub-district

第肆章 · 区域一体化时期的规划　175

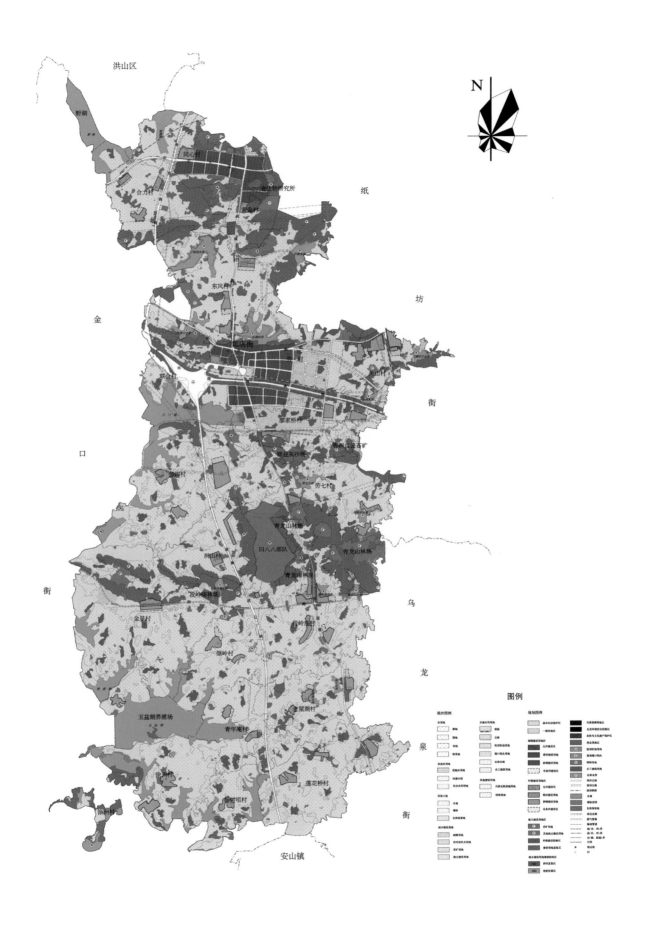

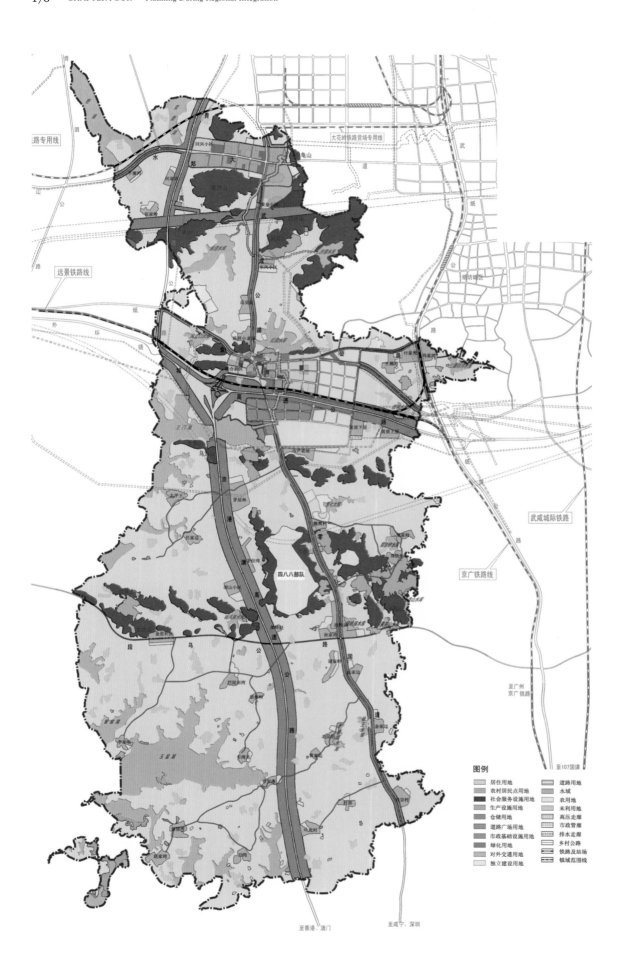

1

江夏区郑店街镇域用地规划图

Land Use Plan for Zhengdianjie Sub-district

2

绿地系统近期建设规划图

Near-term Construction Plan for the Green Space System

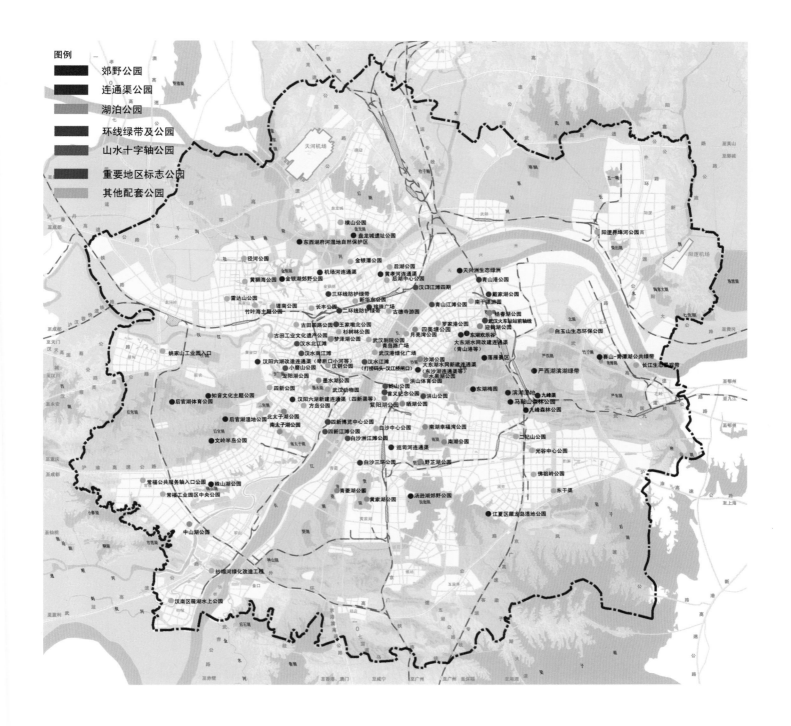

一 近期建设规划

在城市总体规划和土地利用总体规划获批之后，为主动实施"两规"，进一步加强与全市国民经济和社会发展"十二五"规划的空间统筹和衔接，2009年底由武汉市政府组织、武汉市国土资源和规划局牵头，32个市属职能部门和各区共同参与，开展了武汉市近期建设规划的编制工作。

《武汉市近期建设规划》（2009年）以积极承担国家战略、全面提升城市功能为主线，重点打造"辐射全国"的中心城市、"两型社会"的示范城市和"民生幸福"的宜居城市，规划至2015年常住人口达到1040万人，都市发展区新增建设用地196平方公里；同时规划明确了"一核四心、四极九园、一环六楔"的总体发展框架，即重点实施两江四岸核心区改造，加快推进王家墩商务区和四新、杨春湖城市副中心建设，完善提升鲁巷城市副中心功能，着力打造"大光谷"地区、"中国车城""临空经济区""临港产业区"四大产业增长极，积极推进"6+3"新型工业化示范园区建设，全面完成三环线生态隔离带建设，有序推进大东湖、后官湖等六大生态绿楔的实施；形成了工业倍增、服务提档、和谐住区、交通畅达、生态城市、名城保护、设施提升、城乡统筹8个专项行动计划，并制定了包括8大类、38小类、977个项目的近期建设项目库。

与此同步，为了进一步加强对城市建设的引导，武汉市开展了2011年度实施计划的编制工作，进一步将近期建设规划确定的各项建设要求分解落实到年度计划中，确保近期建设规划的有序实施。

Near-term Construction Plan for Road Traffic

Wuhan New Area Conceptual Master Plan (2003)

— Near-term Construction planning

After the approval of the city master plan and the land use master plan, in order to actively implement the "two plans" and further strengthen the spatial coordination and convergence with the "12th Five-Year Plan" for the city's national economic and social development, organized by the Wuhan Municipal Government and led by the Wuhan Natural Resources and Planning Bureau, 32 municipal functional departments and districts participated in the preparation of Wuhan Near-term Construction Plan at the end of 2009.

In Wuhan Near-term Construction Plan (2009), Wuhan has taken the active commitment to the national strategy and the overall improvement of the city's functions as its main line, focusing on building a central city that "radiates the whole country", a model city of "two-oriented society" and a habitable city of "happiness of people's livelihood". It's planned that the permanent population would reach 10.4 million by 2015 and there would be 196 square kilometers of new construction land in the urban development area. The overall development framework of "one core and four centers, four poles and nine parks, one ring and six wedges" has been defined in the Plan, i.e. focusing on the reconstruction of core areas along the two rivers and four banks; accelerating the construction of Wangjiadun Business District and the sub-centers of Sixin and Yangchunhu; improving and upgrading the functions of the sub-center of Luxiang; building four industrial growth poles namely "Greater Optics Valley" "China Auto City" "Airport Economic Zone" and "Port Industrial Zone"; actively promoting the construction of the "6+3" new industrial demonstration park; completing the construction of the ecological isolation zone of the third ring; and orderly promoting the construction of six ecological green wedges including the Great Donghu Lake and Houguan Lake. Eight special action plans have been formed, i.e. industrial multiplication, service promotion, harmonious residential areas, smooth transportation, ecological cities, protection of famous cities, improvement of facilities, and urban and rural overall planning. A library of near-term construction projects including 8 categories, 38 subcategories and 977 items has been drawn up.

At the same time, to further strengthen the guidance on urban construction, Wuhan has carried out the preparation of the 2011 implementation plan, and further decomposed and implemented various construction requirements determined in the near-term construction plan into the annual plan to ensure the orderly implementation of the near-term construction plan.

2 - 以武汉新区为重点的区域发展战略

The regional development strategy focusing on Wuhan New Area

— 武汉新区

21世纪以来，针对三镇发展不平衡的现状，武汉市提出了"三镇均衡发展"的目标，积极创建占地面积368平方公里的武汉新区，并于2003年编制了《武汉新区总体规划（2006~2020年）》。规划将武汉新区界定在长江、汉水、外环线等"两江一路"围合的大汉阳地区，规划人口100万，将其建设成为辐射武汉乃至整个华中地区的现代制造业基地、生产性服务中心、市级文化旅游中心和风貌独特的现代化商住新城，成为展现武汉产业经济及滨江、滨湖自然山水特色的"窗口"。规划形成汉江、四新、沌口等三大组团以及六湖连通的生态网络和"五纵五横"的道路网络。汉江组团对汉阳城区进行整合，完善文化旅游、居住功能，构建文化旅游中心。四新组团形成服务于区域的生产性服务中心和现代居住城市。沌口组团发展现代制造业，建成产业相对集聚、功能互补的工业区，成为武汉新区经济发展的增长点和动力源。

— "中国·光谷"

武汉东湖新技术产业开发区，别称"中国·光谷"，于1988年创建成立，1991年被国务院批准为首批国家级高新区，2009年被国务院批准为全国第二个国家自主创新示范区，规划范围518平方公里。《武汉东湖国家自主创新示范区总体规划（2011~2020年）》要形成推动"两型"社会建设、促进自主创新驱动区域发展、创建全国具有重要典型意义和先行示范作用的"东湖模式"；建成具有活跃的创新经济、和谐的社会人文和绿色的生态环境，以"世界光谷"享誉全球的世界一流科技园区。

该规划依托高新大道、高新三路等快速通道，"集束式"布局服务中心及产业生活组团，打造复合空间主轴，并布局关山、豹澥和未来城三大功能区，以及南北向的严东湖和牛山湖科技生态城，构建光谷综合服务中心、鲁巷城市副中心，花山、左岭、流芳和牛山湖四大支撑中心。

— Wuhan New Area

Since the 21st century, in response to the unbalanced development of the three towns, Wuhan has put forward the goal of "balanced development of the three towns", actively created Wuhan New Area covering 368 square kilometers, and compiled the Wuhan New Area Master Plan in 2003. The Wuhan New Area is defined in the greater Hanyang area surrounded by "two rivers and one road", i.e. the Yangtze River, Hanjiang River and the outer ring road. With a planned population of 1 million, it will be built into a modern manufacturing base, a productive service center, a municipal cultural tourism center and a modern commercial and residential new town with unique features that radiate Wuhan and even the whole central China. It will also become a "window" for Wuhan's industrial economy and the natural landscape of the riverside and lakeside. It's planned to form three clusters, namely Hanjiang River, Sixin and Zhuankou, as well as an ecological network with six lakes connected and a road network with "five vertical and five horizontal lines". The Hanjiang cluster has integrated the urban area of Hanyang and improved cultural tourism and residential functions to construct a cultural tourism center. Sixin has become a productive service center and modern residential city serving the region. Zhuankou has developed the modern manufacturing industry and build an industrial zone with relatively concentrated industries and complementary functions, which has become the growth pole and power source of economic development in Wuhan New Area.

— Optics Valley of China

Wuhan Donghu New Technology Development Zone, also known as "Optics Valley of China", was established in 1988, approved by the State Council as the first batch of national high-tech zones in 1991, and approved by the State Council as the second national independent innovation demonstration zone with a planned area of 518 square kilometers in 2009. Wuhan Donghu National Independent Innovation Demonstration Zone Master Plan (2011—2020) aims to form the "Donghu Model" with typical significance and demonstration role in China which can promote the "two-oriented" society construction and regional development driven by independent innovation, and build a world-class sci-tech park with an active innovative economy, a harmonious social culture and a green ecological environment, with the reputation of "World Optics Valley".

Relying on Gaoxin Avenue, Gaoxin Third Road and other expressways, and the "clustered" service centers and industrial and living clusters, a composite space axis was built; the three functional areas of Guanshan, Baoxie and Future City were arranged; the north-south Yandong Lake and Niushan Lake Science and Technology Eco-city was built; the Optics Valley Comprehensive Service Center, Luxiang Sub-center and four supporting centers of Huashan, Zuoling, Liufang and Niushan Lake were constructed.

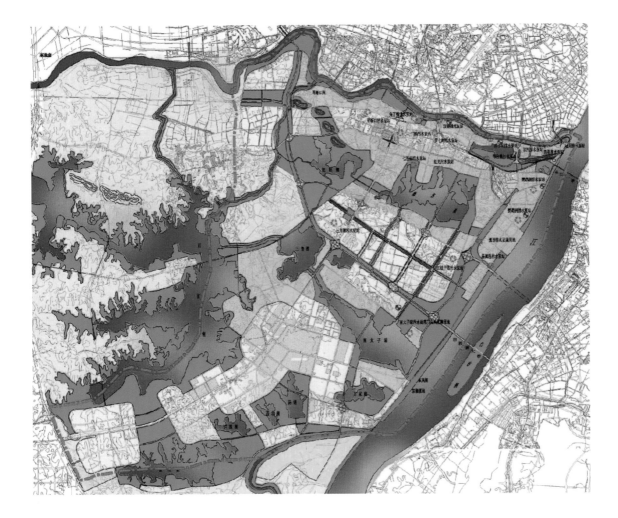

1

武汉新区水环境规划图

Wuhan New Area Water Environment Plan

2

东湖国家自主创新示范区总体规划图

The Comprehensive Planning of National Independent Innovation Demonstration Zone of Donghu Lake

第肆章 · 区域一体化时期的规划

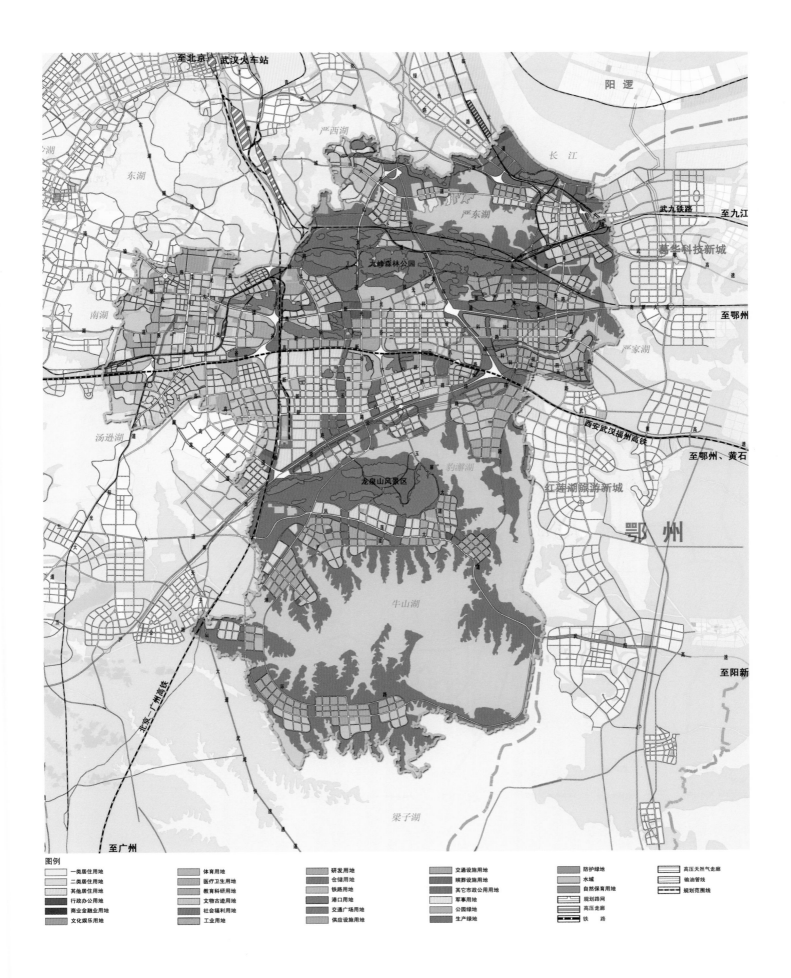

一 王家墩商务区

王家墩军用机场搬迁后，为高效利用王家墩地区的空间资源，引导该地区发展建设，编制了《王家墩商务区规划》（2003年），用地范围7.41平方公里。规划提出重点建设金融商贸区和国际会展中心，形成具有良好环境品质、充沛经济活力、卓越城市形象的新兴城市中心区，使之成为展示21世纪武汉市现代化、国际性城市风貌的重要窗口。以"两纵三横"的主干路网为基础，提出"一心、两轴"和四大功能分区的空间结构。四大功能分区包括中心商务区、全生活城服务区、综合商业区及生活居住区等。规划完成后，商务区在全国率先采取"政府引导、企业运作"开发建设模式，正式开启了市政基础设施建设。

1

东湖国家自主创新示范区规划道路系统图

Road System of the Comprehensive Planning of National Independent Innovation Demonstration Zone of Donghu Lake

2

王家墩商务区用地规划图
（2003年）

Land Use Plan of Wangjiadun Business District (2003)

3

王家墩商务区总平面图
（2003年）

General Plan of Wangjiadun Business District (2003)

— Wangjiadun Business District

After the relocation of Wangjiadun Military Airport, to make efficient use of the space resources in Wangjiadun and guide the development and construction of the area, the Wangjiadun Business District Plan (2003) has been compiled with a land area of 7.41 square kilometers. The Plan proposes to focus on the construction of a financial and business district and an international convention and exhibition center, so as to form a new urban area with good environmental quality, abundant economic vitality and outstanding city image, which will become an important window to show the modern and international features of Wuhan in the 21st century. Based on the main road network of "two vertical and three horizontal lines", the spatial structure of "one center and two axes" and four functional districts has been proposed. The four functional districts are the central business district, the life service district, the comprehensive business district and the residential district. After planning, the business district took the lead in adopting the development and construction mode of "government guidance and enterprise operation" nationwide, officially launching the municipal infrastructure construction.

3 - 以基础设施为重点的城建攻坚战略

The Urban Construction Strategy Focusing on Infrastructure

2000年以来,随着城市快速发展,武汉逐步进入机动化快速发展阶段的多重困难叠加期,主要表现为空间拓展迅猛、轨道起步晚、公交设施薄弱等,机动车拥有量超出2010版总规预测水平,道路服务水平从"轻度拥堵"进入"中度拥堵",拥堵点呈向外扩大态势。交通拥堵这一大"城市病"越来越成为制约武汉发挥核心城市功能、改善居民出行环境的重要因素。在此背景下,武汉市提出了城建攻坚计划,既要克服交通拥堵的痛点,也要找准从快速路和轨道交通双向突破的重点。

— 快速路系统规划

以"1+6"城市空间格局为依托,规划布局了都市发展区"5环18射"和主城区"3环13射"的快速路网络。在当时全国公路干线大建设的背景下,武汉外环线(2007年)和岱黄公路、十升线、金桥大道等出口路、放射线率先建成。同时,城市过江交通问题得到极大改善:白沙洲长江大桥(2000年)、军山长江大桥(2001年)、阳逻长江大桥(2007年)、天兴洲长江大桥(2009年)相继建成通车,进一步形成了环形的骨干交通结构。

2010年开始,大力推进了主城区快速路网建设,实现了主城区30分钟点到点的目标。开启了一环线整治、二环线划圆、三环线提升等工程,实现了主城区城市环线全线贯通,四环线开工建设(预计2020年建成通车),对外出口道路基本形成。主城区快速路系统的形成,有效缓解了城市交通拥堵,大幅改善了交通运行效率,有力支撑了城市空间的快速拓展。

— 轨道交通线网规划

城市轨道交通的规划建设从1980年代开始构想,1990年代形成规划方案,2000年开工建设。2004年,轨道1号线建成通车,武汉轨道交通建设完成发展起步;2010年完成1号线二期工程,全长38.5公里;2012年,轨道2号线一期开通运营,完成了轨道交通跨江连通。

2008年修编完成了《武汉市轨道交通线网规划》,规划提出至2040年形成由3条市域快线和9条市区线构成、总长540公里的线网优化方案,2020年前建成了7条线,总长220公里,形成覆盖三镇中心城区并与主要交通枢纽衔接的轨道交通网络。该项目于同年6月获武汉市政府正式批复。

1

主城区"3环13射"快速路系统图

Expressway System of "3 Rings and 13 Radial Lines" in the Main Urban Areas

2

都市发展区"5环18射"快速路系统图

Expressway System of "5 Rings and 18 Radial Lines" in the Urban Development Area

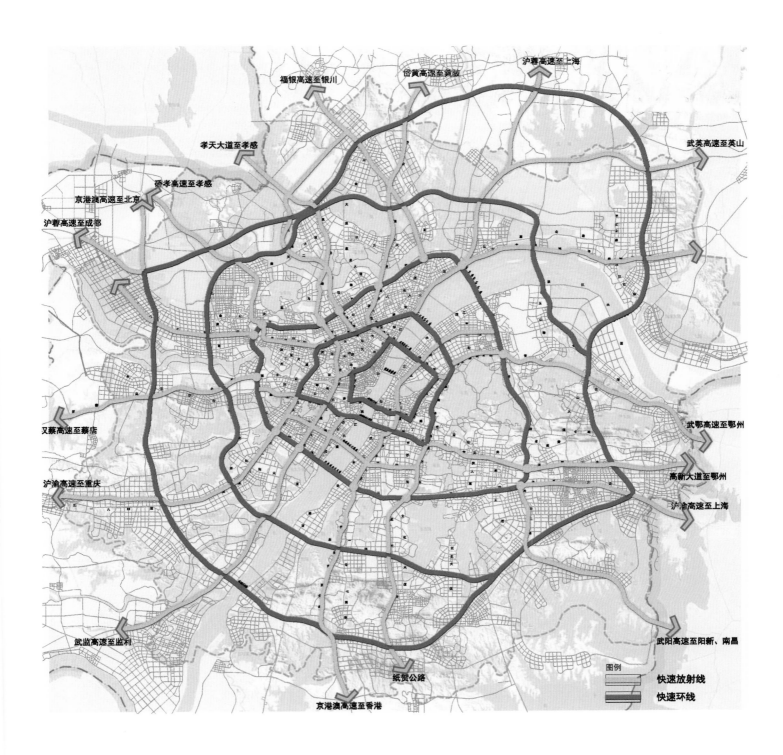

Since 2000, with its rapid urban development, Wuhan has gradually encountered many difficulties in the rapid development of motorization, such as rapid space expansion, lagging rail construction, poor public transport facilities, etc. The number of motor vehicles has exceeded the forecast level of the 2010 master plan; the level of road service has changed from "mild congestion" to "moderate congestion", with congestion points expanding outward. Traffic congestion, a major urban disease, is becoming an important factor that restricts Wuhan from playing its core urban functions and improving residents' travel environment. Under this background, Wuhan has put forward an urban construction plan to solve traffic congestion and find out the key points of solving traffic congestion by expressways and rail traffic.

— Expressway System Planning

Relying on the "1+6" urban spatial pattern, the expressway network of "5 rings and 18 radial lines" in the urban development area and "3 rings and 13 radial lines" in the main urban areas has been laid out. Under the background of national highway construction, Wuhan Outer Ring Road (2007), Daihuang Highway, Shisheng Line, Jinqiao Avenue and other exit roads and radial lines were first built. At the same time, the river-crossing traffic in the city has been greatly improved. Baishazhou Yangtze River Bridge (2000), Junshan Yangtze River Bridge (2001), Yangluo Yangtze River Bridge (2007) and Tianxingzhou Yangtze River Bridge (2009) have been built and opened to traffic, further forming a circular backbone traffic structure.

From 2010, the construction of the expressway network in the main urban areas has been vigorously promoted, achieving the target of 30 minutes' journey from one point to another in the main urban areas. Projects such as the renovation of the First Ring Road, the rounding of the Second Ring Road and the upgrading of the Third Ring Road have been launched. All ring roads in the main urban areas were opened to traffic. The construction of the Fourth Ring Road (expected to be completed and opened to traffic in 2020) has been started and the external exit roads have been basically formed. The formation of the expressway system in the main urban areas has effectively eased urban traffic congestion, greatly improved traffic efficiency and strongly supported the rapid expansion of urban space.

— Rail Transit Network Planning

The planning and construction of urban rail transit began in the 1980s, a planning scheme was formed in the 1990s, and construction began in 2000. In 2004, Line 1 was completed and opened to traffic, and Wuhan's rail transit construction has completed its initial stage. The second phase of Line 1 was completed in 2010, with a total length of 38.5 kilometers. In 2012, the first phase of Line 2 was put into operation, and rail traffic was linked across the river.

In 2008, Wuhan Rail Transit Network Plan was revised and completed. The plan proposes to form a 540-kilometer network by 2040 consisting of three urban expressways and nine urban lines. By 2020, seven lines with a total length of 220 kilometers will be completed to form a rail transit network covering the central urban areas of the three towns and connecting with major transportation hubs. The project was officially approved by the municipal government in June, 2008.

4 – 以产业振兴为重点的工业倍增战略
The Industrial Multiplication Strategy Focusing on Industrial Revitalization

一 武汉市都市工业园规划（2008年）

跨入21世纪以来，武汉老工业基地特有的历史性包袱、体制性缺陷和结构性矛盾等日益突出，制约了工业的发展。为了走出一条老工业基地改造的新路子，武汉市政府做出了发展都市型工业的战略决策。以一环至三环线之间的区域为重点，原则上1个中心城区规划布局1个市级都市工业园区，全市布局8个市级都市工业园区，总规划面积17.07平方公里。都市工业园区建设立足于最大限度地盘活存量用地，大力发展无污染、不扰民和就业、高附加值的都市型工业，注重职住平衡，以硚口汉正街都市工业园区为试点，陆续开展了江岸区堤角、汉阳区洲头和武昌区白沙洲、青山区工人村都市工业园区、洪山区科技都市工业园区等的建设，使之成为国企改革的推进器、优良品牌的加工园、民营中小企业的孵化器、下岗职工再就业基地和工业旅游的新景点。

1
远景年轨道线网规划图
Long-term Rail Transit Network Plan

2
主城区都市工业园分布图
Distribution of Urban Industrial Parks in the Main Urban Areas

― 远城区工业倍增示范园区规划（2010年）

"十二五"时期，世界经济进入深刻调整期，产业转移加速，科技进步加快，新型产业孕育新一轮大发展。为加速工业战略资源向武汉市集中，促进武汉快速成为中西部地区的生产要素核心集聚区和具有强大带动能力的增长极，武汉市委市政府提出了"工业强市"战略，并大力实施工业发展"倍增计划"。这一时期，在主城区产业外迁和"退二进三"的背景下，远城区逐渐成为实施"工业倍增"计划的主战场。为此，《远城区工业倍增示范园区规划》（2010年）以大光谷地区、中国车城、临空经济区和临港产业区等支柱产业集群为4个增长极，采取以示范园建设带动远城区新型工业化的建设思路，在6条城市空间主轴上，集中建设1个新型工业化示范园区，辐射带动其他工业组团，形成特色工业发展板块。最终，规划建设9个新型示范园区和14个一般工业园区。其中，6个新城区各自集中建设一个工业示范园区，其净工业用地不低于20平方公里；汉阳、洪山和青山3个中心城区示范园区和其他一般工业园区规模不低于10平方公里。

― 武汉四大板块综合规划（2014年）

受到历史因素和发展阶段限制，武汉新城建设普遍存在规模偏小、功能偏弱、难以独立成市等问题。为解决上述问题，通过促进工业企业集中布局、新城中心集中打造、基础设施集中配套、强化"产业集聚、空间集中、用地集约"，进一步强化武汉建设国家中心城市的产业"硬实力"，开展了四大板块的综合规划。

武汉西南地区的汽车及零部件产业集群、东南地区的电子信息产业集群、东部地区的钢铁及深加工产业集群已出现明显工业板块发展格局，《武汉四大板块综合规划》（2014年）分别划定大车都、大光谷板块、大临港经济区等工业板块规划范围，同时考虑未来空港地区的快速发展需要，增加划定大临空经济区板块规划范围。四大板块遵循"独立成市、产城联动、城城互动、园园互补"指导方针，激发工业发展要素的聚合效应，打造武汉国家先进制造业中心，促进"产城一体"发展，带动人口外迁集聚，并强化土地节约、集约利用。

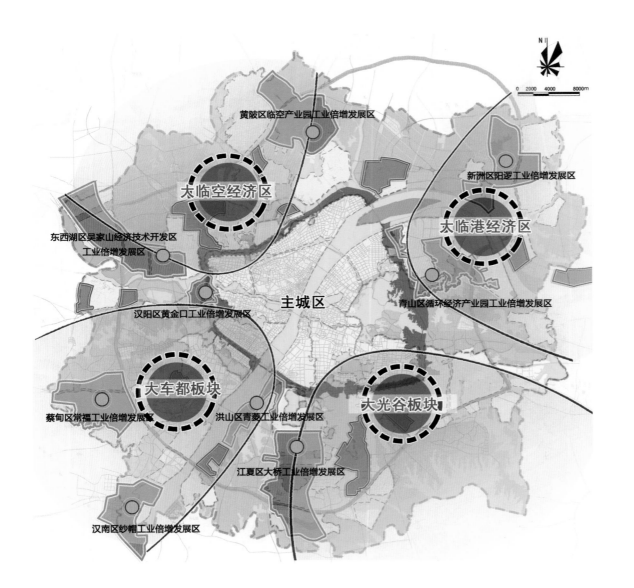

1

武汉市四大工业板块空间布局规划图

Spatial Layout Plan of Four Major Industrial Sectors in Wuhan

2

武汉市工业倍增计划空间发展规划图

Spatial Development Plan for Wuhan Industrial Multiplication Plan

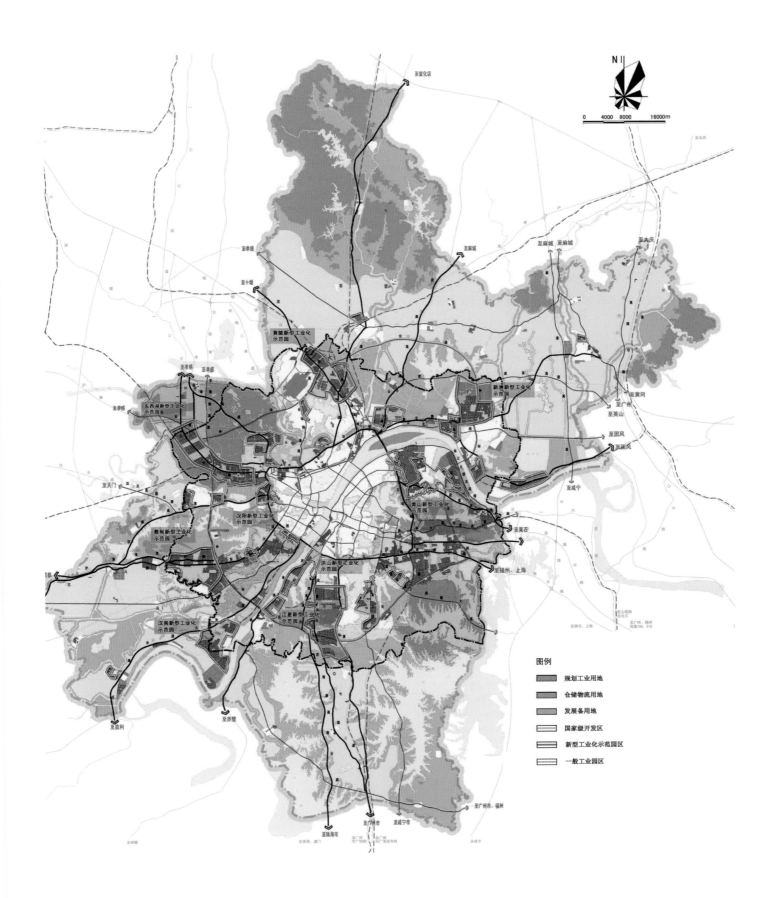

— Wuhan Urban Industrial Park Planning (2008)

Since the beginning of the 21st century, the historical burden, institutional defects and structural contradictions peculiar to Wuhan's old industrial bases have become increasingly prominent, restricting the development of industry. To find a new way to transform the old industrial bases, the municipal government made a strategic decision to develop urban industry. With the area between the First Ring Road and the Third Ring Road as a focus, in principle, a central urban area was planned and laid out with a municipal urban industrial park, and there were 8 municipal urban industrial parks in the whole city, with a total planned area of 17.07 square kilometers. The construction of urban industrial parks is based on maximizing the utilization of inventory, vigorously developing the high value added urban industry with no pollution and no disturbance to the people which brings employment and pays attention to job-housing balance. Taking the Hanzheng Street Urban Industrial Park in Qiaokou as a pilot, construction has been carried out in Dijiao of Jiangan District, Zhoutou of Hanyang District and Baishazhou of Wuchang District; Qingshan District Workers Village Urban Industrial Park and Hongshan District Science and Technology Urban Industrial Park were built. Urban industrial parks have become a propeller for the reform of state-owned enterprises, a processing park with excellent brands, an incubator for private small and medium-sized enterprises, a reemployment base for laid-off workers and a new scenic spot for industrial tourism.

— Plan of Industrial Multiplication Demonstration Zone in the Distant Urban Area (2010)

During the "12th Five-Year Plan" period, the world economy entered a period of profound adjustment, industrial transfer accelerated, scientific and technological progress accelerated, and new industries gave birth to a new round of great development. To speed up the concentration of industrial strategic resources in Wuhan and promote Wuhan to rapidly become a core area gathering production factors and a growth pole with strong driving capacity in the central and western regions, the municipal Party Committee and the municipal government put forward the strategy of "revitalizing the city through industry" and vigorously implemented the "multiplication plan" for industrial development. During this period, under the background of the relocation of industries in the main urban areas and the "shit from the second industry to the third industry", the distant urban areas gradually became the main battlefield for the implementation of the "industrial multiplication" plan. For this reason, Planning of Industrial Multiplication Demonstration Zone in the Distant Urban Area (2010) takes the pillar industrial clusters such as Greater Optics Valley, China Auto City, Airport Economic Zone and Port Industrial Zone as the four growth poles, and adopt the idea of promoting the construction of new industrialization in the distant urban areas through the construction of demonstration parks. On the six main urban spatial axes, a new industrialization demonstration park will be built, influencing and driving other industrial clusters to form a characteristic industrial development plate. Finally, nine new demonstration parks and 14 general industrial parks are planned to be built. Among them, six new urban areas will each focus on the construction of an industrial demonstration park with a net industrial land area of not less than 20 square kilometers. The three central urban demonstration parks of Hanyang, Hongshan and Qingshan and other general industrial parks should have a scale of not less than 10 square kilometers.

— Comprehensive Planning of Four Major Sections in Wuhan (2014)

Limited by historical factors and development stages, the construction of Wuhan New Area has such problems as small scale, weak function and difficulty in building an independent city. To solve these problems, comprehensive planning of four major sectors has been carried out by promoting the centralized layout of industrial enterprises, the centralized construction of new urban centers and the centralized configuration of infrastructure, strengthening "industrial agglomeration, spatial concentration and intensive use of land" and further strengthening Wuhan's industrial "hard power" to build a national central city.

In Wuhan, the automobile and parts industry cluster in the southwest, the electronic information industry cluster in the southeast, and the steel and deep processing industry cluster in the east have shown obvious development patterns of industrial sectors. Comprehensive Planning of Four Major Sections in Wuhan (2014) defines the planning scope of industrial sectors such as Greater Auto City, Greater Optics Valley and Greater Port Economic Zones. Meanwhile, considering the rapid development needs of future airport areas, the Greater Airport Economic Zone has been added into the planning scope. The four major sectors follow the guideline of "independent city, city-industry connection, city-to-city interaction and complementation of industrial parks" to stimulate the aggregation effect of industrial development factors, build Wuhan into a national advanced manufacturing center, promote the "integration of production and city", drive the population to migrate and gather, and strengthen the economical and intensive use of land.

1
—
2
—

1
塔子湖体育中心总平面规划图（2005年）
Master Plan of Tazihu Sports Center (2005)

2
塔子湖体育中心鸟瞰图
A bird's eye view of Tazihu Sports Center

5 - 重大事件保障规划和建设
Planning and Construction to Ensure the Holding of Major Events

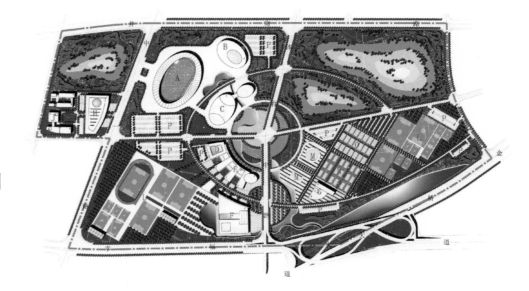

一 六城会——塔子湖体育中心

为举办第六届全国城市运动会，提升城市品位，促进全民健身运动的开展，2005年武汉市规划研究院编制了《塔子湖体育中心总体规划》，用地规模达133.75公顷。按照承办国际单项比赛、全国综合性运动会竞技比赛项目和市民健身娱乐的要求，规划形成集比赛、训练、体育休闲、娱乐等多功能于一体的大型市级体育基地，弥补了汉口地区大型体育设施不足的状况。塔子湖体育中心总体建设规模以4万座体育场、6千座体育馆和4千座游泳跳水馆为主体，辅以足球训练基地、网球中心、篮球俱乐部及全国最大规模的全民健身中心。

— The Sixth National City Games—Tazihu Sports Center

To hold the Sixth National City Games, improve the city's taste and promote the national fitness campaign, in 2005, the Tazihu Sports Center Master Plan was compiled, with a land area of 133.75 hectares. According to the requirements of hosting international individual events, national comprehensive games, competitive events and citizens' fitness and entertainment, a large-scale municipal sports base integrating competition, training, sports and leisure, entertainment and other functions was planned to make up for the lack of large-scale sports facilities in Hankou. Tazihu Sports Center is mainly composed of a 40,000-seat stadium, a 6,000-seat gymnasium and a 4,000-seat natatorium, supplemented by a football training base, a tennis center, a basketball club and the largest national fitness center in China.

一 八艺节——琴台文化艺术中心

2007年为高水平举办全国第八届文化艺术节，武汉市在梅子山下月湖畔兴建月湖文化艺术区，承担活动主会场功能，《月湖文化艺术区规划》（2003年）分为琴台古迹区、文化艺术中心等8个景区，将月湖、琴台、梅子山连为一体。该规划设计融入了俞伯牙与钟子期的典故，将"高山流水觅知音"的文化寓意于琴台大剧院的建筑形态中，深得专家和市民的好评，成为反映武汉文化艺术的标志性景观之一。

月湖文化艺术中心由琴台大剧院、音乐厅和公园组成。琴台大剧院建筑面积65650平方米、1800座，另有400座的多功能厅。音乐厅建筑面积36858平方米，设1602座交响乐厅，415座室内乐厅。公园51.5万平方米，是全国一流的艺术殿堂。

一 辛亥革命百年庆典——
武昌古城、武汉大道、东沙湖连通工程

2011年是辛亥革命100周年，这是继我国改革开放30周年、新中国成立60周年后举办的又一重要纪念活动。武汉市作为辛亥革命的首义地，成为了百年庆典活动的重要分会场。《武昌古城保护与复兴规划》（2008）通过对庆典活动的主动策划与安排，对辛亥遗址、剧院场馆、主要景点、交通枢纽、酒店宾馆等地点进行了详细的勘察与梳理，制订了迎送宾客路线、庆典活动路线、遗址参观路线、城市建设成就展示和民俗文化路线等路线，以武昌古城范围为核心、以武汉大道沿线为纽带、以东沙湖连通工程为亮点，开展了27处辛亥遗址、25条通道、15处门户地段和31处主要景点的综合整治工作，为顺利、成功举办该庆典奠定了较好的基础，有力地提升了城市环境面貌。

1	3
2	

1 月湖文化艺术区总平面规划图（2003年）
Master Plan of Yuehu Culture and Arts Zone (2003)

2 月湖文化艺术区鸟瞰图
A bird's eye view of Yuehu Culture and Arts Zone

3 武汉大道夜景鸟瞰图（2008年）
A bird's eye view of Wuhan Avenue at night (2008)

— The Eighth Culture & Arts Festival — Qintai Culture and Arts Center

To hold the 8th National Culture and Arts Festival at a high level, Wuhan has built a Yuehu Culture and Arts Zone by Yuehu Lake at the foot of Meizi Mountain as the main venue for the event. Eight scenic spots were planned in Planning of Yuehu Culture and Arts Zone (2003), including Qintai Historic Site Zone and Culture and Arts Center, integrating Yuehu Lake, Qintai and Meizi Mountain. The planning and design drew inspiration from the story of Yu Boya and Zhong Ziqi, and incorporated the cultural implication of "looking for bosom friends in high mountains and flowing water" into the architectural form of Qintai Grand Theater, which has won high praise from experts and citizens and has become a landmark landscape reflecting Wuhan's culture and art.

Yuehu Culture and Arts Center consists of Qintai Grand Theater, Concert Hall and Park. Qintai Grand Theater has a floor area of 65,650 square meters, with 1,800 seats and 400-seat multi-function hall. The concert hall has a floor area of 36,858 square meters, with 1,602-seat symphony hall and 415-seat chamber music hall. The park covers 515,000 square meters and is the first-class art hall in China.

— The centennial celebration of the Revolution of 1911 — Wuchang Ancient City, Wuhan Avenue and connection of Donghu Lake and Shahu Lake

The year 2011 marks the 100th anniversary of the 1911 Revolution of 1911, which is another important commemorative activity held after the 30th anniversary of China's reform and opening-up and the 60th anniversary of the founding of the People's Republic of China. As the birthplace of the Revolution of 1911, Wuhan is an important sub-venue for the centennial celebration. Through active planning and arrangement of the celebration activities, detailed investigation and sorting have been carried out for Xinhai Revolution ruins, theatres and venues, major scenic spots, transportation hubs, hotels, guesthouses and other locations. Planning for Conservation and Revival of Wuchang Ancient City (2008) draws up a route for welcoming and sending guests, a route for celebration activities, a route for visiting the ruins, a route for displaying achievements in urban construction and folk culture, and a route for business investigation. With the scope of Wuchang Ancient City as the core, Wuhan Avenue as the link, and the Donghu Lake and Shahu Lake connection project as the highlight, the comprehensive renovation has been carried out for 27 Xinhai Revolution ruins, 25 passages, 15 gateway sections and 31 major scenic spots, laying a good foundation for the smooth and successful holding of the event and effectively improving the urban environment.

1　楚河汉街实景
Photo of Chu River and Han Street

2　东、沙湖连通工程规划总平面图（2009 年）
Master Plan of Donghu Lake and Shahu Lake Connection Project (2009)

3　首义板块鸟瞰图（2008 年）
A bird's eye view of the Shouyi area (2008)

第肆章 · 区域一体化时期的规划

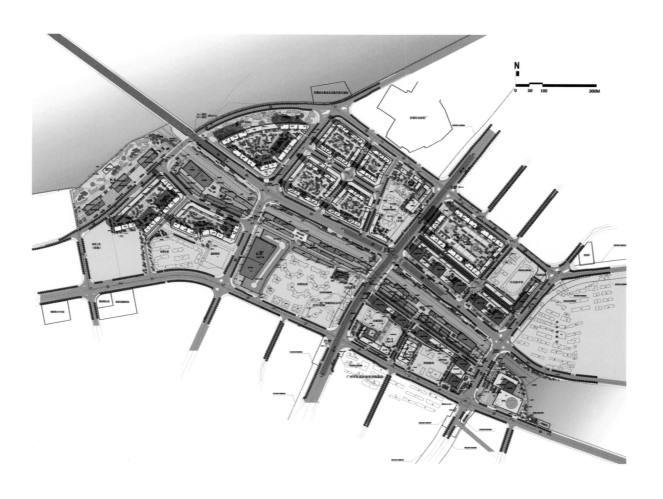

二、新时代转型期的规划
Planning During Transformation in the New Era

2012 - present

2011年，中国城镇化率达到50%，这一年，武汉市第十二次党代会提出了"建设国家中心城市、复兴大武汉"的目标，这是武汉新的发展起点。2015年，中央城市工作会议提出城市工作以人民为中心、"一个尊重、五个统筹"新要求，这些为做好新时期城市工作指明了方向，加快转变城市发展方式，不断提升城市环境质量、人民生活质量、城市竞争力，走出一条具有中国特色、武汉特点的城市发展之路。2018年4月，习近平总书记在视察湖北期间指出"推动长江经济带发展要正确处理好的'五大关系'"，"提出了'四个切实'要求"，对长江经济带高质量发展进行了系统部署。同时，国家多重战略聚焦武汉，"一带一路"、长江经济带、长江中游城市群等战略叠加，武汉市委市政府抢抓机遇，提出了建设"三化"大武汉和国家中心城市，全面复兴大武汉的战略目标。2018年，武汉在国内外城市排名中的位次有了很大提升，在经济日报中国城市综合实力排行榜中，武汉成为"中国第五城"，也就是"北上广深武"；在全球化与世界城市研究网络发布的全球城市体系排名中，武汉被列为国际二线城市，与柏林、费城等著名城市处在同一水平线上。

综合国内外形势变化，武汉既需要对接国家"两个一百年"目标，勇担国家赋予的重要使命，继续乘风破浪大步向前，又需要实现经济、技术、社会、空间等发展模式上的转型，更重要的是探索能够传承荆楚文化、符合生态文明建设要求的城市发展方式，走出具有武汉特色的现代化城市发展新路子。

In 2011, China's urbanization rate reached 50%, and the goal of "building a national central city and rejuvenating Wuhan" was also put forward at the 12th Party Congress of Wuhan in the same year, which was a new starting point for Wuhan's development. In 2015, the new requirement of "one respect and five overall plans" that urban work should be people-centered was put forward at the Central Urban Work Conference. These have set the direction for urban work in the new era, accelerated the transformation of urban development mode, continuously improved the quality of urban environment and people's life and the competitiveness of the city, and led to the road of urban development with Chinese and Wuhan characteristics. In April 2018, General Secretary Xi Jinping pointed out during his visit to Hubei that "to promote the development of the Yangtze River Economic Belt, we must properly handle the "five major relationships" and "put forward the 'four practical' requirements", and systematically deployed the high-quality development of the Yangtze River Economic Belt. Meanwhile, the national multiple strategies are focused on Wuhan; "the Belt and Road Initiative", the Yangtze River Economic Belt, the city cluster in the middle reaches of the Yangtze River and other strategies are superimposed. The Municipal Party Committee and the Municipal Government seized the opportunity and put forward the strategic goal of building a modern, international and ecological Wuhan and a national central city and comprehensively rejuvenating Wuhan. In 2018, Wuhan's position in the ranking of cities at home and abroad has been greatly improved. In the ranking of cities in China with comprehensive strength in Economic Daily, Wuhan has become "the fifth city in China" after Beijing, Shanghai, Guangzhou and Shenzhen. In the global urban system ranking released by the Globalization and World Cities Research Network, Wuhan is listed as a second-tier international city at the same level as famous cities such as Berlin and Philadelphia.

In view of the changes in the internal and external situations, Wuhan needs not only to meet the China's Two Centenary Goals, bravely shoulder the important mission entrusted by China, and continue to make great strides forward through the wind and waves, but also to realize the transformation of economic, technological, social and spatial development modes. More importantly, it needs to explore a new way of urban development that can inherit Jingchu Culture and satisfy the requirements of ecological civilization construction, and find a new way of modern urban development with Wuhan characteristics.

1 – 立足百年的 2049 远景发展战略研究

Research on the Long-Range 2049 Long-Term Development Strategy

2049 是新中国成立百年，也是武汉解放百年，一个城市发展需要"百年大计"，才能避免走弯路、错路。《武汉 2049 远景发展战略规划》（2016 年）编制目的并不是请谁批准，更没有想在一定范围内把武汉做大、当龙头的奢望，而是更为脚踏实地地为武汉谋划未来的发展方向。"武汉 2049"核心在于解决三大问题：

2049 marks the centennial of the founding of People's Republic of China and the liberation of Wuhan. The development of a city needs a "long-range plan" to avoid detours and mistakes. Wuhan 2049 Long-term Development Strategic Planning (2016) is not formulated to ask for approval, or build Wuhan into a large and leading city within a certain range, but to plan the future development direction for Wuhan in a more down-to-earth way. The core of "Wuhan 2049" lies in solving three major problems:

第一，武汉的目标是什么？

工业化时期，城市发展的目标更多聚焦经济增长的速度和规模，工业倍增和城建攻坚成为自然选择，进入后工业化时期，城市发展的目标更加多元，从关注经济增长转向关注生态、社会、人文等。在"物质主义"向"后物质主义"的时代转变下，2049 年的武汉，应该成为一个更具竞争力、更可持续发展的世界城市。

第二，明确武汉不能做什么？

我们可能很难准确地预见 2049 年具体的发展情况，但江、湖、山体不能够再填挖，城市不能再摊大饼、产业发展要更集约等应是城市发展遵守的基本底线。

第三，确定武汉该做什么？

也就是总目标指引下的五个城市的可持续目标。要通过生态安全、生态底线、蓝绿网络与低碳发展构建绿色的城市；要通过打造中部地区的国际交通枢纽、物流运营枢纽来建设高效的城市；通过提供优质的居住环境来建设宜居的城市；通过营造地域文化精神、空间场所来建设包容的城市；通过"主城区"优化核心职能、"四个次区域"引领城市外围、轴线带动"1+8"城市圈等构建活力的城市。

Firstly, what's the goal of Wuhan?

During the industrialization, urban development focused more on the speed and scale of economic growth. Industrial multiplication and breakthroughs in urban construction were the natural choices. In the period of post industrialization, the goals of urban development were more diversified, from economic growth to ecology, society and humanity. In the transformation from "materialism" to "post-materialism", Wuhan in 2049 will be a more competitive and sustainable global city.

Secondly, what can't Wuhan do?

It may be difficult for us to accurately predict the specific development in 2049, but rivers, lakes and mountains can no longer be filled and excavated, cities can no longer spread disorderly, and industrial development should be more intensive; these are the bottom lines to follow.

Thirdly, what should Wuhan do?

It's about the sustainable goal of the five cities under the guidance of the general objective. It's necessary to build a green city through ecological security, ecological bottom line, blue-green network and low-carbon development; to construct an efficient city by building an international transportation hub and logistics operation hub in the central region; to forge a livable city by providing high-quality dwelling environment; to build an inclusive city by creating regional cultural spirit and spatial place; and to optimize the core functions of the "main urban area", "four sub-regions" to lead the outskirts of the city, and to drive the "1+8" city circle axially to build a vibrant city.

2 – 新时代国土空间规划的探索

Exploration of Land and Space Planning in the New Ara

— "1331"引导的 2035 版城市总体规划

全球化时代，国与国之间的竞争，本质上是城市与城市之间的竞争，也是以城市群为特征的区域之间的竞争。

十八大以来，我国的区域发展战略经历了从板块到轴带的转变，从西部开发、东北振兴、中部崛起、东部率先等板块发展战略，转变到"一带一路"、京津冀协同、长江经济带等轴带发展战略，聚焦于沿海沿江沿线经济带为主的纵向横向经济轴带。2015 年中央城市工作会议明确提出，要以城市群为主体形态，科学规划城市空间布局，实现紧凑集约、高效绿色发展。

近年来，武汉奋起直追，发展成就有目共睹，在 2018 年的全球城市体系排名中，武汉被列为国际二线城市，与柏林、费城等著名城市处在同一水平线上。城市定位逐渐由区域中心向国家中心升级，城市空间结构逐渐呈现出"主城—主城 + 卫星城—都市发展区—大都市区"的发展态势。立足于国家战略要求和区域责任，对接"两个一百年"奋斗目标，武汉启动了 2035 版《武汉市城市总体规划（2017～2035 年）》的编制工作。

规划确定武汉 2035 年城市发展目标为"创新引领的全球城市，江风湖韵的美丽武汉"。在此基础上，将城市性质确定为"国家中心城市，全国重要的科技创新中心、现代服务中心、先进制造中心和综合交通中心，国际滨水文化名城"。规划预测 2035 年全市常住人口规模达 1660 万人，并按 2000 万管理服务人口配置基础设施和公共服务设施。

规划构建"1331"武汉全域空间格局。"1"为 1 个主城，突出现代服务和环境品质提升，实施"两降两增两保"策略，即：降低人口、建筑密度，增加绿色开放空间、公共服务设施，保护历史文化街区、山体湖泊及周边环境。"3"为 3 个副城，突出战略功能和美丽城市打造，分别是光谷、车都、临空副城，承载国家中心城市核心职能，每一个副城按照"大城市"标准，建设功能突出、配套完善的综合性城市。第二个"3"为 3 个新城组群，突出产城融合和宜居宜业发展，分别为东部、南部、西部新城组群，承载国家中心城市重要职能，每一组群均按照"中等城市"标准建设产城融合新城。后一个"1"为长江新城，面向未来，以超前理念、世界眼光，强化未来产业、特色资源禀赋和发展潜力，注重发展弹性，打造践行新发展理念的未来城市样板。

对外围非集中建设区，构建"功能小镇 + 生态村庄 + 郊野公园"的功能体系。依托山水资源和区位优势，推进功能小镇建设，形成田园化生态村庄体系，构建复合型郊野公园集群。

— 中心城区、新城区分区规划

为保障武汉市新一轮城市总体规划战略性内容落地，充分吸纳各区发展诉求，2016 年 10 月，在武汉市委市政府的综合统筹下，市国土资源和规划局会同各中心城区政府开展分区规划大纲编制工作，会同各新城区政府、开发区管委会开展新城区分区规划编制工作。两项工作作为国土空间规划改革过渡期控规编制和指导建设的上位依据，也将作为下一步开展区级国土空间规划编制工作的基础。

在规划内容上，中心城区分区规划大纲落实了总规结构性和强制性内容，同时结合各区特色，对接党的十九大精神和市区"两会"要求，增加了各区的发展重难点问题、城市设计等内容的研究，明确了各区目标定位、空间结构、功能特色、重大项目、图斑处置等核心问题，并指导了下位控规升级版工作，有效地服务了各区经济、社会的发展。新城区分区规划重点关注了非集中建设区规划研究，探索郊野公园、乡村振兴等创新规划的发展策略，进一步深化了各区综合交通、城市设计等重点专项，并加强与土地利用总体规划衔接，以实现支撑总规、指导控规的目标。

— The 2035 Urban Master Plan guided by "1331"

According to the Plan, the urban development goal of Wuhan is to become "a global city led by innovation, a beautiful Wuhan with beautiful scenery" by 2035. On this basis, the nature of the city is determined as "national central city, national important center for scientific and technological innovation, modern service center, advanced manufacturing center, integrated transport center, and international well-known waterfront cultural city". It is predicted that by 2035, the city's permanent population will reach 16.6 million, and infrastructure and public service facilities will be allocated to manage and serve a population of 20 million.

The "1331" Wuhan overall spatial layout has been planned and built. The first "1" means one main city. It highlights the improvement of modern services and environmental quality and implements the strategy of "two reductions, two increases and two protections", namely, reducing the density of population and buildings; increasing green open spaces and public service facilities; and protecting historical and cultural blocks, mountains and lakes and the surrounding environment. "3" means three sub-cities. It highlights the strategic functions and constitution of a beautiful city. There are Optics Valley, Capital of Vehicles and Airport sub-cities, carrying the core functions of the national central city. Each sub-city follows the "metropolis" standard, building a comprehensive city with prominent functions and complete supporting facilities. The second "3" is the three new city clusters, highlighting the city-industry integration and livable and industrial development. There are the eastern, southern and western new city clusters, bearing the important functions of the national central city. Each cluster is constructed in accordance with the "medium-sized city" standard, building

第肆章 · 区域一体化时期的规划

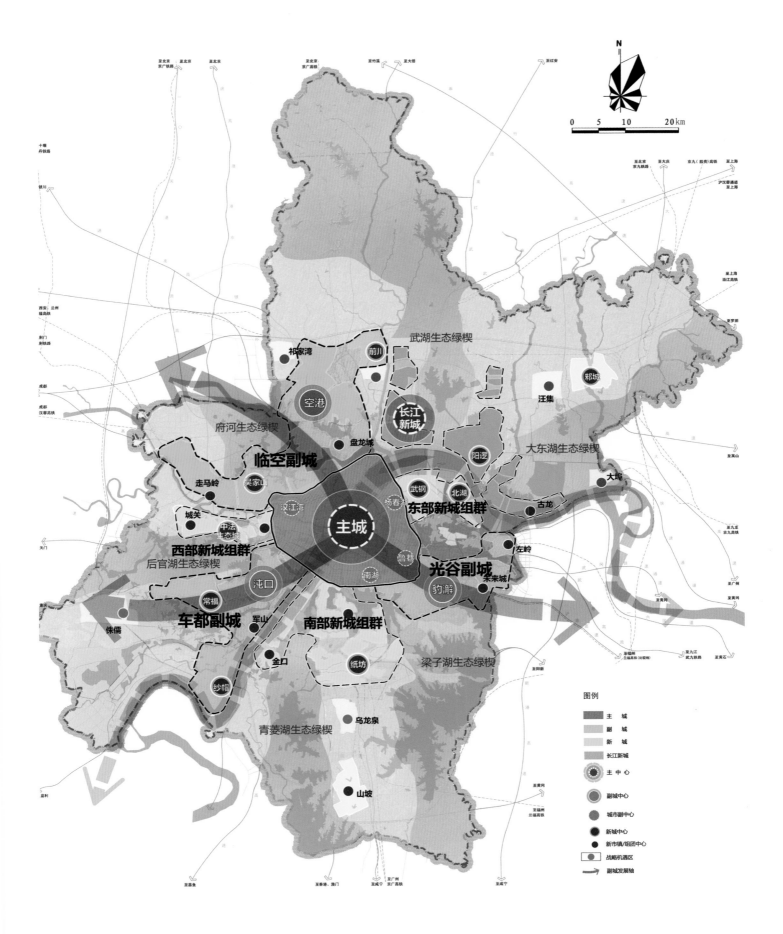

2035 版城市总体规划·"1331"
的空间布局图

2035 Urban Master Plan —
Spatial Layout of "1331"

a new city of city-industry integration. The last "1" means new urban area of the Yangtze River. The future industries, characteristic resources endowments and development potential should be strengthened with a future orientation, advanced ideas and global vision. Attention should be paid to the flexibility of development and a model future city that implements the new development ideas should be created.

For the peripheral non-centralized construction area, the functional system of "functional town + ecological village + country park" should be constructed. Relying on landscape resources and geographical advantages, the construction of functional towns should be promoted to form an idyllic ecological village system and build a compound country park cluster.

In the era of globalization, the competition between countries is essentially the competition between cities and regions characterized by city clusters.

Since the 18th National Congress of the Communist Party of China, China's regional development strategy has undergone a transformation from plate to axial belt, from plate development strategies such as Great Western Development, Revitalization of Northeast China, Rise of Central China, and priority to development of eastern China, to axial development strategies such as "The Belt and Road Initiative", Beijing-Tianjin-Hebei Cooperation, and Yangtze River Economic Belt, focusing on the longitudinal and horizontal economic axial belt dominated by the economic belt along the coast, river and railway. It was clearly proposed in the 2015 Central Urban Work Conference that city clusters should be taken as the main form so as to scientifically plan the urban spatial layout and realize compact, intensive, efficient and green development.

In recent years, Wuhan has been catching up and its development achievements are obvious to all. In the 2018 global urban system ranking, Wuhan was listed as a second-tier international city at the same level as Berlin, Philadelphia and other famous cities. The urban positioning has been gradually upgraded from a regional center to a national one, and the urban spatial structure has gradually shown a development trend of "main urban area — main urban area + satellite city — urban development area — metropolitan area". Based on the national strategic requirements

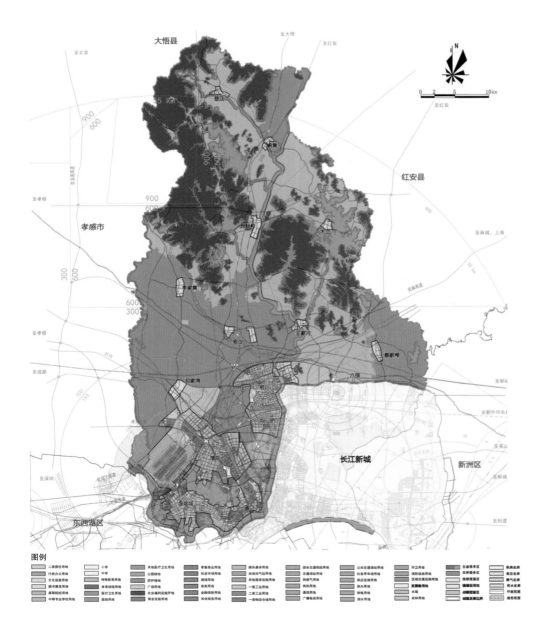

and regional responsibilities and in line with the Two Centenary Goals, Wuhan started the compilation of version 2035 Wuhan City Master Plan (2017—2035).

— Zoning Plans for Central and New Urban Areas

To ensure that the implementation of a new round of strategic urban planning in Wuhan and the absorption of development demands from all districts, in October 2016, under the comprehensive coordination of the Municipal Party Committee and the Municipal Government, the Wuhan Land and Resources Planning Bureau worked with the governments of the central urban areas to prepare the outline of the zoning plan, and with the governments of new urban areas and the development zone administrative committee to prepare the zoning plan of the new areas. The two tasks will serve as the clear basis for the preparation and guidance of the regulatory plan for the transitional period of land and space planning reform, and will also serve as the basis for the next step in the preparation of district-level land and space planning.

In terms of planning content, the outline of the central urban area zoning plan has been implemented along with the structural and mandatory contents of the general rules, combined with the characteristics of each area, in line with the spirits of the 19th National Congress of the Communist Party of China and the requirements of the "Two Sessions" in the urban district, enhanced the research on the important and difficult issues in the development of each district, urban design and other contents, clarified the core issues such as the target orientation, spatial structure, functional characteristics, major projects, parcels disposal and so on of each district, and guided the work of upgrading the subordinate regulatory plan and effectively promoted the economic and social development of each district. The new area zoning plan focuses on the planning and research of non-centralized construction areas, explores the development strategies of innovative planning such as country parks and rural revitalization, further deepens key projects such as comprehensive transportation and urban design in various districts, and strengthens the connection with the overall land use planning to achieve the goals of supporting the overall planning and guiding the regulatory plan.

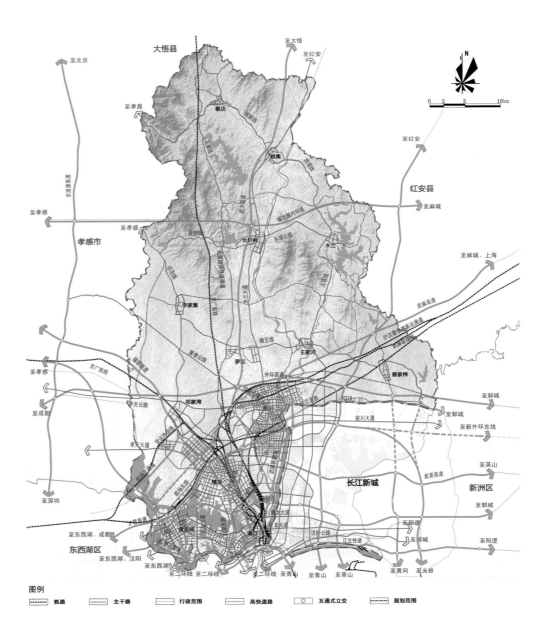

1

黄陂区分区规划·城乡空间布局规划图

Zoning Plan for Huangpi District: Urban and Rural Spatial Layout Plan

2

黄陂区分区规划·交通体系规划图

Zoning Plan for Huangpi District: Traffic System Planning

一 主城区控制性详细规划升级版

按照武汉市建设国家中心城市、打造"城市、经济、民生"三个升级版的总体目标，为全面提升城市功能品质，推进规划编制和管理的精细化，武汉市国土资源和规划局组织开展了《主城区控制性详细规划导则升级版》（2017年）的编制工作。

规划重点升级公共服务设施、绿化、交通和市政等公益性设施，深化落实总体规划、分区规划战略设想，专项规划突出了以下8个方面的内容。一是建立了"立体化"的控规管控体系，由地面单维管控向地面、地上和地下三维立体管控升级；二是提出了"分区化"的控规升级模式，管理模式由均质化向动静分区差异化升级；三是制定了"精细化"的控规组织和编制方式，全面提升基础数据精度、设施配置单元及升级管控要素；四是构建了"动态化"的控规评估机制，结合大数据信息平台，评估方法由静态问题分析向动态综合评估升级；五是突出了"人本化"的控规编制理念，细分了设施配置的种类和标准；六是强化了"品质化"的控规提档重点，加强了公共空间、道路网密度、社区生活圈和景观环境等4个方面的升级；七是提升了"实施化"的控规管控策略，划定新管理单元，测算拆建比，提出升级实施策略；八是完善了"规范化"的控规管理规定，优化控规编审、维护、修改程序，明确控规与专项规划、规划管理的工作关系。

一 国土空间规划的探索

建立国土空间规划体系是生态文明体制改革的重要环节，是城市治理体系和治理能力现代化的重要举措。国土空间规划编制，需要把握规划责权、模式、内容和技术的转变。一是要厘清责权边界，"一级政府、一级事权、一级规划"是国土空间规划责权边界确定的基本逻辑。从规划编制上，要围绕地方政府事权，制定解决地方面临问题和实际需求的空间规划和空间政策；从规划审批上，要对应上级政府事权，提炼形成精简、刚性

1 2

1
控规升级版动静分区图
（2017年）

Upgraded Dynamic and Static Areas of the Regulatory Plan (2017)

2
品质化提升绿化及公园图
（2017年）

Quality-oriented Improvement of Greening and Parks (2017)

的空间规划报批内容；二是要适应模式转化，以促进高质量发展为目标导向，空间规划逐步由"建设型"向"治理型"转变，在生态文明建设和人民美好生活需求保障的指引下，更加注重生态修补、空间织补等，强调发展质量和发展效益；三是把握内容重点，坚持"守住底线"与"高质量发展"规划内容相结合，科学布局生产空间、生活空间、生态空间，处理好战略定位、空间格局、要素配置关系，使空间布局同战略定位一致、要素配置同战略定位统一，健全用途管制；四是创新规划技术，形成更加科学的国土空间规划的"观察—分析—选择"过程。

规划遵循"共抓长江大保护"的基本思想，将"生态优先、绿色发展"的理念作为规划主线，通过锚固城市生态框架严守生态空间，加强创新驱动，构建绿色发展体系，构建长江沿线3~5公里腹地分层次的空间管控体系。突出带动区域发展的责任，适应武汉当前城市区域化、区域一体化发展的趋势，以1小时交通通勤距离为基础（60~80公里半径），综合考虑经济、人口、生态、设施等要素，构建武汉大都市区，加强武鄂、汉孝、武咸、武仙洪等临界地区的空间、功能、交通、生态等协调一体。

在统一现状底图方面，做实做细"三调"工作，在开展"双评价"和规划实施评价的基础上，构建统一空间管控体系。促进"全域"管控，着力解决现行空间类规划"重城轻乡"的问题，重点提升非集中建设区管控短板，以乡村全面振兴为重点，构建农业农村地区空间规划、农村产权、乡村功能和空间治理的"编管合一"体系。加强"全要素"管控，按照系统治理的思路，制订各类用地之间的转换规则，强化对非耕农地、生态用地的用途管控，实现全类型资源要素的用途管控，构建"山水林田湖草"生命共同体。坚持"全过程"管控，将自然资源的保护、修复和建设的全过程结合，从"单纯划线"走向"主动实施"。突出"全维度"管控，围绕国土空间用途管制制度，构建"分区、用途、指标、名录"互相支撑的多手段综合管控体系。

— Upgraded Version of the Detailed Regulatory Plan for Main Urban Areas

According to Wuhan's overall goal of building a national central city and three upgraded versions of "city, economy and people's livelihood" and in order to comprehensively improve the functional quality of the city and promote the refinement of planning formulation and management, Wuhan Land Resources and Planning Bureau organized and carried out the compilation of the Upgraded Guidelines on Detailed Regulatory Plan for Main Urban Areas (2017).

The plan attaches great importance on upgrading public welfare facilities for public service, greening, transportation and municipal administration and so on, deepening the implementation of the strategic assumption of the master plan and zoning plan. The special plan highlights eight aspects. Firstly, a "three-dimensional" control system has been established to upgrade from one-dimensional ground control to three-dimensional ground, aboveground and underground one. Secondly, the upgraded mode of "zoning" control has been put forward with the management mode upgraded from homogenization to dynamic and static zoning. Thirdly, the "refined" organization and compilation method of the regulatory plan has been formulated to comprehensively improve the accuracy of basic data, configure the facility units and upgrade the control elements. Fourthly, the "dynamic" regulatory plan evaluation mechanism has been constructed. Combined with the big data information platform, the evaluation method has been upgraded from static problem analysis to dynamically comprehensive assessment. Fifthly, the "people-oriented" compilation philosophy of the regulatory plan has been highlighted and the types and standards of facility configuration have been specified. Sixthly, the upgrading focus of "quality orientation" for the regulatory plan has been stressed and the upgrading has been enhanced in the four aspects of public space, road network density, community life circle, and landscape environment. Seventhly, the "implementation" control strategy of the regulatory plan has been improved to designate the new administrative unit, measure and calculate the demolition and construction ratio and propose the upgrading implementation strategy. Eighthly, the "standard" administrative provisions of the regulatory plan have been perfected to optimize the procedures for compilation, maintenance and modification of the regulatory plan and clarify the relationship between the regulatory plan and special plan and planning management.

— Exploration of Land and Space Planning

The establishment of a land and space planning system is an important link in the reform of the ecological civilization system and a significant measure to modernize the urban governance system and governance capacity. For the compilation of land and space plans, it's necessary to grasp the transformation of planning responsibility and authority, modes, contents and technologies. Firstly, it is necessary to clarify the boundary of responsibility and authority. "First-level government, administrative power and planning" is the basic logic for determining the boundary of responsibility and authority in land and space planning. In planning formulation, it is necessary to formulate space plan and policy around the local governmental administrative power to solve the problems and actual needs faced by the local government. In planning approval, it is necessary to refine and form a concise and rigid content of space planning approval corresponding to the administrative power of the higher-level government. Secondly, the model

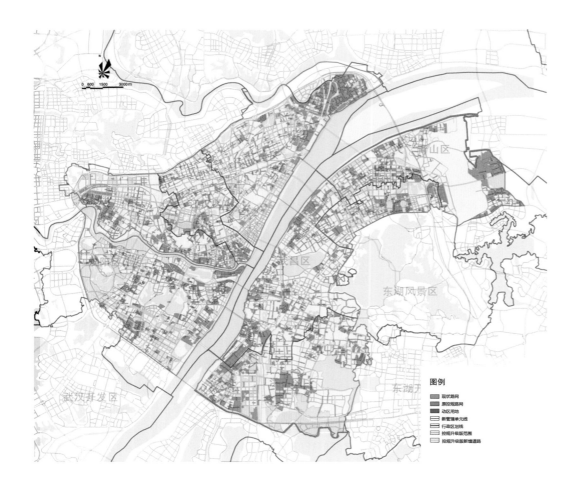

transformation should be adapted so as to promote the goal orientation of high-quality development. The space planning should be changed from "construction" to "governance" gradually. Under the guidance of building an ecological civilization and meeting the people's needs for a better life, attention should be paid to the ecological remedy and spatial darning in order to emphasize the development quality and outcome. Thirdly, the key contents should be grasped and persistence in "holding the bottom line" and planning content of "high-quality development" should be combined in order to scientifically arrange the space for production, life and ecology, handle the relationship between the strategic positioning, spatial pattern and element allocation, accord spatial arrangement and strategic orientation, unify element allocation and strategic positioning and improve use control. Fourthly, the planning technology should be innovated to form a more scientific process of "observation-analysis-selection" for land and space planning.

Following the basic idea of "joint efforts to protect the Yangtze River" and taking the philosophy of "ecological priority and green development" as the main line of planning, we will strictly adhere to the ecological space by anchoring the ecological framework of the city, strengthen innovation-driven construction of a green development system, and build a hierarchical spatial control system in the hinterland 3～5 kilometers along the Yangtze River. The responsibility to promote regional development is highlighted to adapt to the current trend of Wuhan's urban regionalization and regional integration. Based on the one-hour traffic commuting distance (a radius of 60～80 kilometers), considerations are given to the factors such as economy, population, ecology, facility and so on comprehensively so as to build the Wuhan City Circle and enhance the coordination and integration of space, functions, transportation, ecology and so on in critical areas such as Wuhan and Ezhou, Wuhan and Xiaogan, Wuhan and Xianning, Wuhan, Xiantao and Honghu.

As regards the base map of current situation, the "Third National Land Survey" should be carried out in a meticulous and down-to-earth way to build a unified spatial control and management system on the basis of the "double evaluation" and planning implementation evaluation. "All-round" control and management should be promoted to strive to solve the problems in "preferring cities to villages" in the current space planning. Efforts should be made to eliminate the weakness in the management and control on the non-centralized construction area, focus on the overall revitalization of villages and build a "combination of management and organization" system for space planning, rural property rights, rural functions and spatial governance in agricultural and rural areas. Control on "all factors" should be enhanced according to the idea of systematical governance, to formulate conversion rules between various types of lands, strengthen the use control of non-cultivated farmland and ecological land, realize the use control of all types of resource factors, and build a life community of mountains, rivers, forests, fields, lakes and grasses. Control on the "whole process" should be maintained to combine the entire process of natural resources protection, restoration and construction and change from "simple arrangement" to "active implementation". The "all-dimensional" control should be highlighted for the use control system of the land and space so as to set up the comprehensive control system of multiple means with mutual support of "zoning, uses, indicators and directories".

1

道路网升级图（2017年）

Upgraded Road Network (2017)

2

规划邻里中心升级图（2017年）

Upgraded Planning of Neighborhood Centers (2017)

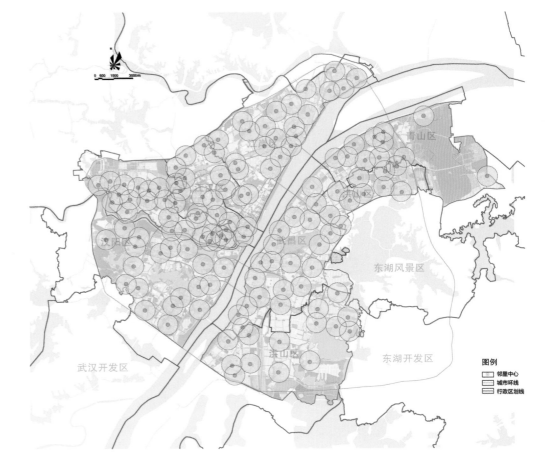

3 - 生态文明建设战略引领下的空间保护规划
Space Protection Planning Under the Guidance of the Ecological Civilization Construction Strategy

党的十八大、十八届三中全会将生态文明建设作为国家战略，强调要"划定生态保护红线""用制度保护生态环境""要实施重大生态修复工程，增强生态产品生产能力""优化国土空间开发格局""共抓长江大保护，实现高质量发展"。随后，国家又出台了生态文明体制改革总体方案，以及加快推进生态文明建设的意见，标志着我国生态文明建设的顶层设计和实施路径已经明确。

从这些年武汉的城市发展和建设来看，总体上"摊大饼"的状况还没有得到完全遏制，中心城区和新城区的边界不是在强化，而是在不断模糊。为此，迫切需要更加关注城市生态空间的规划、管控和实施。这一阶段，武汉从关注全域生态资源的规划和管控起步，建立了一套"生态框架+划线控制+法规保障"的全链条路径；从关注局域生态资源的修复和实施出发，让"轴、楔、环、廊"的生态框架逐步落地；从关注流域生态资源的跨界保护开始，聚焦到长江大保护和武汉城市圈等更大区域范围。

一 关注全域生态资源的规划和管控

● 全域生态框架规划（2012年）

武汉作为中部地区的中心城市、国家"两型社会"建设综合配套改革试验区，在2010年版《武汉市城市总体规划（2010~2020年）》编制完成后，即开展了持续的生态空间相关规划研究，对生态空间规划体系构建、以基本生态控制线为核心的生态空间管控模式及相关政策法规等，均进行了积极的探索和实践，为推进武汉生态文明建设、促进特大城市集约有序发展奠定了坚实基础。

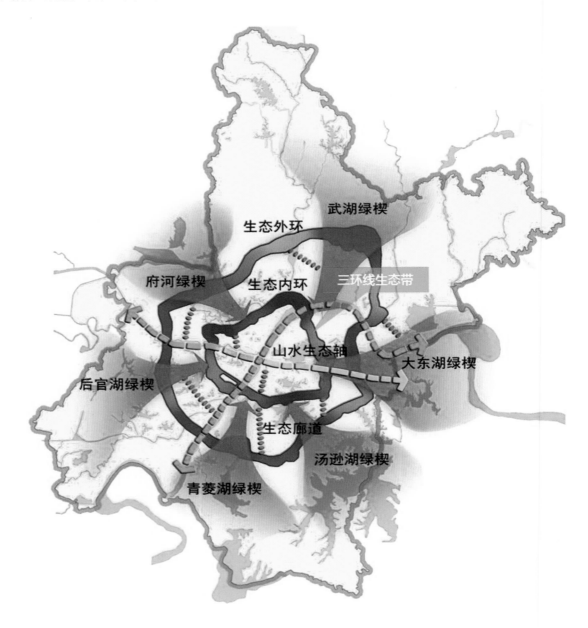

武汉市生态框架体系
Wuhan's Ecological Framework System

武汉城市生态空间规划管控的总体思路是：从规划编制、管控模式和政策法规制定三个环节，提出一套完整的城市生态空间管控的行动体系。

在规划编制上，构建从单纯划线到功能管控的全域生态空间规划体系；在具体做法上，全面对接各行业的专项规划，整合各部门对各类生态资源的保护要求，系统梳理山、水、自然保护区与水源保护区等资源型生态要素以及生态廊道、生态绿楔核心区等结构型生态要素，采取"分层叠加"的方式，实现了基本生态控制线的市域全覆盖，明确了全市生态底线区、生态发展区、弹性控制区的范围，确保了生态资源的"应保尽保"，完成了总体规划确定的生态框架"落地"。

● "两线三区"划定基本生态控制线（2014年）

在管控模式上，《武汉市城市开发边界划定》（2014）构建以基本生态控制线为核心的"两线三区"管控模式。即划定城市增长边界（UGB）和生态底线"两线"，形成集中建设区、生态发展区和生态底线区——"三区"，通过城市增长边界（UGB）反向确定基本生态控制线范围。针对生态空间的不同区位，进一步制定不同的分区管控策略。首先，严格准入，实现新增建设项目的有效管控，即按照生态底线区和生态发展区分别制定相应的准入要求，明确准入建设项目类型、相关建设控制指标，严格规定准入程序，严禁不符合准入条件的建设项目进入。其次，以"建"促"保"，实现生态功能的整体提升，推进郊野公园、绿道等重点生态工程建设，将生态公园建设与新农村建设、农业产业化发展捆绑结合。最后，分类解决，对既有建设项目进行妥善处置，提出"保留、整改或置换用地、迁移"三大类共七种具体的处置方式。

● "政府令、决定、条例"三步走（2012～2016年）

在法规体系上，建立从地方性法规到实施细则的生态空间政策法规体系保障。按照"政府令—人大决定—条例"三步走计划推进生态空间的地方性立法工作。2016年10月1日正式施行《武汉市基本生态控制线管理条例》，成为全国首个针对全市域生态空间的地方性法规。

The 18th National Congress of the Communist Party of China and the 3rd Plenary Session of the 18th Central Committee of the Communist Party of China took the construction of ecological civilization as a national strategy, stressing the need to "draw a boundary line for ecological protection", "protect the ecological environment with systems", "implement major ecological restoration projects to enhance the production capacity of ecological products", "optimize the pattern of land and space development", and "jointly protect the Yangtze River and realize high-quality development". Later, China has issued an overall plan for the reform of the ecological civilization system and put forward opinions on accelerating the construction of ecological civilization, which indicates that the top-level design and implementation path for the construction of ecological civilization have been made clear in China.

From the urban development and construction of Wuhan in recent years, the "urban sprawl" has not been completely curbed, and the boundary between the central urban areas and the new urban areas has been blurred instead of strengthened. To this end, it is urgent to pay more attention to the planning, control and implementation of urban ecological space. At this stage, Wuhan established a set of full-chain path of "ecological framework + layout control + legal protection" from the planning and control of regional ecological resources, gradually implemented the ecological framework of "axis, wedge, ring and corridor" from the restoration and implementation of local ecological resources, and focused on the Yangtze River protection and Wuhan City Circle and other larger regions from the cross-border protection of ecological resources in the river basin.

— Planning and control focusing on regional ecological resources

● *Regional Ecological Framework Plan (2012)*

As a central city in the central region and an experimental zone for the comprehensive supplementary reform of national construction of a "two-oriented" society, Wuhan has carried out continuous ecological space-related planning research after the preparation of the Version 2010 of Wuhan City Master Plan (2010—2020). It has actively explored and practiced the construction of the ecological space planning system, the ecological space control mode with the basic ecological control line as the core, and relevant policies and regulations and so on, laying a solid foundation for promoting the construction of Wuhan's ecological civilization and the intensive and orderly development of the metropolis.

The general idea of Wuhan's urban ecological space planning and control is to put forward a complete action system for urban ecological space control from the three aspects of planning compilation, control mode and formulation of policies and regulations.

For planning compilation, it has established a regional ecological and spatial planning system from simple arrangement to functional control. In terms of specific measures, it has fully integrated the special plans of various industries, integrated the protection requirements of various departments for various ecological resources, systematically combed the resource-based ecological elements such as mountain, water, nature reserves and water source conversation areas, and structural ecological elements such as ecological corridors and ecological green wedge core areas, and adopted the "layered superposition" method to realize the full coverage of the basic ecological control line in the urban area, define the scope of the urban ecological bottom line area, ecological development area and flexible control area, ensure the "maximum protection" of ecological resources and complete the "implementation" of the ecological framework determined by the master plan.

● *"Three Areas by Two Lines" delimits the basic ecological control line (2014)*

In terms of management and control mode, the management and control mode of "Three Areas by Two Lines" with the core of the basic ecological control line is constructed. That is, the "Two Lines" of the urban growth boundary (UGB)

and ecological bottom line will be defined to form "Three Areas" of concentrated construction area, ecological development area and ecological bottom line area, and the scope of basic ecological control line will be reversely determined the UGB. According to the different locations of ecological spaces, different zoning control strategies will be further developed. First of all, strict access is required to effectively control new construction projects, i.e. corresponding access requirements are formulated in accordance with the ecological bottom line area and the ecological development area. The types of construction projects are admitted and relevant construction control indicators are defined clearly. Access procedures are strictly stipulated and construction projects that do not meet the access conditions are strictly prohibited from entry. Secondly, the protection can be promoted by construction so as to realize the overall improvement of ecological functions, advance the construction of country parks, greenway and other key ecological projects, and combine the construction of ecological parks with the new rural construction and the development of agricultural industrialization. In the end, the solution by classification is adopted, the existing construction projects are properly disposed, and seven specific disposal methods are put forward in three categories of "reservation, rectification or replacement of land, and relocation".

● *Three steps of "government Directive, Decision and Regulation" (2012—2016)*

In regard to the system of law and regulations, the system of policies and regulations for protection of ecological spaces varying from the local laws and regulations to the implementing regulations is established. The local legislation of ecological space is advanced as planned on the basis of the three steps comprised of "government directives, decisions by the National People's Congress, and regulations". On October 1, 2016, the Regulations of Wuhan on the Administration of Basic Ecological Control Line was officially implemented, becoming the first local regulation on the ecological space in the entire city.

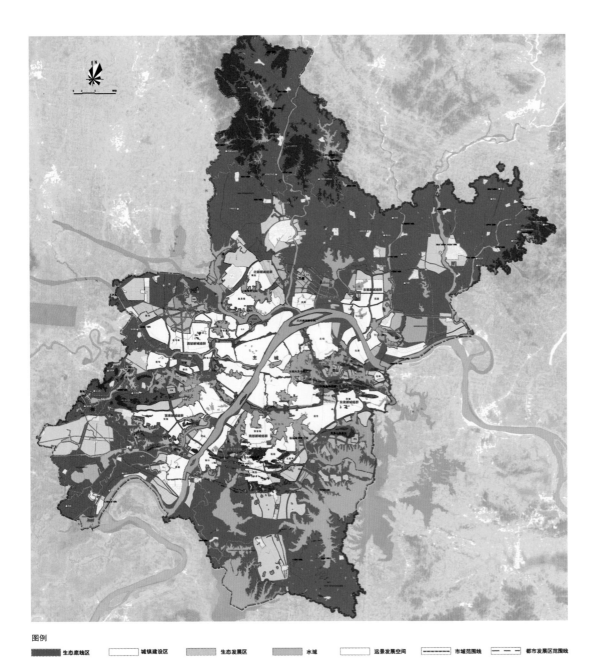

1

基本生态控制线范围图

Scope of the Basic Ecological Control Line

2

三环线绿化带"一环串多珠"生态景观结构（2014年）

Ecological Landscape Structure of "One Ring with Multiple Beads" in the Green Belt of the Third Ring Road (2014)

一 关注局域生态资源的修复和实施

● 环：
三环线城市生态带实施规划及园博园概念规划（2011~2014年）

三环线城市生态带作为武汉市生态内环，是承担主城与新城的隔离功能、防止城市无序蔓延的重要屏障。2011年，按照市委市政府"划定主城和新城组群生态隔离带，界定城市增长边界，防止城市无序蔓延"的要求，同时发挥环城绿带对周边地区发展的积极意义，塑造城市良好门户景观，武汉市国土资源和规划局开展了《三环线沿线综合规划》，明确了三环线生态带"一环多珠"空间布局。

市国土资源和规划局2014年又开展了《三环线城市生态带实施规划》的编制工作，划定了三环线城市生态带范围，对沿线91公里长绿化用地的保护开发做出针对性的指引，锁定了绿化用地边界，并对"森林带、生态区、33珠"提出建设指引。"森林带"着重塑造沿线较密集、大尺度、连续流动的绿化空间感受；"33珠"又分为6个生态郊野公园以及27个城市公园两类。其中"6大珠"强调原生态的景观感受，满足人们亲近大自然的愿望；"27小珠"强调日常文化活动场所的设计，满足居民休闲活动的需要。

该规划完成后，三环线沿线各区政府及相关部门启动实施了一系列重点建设工程，生态带建设完成90%，沿线串接公园完成约40%。最著名的莫过于《园博园概念规划（2011~2014年）》及实施设计。2012年5月10日，武汉市政府凭借将城市最大的垃圾场作为园博会主会场的大胆创意，获得了第十届中国国际园林博览会的承办权。该届园博会以"绿色连接你我，园林融入生活"为主题，举办一届"园林、彰显城市魅力、提升城市品质、融入城市生活"的创新示范型园博会。该规划期望以园林来治愈这片土地，强调自然、人和城市的共生、平衡。规划着力点一是强化垃圾场治理安全与掇山造景并重，通过对垃圾场土壤的样本分析，确定采用好氧修复技术和封场技术相结合的方式，对基地进行生态治理；二是以"慢排缓释"和"源头分散"为设计理念，优先利用雨水花园、植草沟、下沉式绿地等措施来组织排水，最大限度地实现雨水在城区的积存、渗透和净化，打造"海绵示范园区"；三是整体承续荆楚文化文脉，强调本土文化，展示"最湖北、最武汉"的文化元素，体现园博会的开放与兼容，创建大都会新景；四是以"节约型园林"的理念贯穿规划实施的全过程，通过"新理念、新工艺、新材料"的综合运用，充分阐释了节约型园林的内涵。

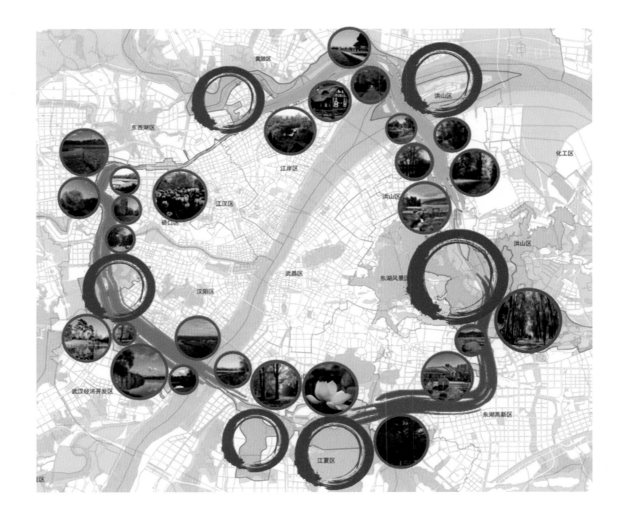

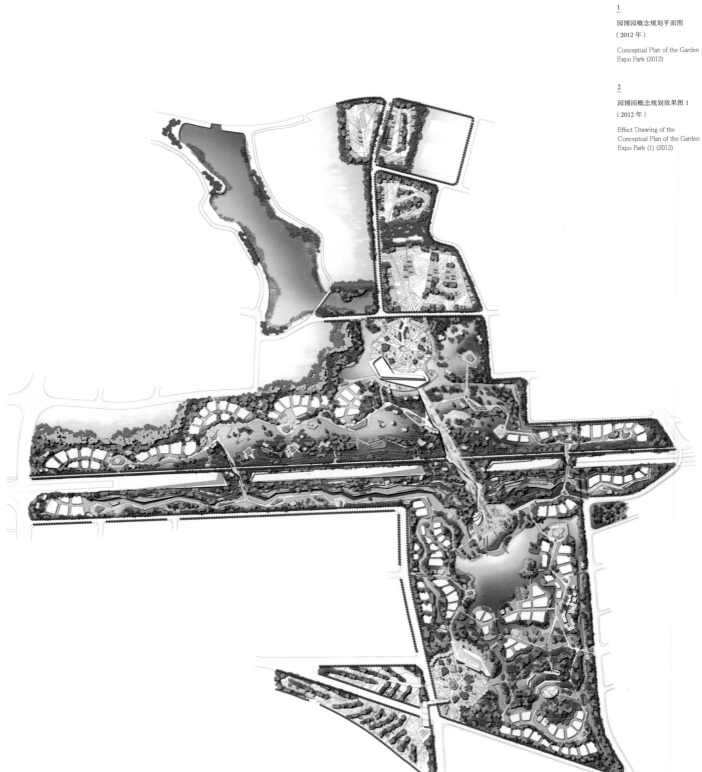

1

园博园概念规划平面图
（2012 年）

Conceptual Plan of the Garden Expo Park (2012)

2

园博园概念规划效果图 1
（2012 年）

Effect Drawing of the Conceptual Plan of the Garden Expo Park (1) (2012)

— Restoration and Implementation Focusing on Local Ecological Resources

● *RING:*
Implementation Plan of Urban Ecological Belt of the Third Ring Road and Conceptual Plan of Garden Expo Park (2011—2014)

As the inner ecological ring of Wuhan, the urban ecological belt of the Third Ring Road is an important barrier to isolate the main urban areas from the new urban areas and prevent urban sprawl. In 2011, in accordance with the requirements of the municipal Party committee and the municipal government to "delimit the ecological isolation zone between the main urban area and the new city cluster, define the urban growth boundary, and prevent urban sprawl", and meanwhile giving full play to the positive significance of the surrounding green belt for the development of surrounding areas, and shaping a good gateway landscape for the city, the Wuhan Land Resources and Planning Bureau launched the Comprehensive Plan along The Third Ring Road, and defined the spatial layout of the "one ring and many beads" of the ecological belt along the Third Ring Road.

Wuhan Land Resources and Planning Bureau began the preparation of the Implementation Plan for the Urban Ecological Belt of the Third Ring Road in 2014. It delineated the urban ecological belt of the Third Ring Road, provided targeted guidelines for the protection and development of the 91-kilometer-long green land along the line, locked the boundary of the green land, and proposed guidelines for construction of the "forest belt, ecological zone and 33 beads". The "forest belt" focused on shaping a dense, large-scale and continuous flow of green space along the line. "33 Beads" consisted of 6 ecological country parks and 27 city parks. Among them, "6 Big Beads" emphasized the original ecological landscape experience and satisfied people's desire to get close to nature. "27 Small Beads" emphasized the design of daily cultural activities places to meet the residents' needs of leisure activities.

After planning, the district governments and relevant departments along the Third Ring Road started to implement a series of key construction projects. 90% of the ecological belt construction was completed and about 40% of the parks connected in series along the route were completed. The most famous was The Conceptual Plan of the Garden Expo Park and the implementation design. On May 10, 2012, the Wuhan Municipal Government won the right to host the 10th China International Garden Expo with its bold idea of using the city's largest garbage dump as the main venue of the Expo. With the theme of "Green Joins You and Me, Garden Blends into Life", this Expo would be innovative and exemplary with the idea that "garden highlights city charm, promotes city quality and is blended into city life". The plan aimed to heal the land with gardens, emphasizing the symbiosis and balance between nature, people and the city. The first was to attach equal importance to strengthening the safety of waste yard management and making rockery and landscaping. Through the analysis of soil samples from waste yard, the combination of aerobic remediation technology and field closure technology was adopted to conduct ecological management of the base. The second was to take "slow discharge and slow release" and "scattered sources" as the design concepts, and give priority to the use of rainwater gardens, grass planting ditches, sunken green spaces and other measures to organize drainage, thus maximizing the accumulation, infiltration and purification of rainwater in urban areas, and building a "sponge demonstration park". The third was to carry on the whole Jingchu culture, emphasize the local culture, display the most representative elements of Hubei and Wuhan culture, embody the openness and compatibility of the Expo, and create a new metropolis. The fourth was to put the concept of "conservation-oriented garden" through the whole process of planning and implementation. Through the comprehensive application of "new ideas, new processes and new materials", the connotation of conservation-oriented garden was fully explained.

212 CHAPTER FOUR • Planning During Regional Integration

● 楔：
绿楔导则、后官湖绿道、东湖绿心概念规划及绿道实施规划
（2013~2018年）

根据《关于加强基本生态控制线管理的实施意见》，为深化完善基本生态控制线规划体系，积极推进生态绿楔控规导则编制工作，细化、落实基本生态控制线总体管控要求，实现都市发展区内控规全覆盖，武汉市自2013年以来开展了后官湖、青菱湖、武湖、汤逊湖、府河、大东湖等绿楔的控规编制工作，明确绿楔功能和空间结构，划定生态实施单元，提出单元内村庄发展、用地功能指引，做好"五线"及各类生态要素的规划管控。

特别是东湖绿道的建设，既是体现"让城市安静下来"城建理念转变的重要载体，也是主动实施大东湖绿楔的保护工程。《东湖生态绿心概念规划》（2017年）秉承"公众参与观、国际化视野、大区域视角、一体化思维、全过程规划"的规划理念，避开就绿道论绿道的狭隘规划观，全方位打造世界级东湖绿道。该规划通过打破传统的层级与尽端式绿道网建构方式，以大区域生态格局为根本，采取直接串联方式构建贯通城市重要功能区的"6+3"区域绿道体系，构成东湖"D+H"绿道结构，旨在建设一个以区域整体发展为背景的东湖绿道，实现东湖与城市的良性互动。

规划按"建设先易后难，串联现状重要景点、道路交通支撑及尊重公众调查意愿"等原则，明确建设时序，其中环郭郑湖梨园广场—磨山北门—枫多山——棵树段、磨山核心景区及环团湖落雁景区为一期启动段，全长28.7公里。结合湖中道、湖山道、磨山道、郊野道不同路段的地段特征，通过断面改造、主题区域设计，实现"漫步湖边、畅游湖中、走进森林、登上山顶"的市民体验。

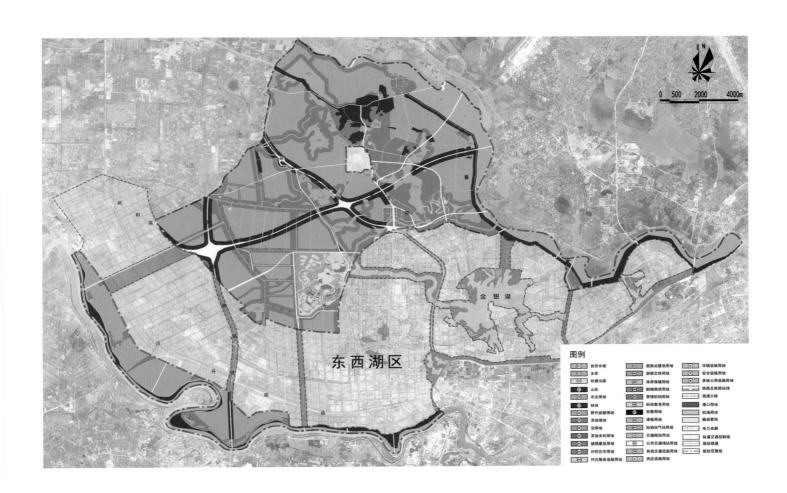

1–2 园博园概念规划效果图（2012年）
Effect Drawing of the Conceptual Plan of the Garden Expo Park (2012)

3 生态绿楔控规导则东西湖府河用地规划图
Guidelines for the Control of Ecological Green Wedges — Land Use Plan of Fu River in Dongxihu District

CHAPTER FOUR • Planning During Regional Integration

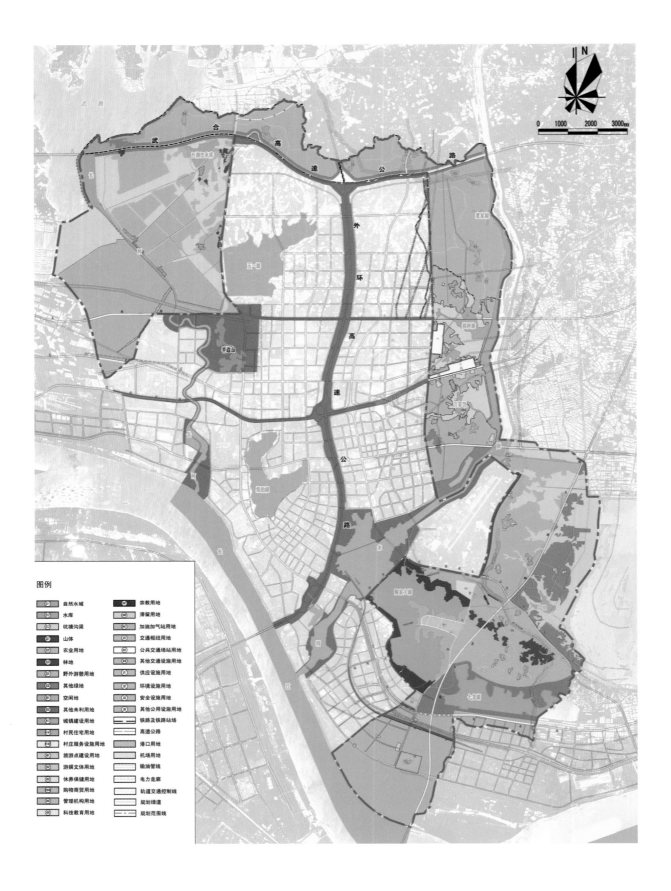

- **WEDGES:**

Green Wedge Guidelines, Houguan Lake Greenway, Donghu Oasis Conceptual Plan and Greenway Implementation Plan (2013~2018)

According to the Implementation Opinions on Strengthening the Management of Basic Ecological Control Lines, in order to deepen and improve the planning system of basic ecological control lines, actively promote the preparation of guidelines for ecological green wedge control rules, refine and implement the overall control requirements of basic ecological control lines, and achieve full coverage of internal control rules in urban development areas, Wuhan had carried out the preparation of control rules for green wedges such as Houguan Lake, Qingling Lake, Wuhu Lake, Tangxun Lake, Fu River and Dadong Lake since 2013, clarifying the function and spatial structure of green wedges, delineating ecological implementation units, proposing functional guidelines for village development and land use within the units, and accomplishing the plan and control of the "Five Lines" and various ecological elements.

The construction of Donghu Greenway, in particular, was not only an important carrier to reflect the transformation of the urban construction concept of "quieting down the city", but also an initiative to implement the protection project of the Green Wedge in Dadong Lake. Donghu Ecological Oasis Conceptual Plan (2017) adhered to the plan concept of "public participation, internationalized vision, large regional perspective, integrated thinking and whole-process planning", avoided the narrow planning concept of greenways, and created a world-class Donghu greenway in all directions. The plan broke the traditional hierarchical and end-to-end greenway network construction mode, took the large-area ecological pattern as the foundation, and adopted the direct series mode to construct a "6+3" regional greenway system that ran through the important functional areas of the city, and formed the "D+H" greenway structure of Donghu. Its aim was to build the Donghu Greenway in the background of the overall regional development, and realize the benign interaction between Donghu and the city.

According to the principle of "building the easier first, connecting the important scenic spots, road traffic support, and respecting the public will", the plan clearly defined the construction time sequence. The first phase of the construction started from the Liyuan Square around Guozheng Lake—North Gate of Moshan Mountain—Fengduo Mountain—One Tree Section, the Core Scenic Spot of Moshan Mountain and Luoyan Scenic Spot around Tuanhu Lake, with a total length of 28.7 kilometers. According to the regional characteristics of different sections of lake middle road, lake mountain road, Moshan road and country road, the citizens' experience of "walking along the lake, boating in the lake, walking into the forest and climbing to the top of the mountain" could be realized through section reconstruction and theme area design.

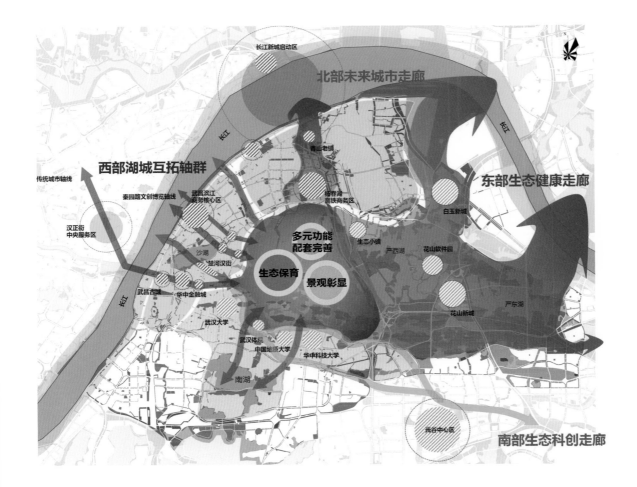

1

生态绿楔控规导则新洲区用地规划图

Guidelines for the Control of Ecological Green Wedges — Land Use Plan in Xinzhou District

2

东湖城市生态绿心空间拓展结构图

Spatial Expansion Structure of Donghu Urban Ecological Oasis

CHAPTER FOUR • Planning During Regional Integration

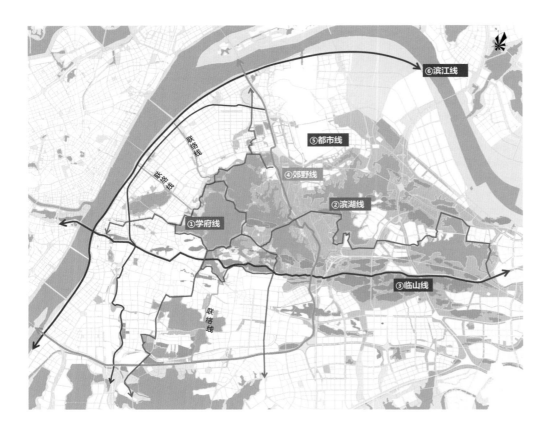

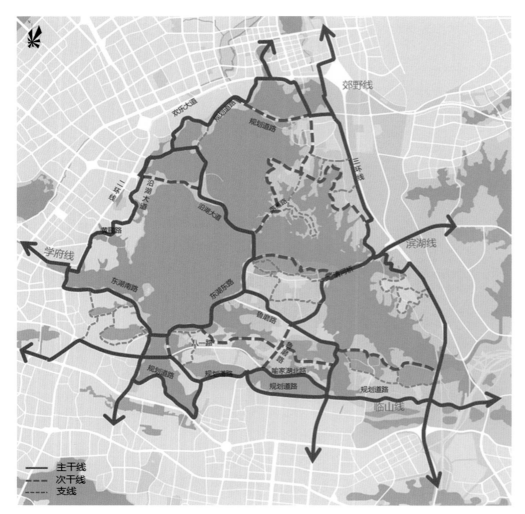

1	3
2	4

1
大江南绿道体系规划图
Greater Jiangnan Greenway System Plan

2
东湖绿道体系规划图
Donghu Greenway System Plan

3
梅园全景广场效果图
Effect Drawing of Meiyuan Panoramic Square

4
东湖城市生态绿心效果图
Effect Drawing of Donghu Urban Ecological Oasis

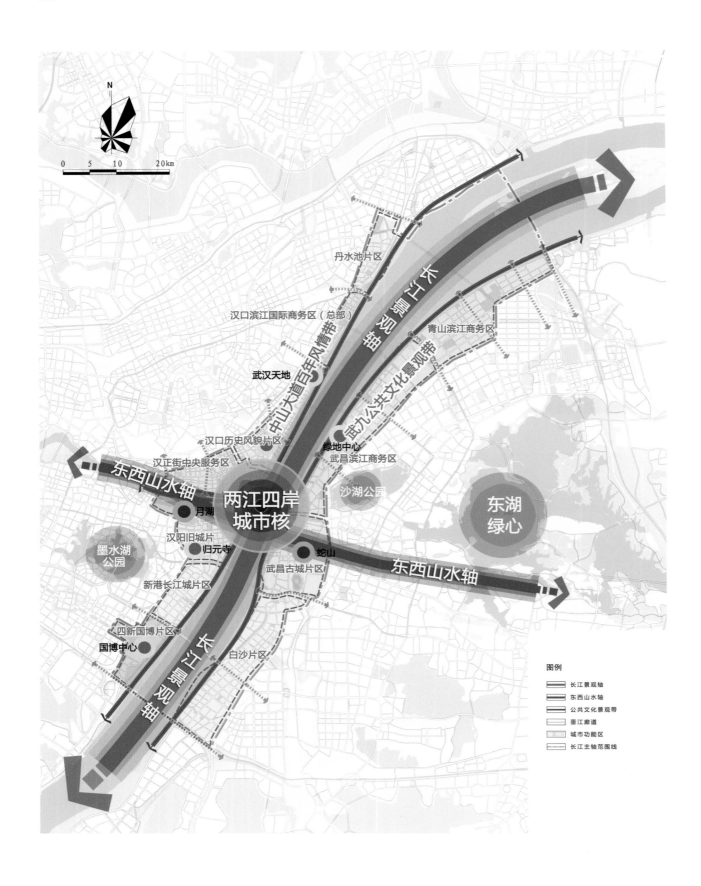

1

长江主轴双修·主轴结构图
(2018年)

Double Maintenance of the
Yangtze River Main Axis —
Structure of the Main Axis (2018)

2

通顺河规划图

Tongshun River Plan

● 轴：

长江主轴双修实施性规划（2018年）

为深化贯彻国家"城市双修"的工作部署要求，落实市委市政府有关"长江主轴"建设的目标任务，积极应对转型发展要求，从而有效解决武汉"城市病"问题，提升城市治理水平，武汉市2018年开展了《长江主轴（核心区）城市双修总体规划》编制工作。规划聚焦"环境、文化、功能"3个关键词，通过"城市双修"，将核心区建设成为展示武汉城市形象和文化风貌的卓越示范区，着力打造"人水和谐"的活力空间、长江文明的卓越之心和宜居漫行的精致城区，开展了1个总体规划以及产业、历史街区、慢行系统、蓝绿空间等4个专项规划。

● 廊：

一河一策（2018年）

全面推行"河长制"是落实绿色发展理念、推进生态文明建设的内在要求。为此，武汉市结合实际制定了《关于全面推行河长制的实施意见》以及《武汉市全面推行河湖长制三年行动计划（2018~2020年）》，重点对汉江、府澴河、通顺河、举水、汉北河、长江、倒水、滠水、东荆河、马影河等10条跨界河流，制定了规划方案，提出治理保护对策措施，明确责任分工，切实落实截污、清淤、河容整治、两岸绿化、环境卫生、排污总量控制、日常管控等工作，力争经过3~5年的努力，基本实现"水清、岸绿、河畅、水动、景美"的美好愿景。

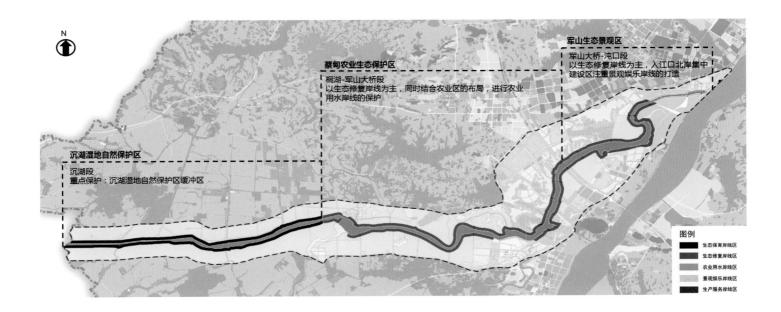

● AXIS:

Implementation Plan for Double Maintenance of the Main Axes in Yangtze River (2018)

In order to deepen and implement of the national deployment requirement for "double maintenance", implement the objectives and tasks of the municipal Party committee and the municipal government regarding the construction of the "Yangtze River main axis" and actively respond to the requirements of transformation and development, so as to effectively solve the problem of Wuhan's "urban diseases" and improve the level of urban governance, Wuhan launched the preparation of the Master Plan for Double Maintenance of Cities (Core Areas) on the Yangtze River Main Axis in 2018. The plan focused on the three key words of "environment, culture and function". Through the "urban double maintenance", the core area would be built into an excellent demonstration area to show Wuhan's urban image and cultural features. Efforts would be made to create a harmonious vitality space between human and water, an outstanding civilization heart of Yangtze River and a fine livable and slow-moving urban area. One master plan and four special plans including industry, historical blocks, slow-moving system and blue-green space had been carried out.

● CORRIDOR:

One Policy for One River (2018)

The full implementation of the river chief system was the inherent requirement of implementing the concept of green development and promoting the construction of ecological civilization. For this reason, Wuhan City formulated the Implementation Opinions on the Full Implementation of the River Chief System and the Three-Year Action Plan for the Full Implementation of the River and Lake Chief System in Wuhan City (2018—2020), focusing on making planning schemes and proposing control and protection countermeasures for 10 trans-boundary rivers including the Hanjiang River, Fuhuan River, Tongshun River, Jushui River, Hanbei River, Yangtze River, Daoshui River, Sheshui River, Dongjing River and Maying River. Wuhan would clarify the division of responsibilities, effectively implement pollution control, dredging, river regulation, greening on both sides of the river, environmental health, total sewage control, daily control, etc., and strive to basically realize the vision of "clear water, green banks, smooth rivers, dynamic water and beautiful scenery" after 3～5 years' efforts.

一 关注流域生态资源的跨界保护

● 长江生态大保护滨江带规划（2018年）

为深入贯彻习近平总书记视察湖北关于"共抓大保护、不搞大开发"的重要讲话精神，落实自然资源部陆昊部长调研武汉长江大保护的有关指示要求，加快推进武汉长江大保护工作，促进长江沿线生态环境保护和资源合理利用，2018年武汉市开展了《武汉长江生态大保护沿江带空间规划》编制工作。

长江大保护的目标是助力武汉市打造世界级城市长江主轴景观带，建设长江大保护与绿色发展示范区，使长江武汉段成为世界大河治理最新成就集中展示地、长江大保护及绿色发展典范。在此基础上，提出了"主城提升、新城重构、郊野保育、江湖连通、协同管控"的武汉模式。

同时，规划还提出了固化沿江生态框架、修复沿江带生态要素两大举措。按照武汉长江作为排水骨干通道、取水水源地、内河航运水道、景观生态廊道的功能定位要求，规划将长江段岸线按生产、生活、生态三大功能，划分为生态保育、生态修复、农业生产、景观娱乐、生产服务等5大岸线区，分类提出了岸线保护控制要求。在长江左右岸，各建设一条近100公里长的滨江绿道，打造全球独一无二的"双百里"江滩绿道。

● 武汉城市圈禁限建区空间管控规划研究（2017年）

在当前湖北生态省建设、长江经济带生态保护等背景下，武汉城市圈内城乡建设和城乡安全协调发展的诉求日益强烈，要求进一步完善禁限建区管控内容和相关机制。

这是国内首次从区域尺度开展的禁限建管控研究，按照"省级引领、市县参与、省市联动"的工作模式，采取区域性空间管制和城市层级禁限建规划两级编制模式，制定指导各市县禁限建管控的区域整体空间保护规划，以及落实具体空间发展和保护的市县禁限建管控规划。利用GIS技术构建了区域尺度的生态框架，并结合管控事权，将传统的禁建区、限建区、适建区进一步细分为一级禁建区、二级禁建区、限建区、适建区四大分区，同时科学制订了不同分区的管控任务和管控要素，在确保规划管控分工明确、传导有力的同时，为武汉城市圈内各城市禁限建规划的编制提供了统一的技术标准。

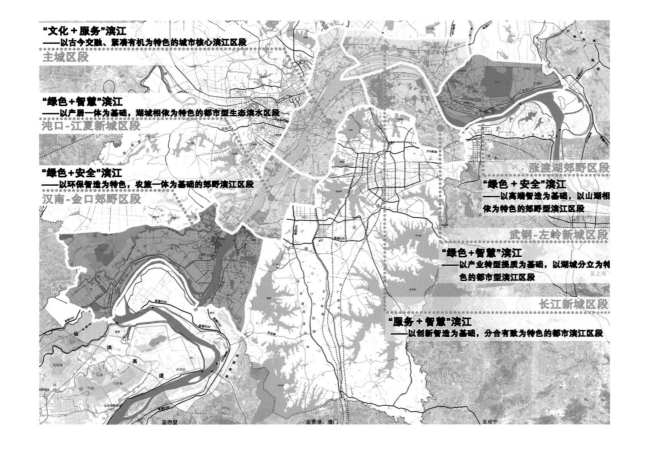

1
功能空间结构图
Functional Space Structure

2
岸线设施规划图
Shoreline facility planning

3
生态保护和修复规划图
Ecological Protection and Restoration Plan

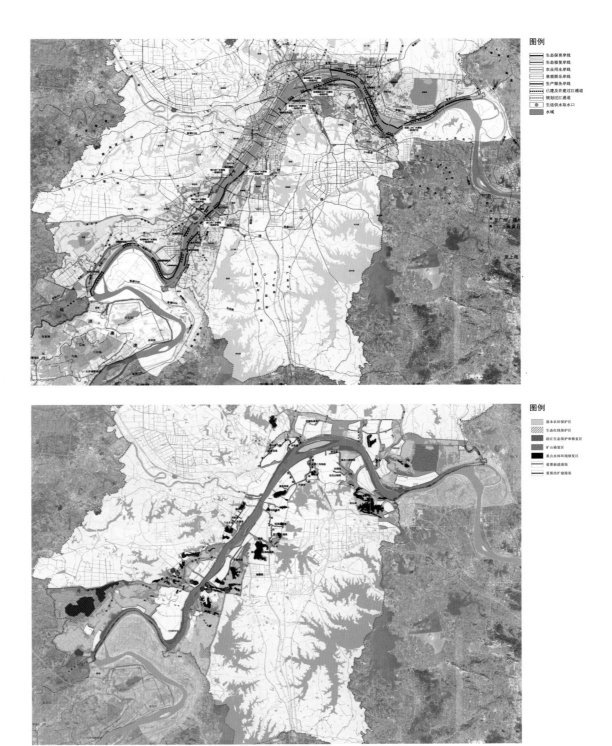

— Trans-Boundary Protection Focusing on Ecological Resources in the River Basin

● *Plan for Great Ecological Protection along the Yangtze River (2018)*

To thoroughly implement the spirit of the important speech made by General Secretary Xi Jinping during his inspection visit to Hubei Province on "jointly promoting great protection and avoiding large-scale development", implement the relevant instructions of Lu Hao, Ministor of Ministry of Natural Resources. on investigating the great protection of the Yangtze River in Wuhan, accelerate the great protection of the Yangtze River in Wuhan, and promote the protection of the ecological environment and rational utilization of resources along the Yangtze River, Wuhan launched the preparation of Spatial Plan for Great Ecological Protection along the Yangtze River in Wuhan in 2018.

The great protection of the Yangtze River aimed to help Wuhan build a world-class urban landscape belt along the Yangtze River main axis, and a demonstration area for the protection and green development of the Yangtze River, so that the Wuhan section of the Yangtze River would become a showcase for the latest achievements in the world's river management, and a model for the protection and green development of the Yangtze River. On this basis, the Wuhan mode of "main urban area upgrading, new urban area reconstruction, countryside conservation, river lakes connection and coordinated management and control" was proposed.

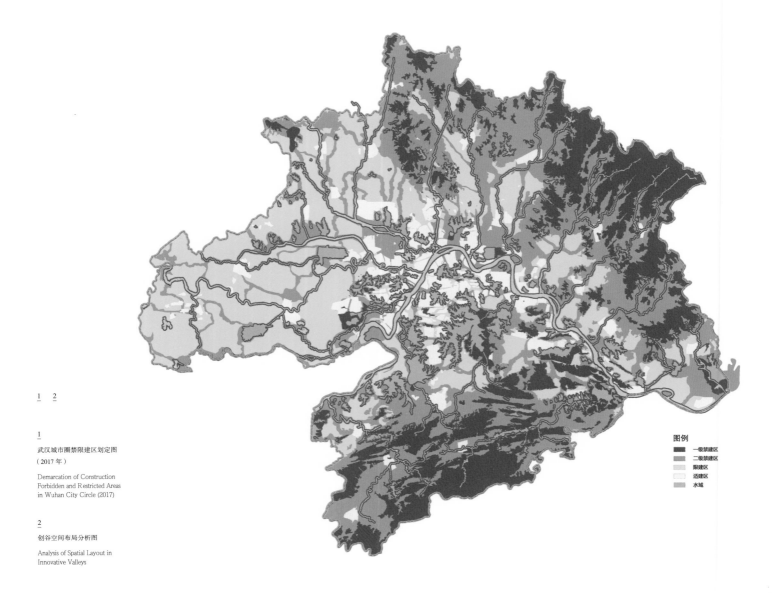

1

武汉城市圈禁限建区划定图
（2017 年）

Demarcation of Construction Forbidden and Restricted Areas in Wuhan City Circle (2017)

2

创谷空间布局分析图

Analysis of Spatial Layout in Innovative Valleys

Meanwhile, the Plan raised two major measures: solidifying the ecological framework along the river and restoring the ecological elements along the river. According to the functional positioning requirements of Yangtze River in Wuhan as the main drainage channel, water source, inland navigation channel and landscape ecological corridor, the Plan divided the shoreline of the Yangtze River into five major shoreline areas, including ecological conservation, ecological restoration, agricultural production, landscape entertainment and production services, according to the three major functions of production, life and ecology, and put forward the requirements for shoreline protection and control. On the left and right banks of the Yangtze River, it was designed to build a riverside greenway of nearly 100 kilometers each to create the unique "double-hundred-mile" riverside greenway in the world.

● *Study on the Spatial Control Plan of Construction Forbidden and Restricted Areas in Wuhan City Circle (2017)*

Under the background of the construction of Hubei ecological province and the ecological protection of the Yangtze River Economic Belt, the demand for coordinated development of urban and rural construction and security in Wuhan City Circle was increasingly strong, requiring further improvement of the control and related mechanisms of construction forbidden and restricted areas.

This was the first time in China to carry out the study on control of forbidden and restricted areas at the regional level. According to the working model of "provincial guidance, participation by cities and counties, and linkage between provinces and cities", the two-level preparation model of regional spatial control and city construction ban and control was adopted to formulate the overall regional spatial protection plans to guide the control of forbidden and restricted construction of all cities and counties, and to implement the control plans of forbidden and restricted construction of cities and counties for specific spatial development and protection. The GIS technology was adopted to construct an ecological framework on a regional scale. Combined with the management and control powers, the traditional forbidden zone, restricted zone and suitable zone were further subdivided into four sub-zones: primary forbidden zone, secondary forbidden zone, restricted zone and suitable zone. Meanwhile, the management and control tasks and elements of different sub-zones were scientifically formulated, thus ensuring clear division of planning and control and strong transmission, and providing unified technical standard for the preparation of the forbidden and restricted planning of cities in Wuhan City Circle.

4 - 面向国家产业创新中心的园区规划（2016~2018年）

Park Plan for National Industrial Innovation Center (2016—2018)

中国经济发展进入新常态，是综合分析世界经济长周期和我国发展阶段性特征及其相互作用做出的重大判断。走传统工业化道路所倚重的后发优势等也在逐步削弱，不足以支撑武汉追赶先进城市。因此，武汉需要不断推进创新驱动发展，抓住科技产业变革的机遇，创新领军企业、战略性新兴产业等。

近年来，武汉市自主创新能力显著增强，在经济社会发展和产业结构调整中的重要作用进一步增强，重点打造光电子信息和生物健康产业两个战略性新兴产业集群，将其作为主要支柱性高新技术产业。与此同时，武汉市启动了"创谷计划"，稳步推进众创空间建设，制定了《关于加快发展众创空间支持大众创新创业的实施意见》，认定国家级众创空间14个，省级众创空间50个，市级众创空间10个，鲁巷、街道口等高校密集、商业发达区域已初步形成众创街区。2016年，规划提出了"园区、校区、城区、湖区"四种类型的创新创业集聚区，提出了江岸区幸福创新生态半岛等42个"创谷"，启动实施了南太子湖创新谷等13个"创谷"建设。

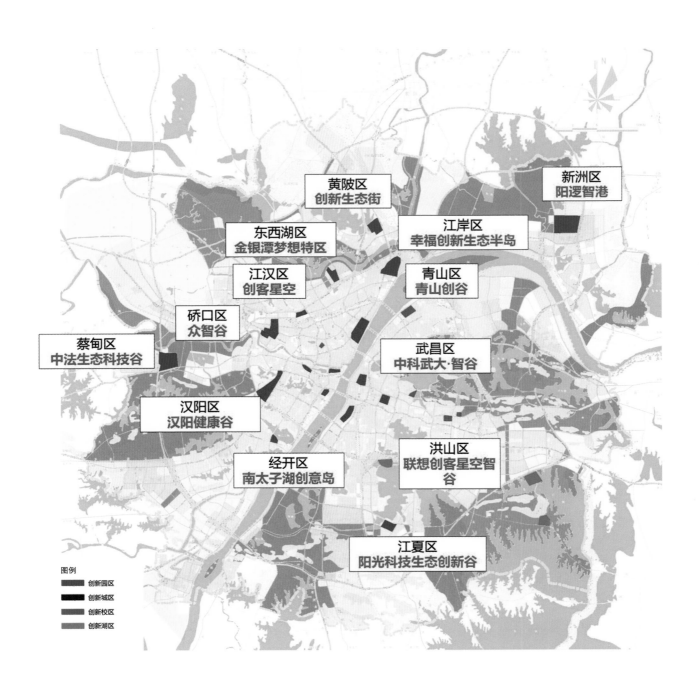

224 CHAPTER FOUR • Planning During Regional Integration

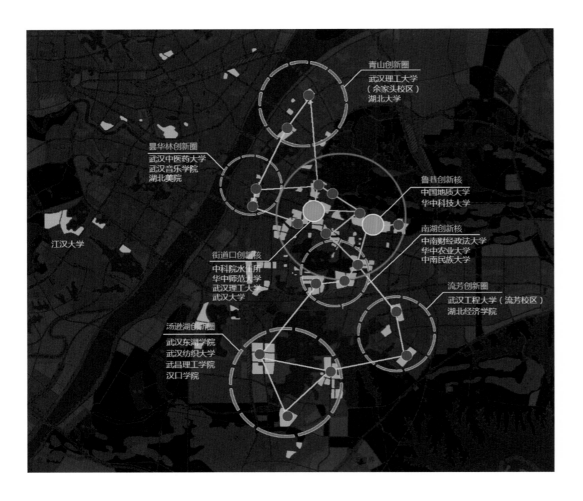

1
创新校区分析图
Analysis of Innovative Campuses

2
创新城区分析图
Analysis of Innovative Urban Areas

3
创新园区分析图
Analysis of Innovative Parks

China's economic development has entered a new normal, which was a major judgment based on the comprehensive analysis of the long cycle of the world economy and the phased characteristics of China's development and their interactions. The late-developing advantage that we rely on in traditional industrialization was also being gradually weakened, which is not enough to support us to catch up with advanced cities. Therefore, we needed to continuously promote innovation-driven development, seize the opportunity of technological industry transformation, and innovate in leading enterprises and strategic emerging industries.

In recent years, Wuhan's independent innovation capability has been significantly enhanced, and its important role in economic and social development and industrial structure adjustment has been further enhanced. Two strategic emerging industrial clusters i.e. optoelectronic information and bio-health industry have been built as the main pillar of high-tech industries. Meanwhile, Wuhan launched the "valley innovation plan" to steadily promote the construction of public innovation space, and formulated the Implementation Opinions on Accelerating the Development of Public Innovation Space to Support Public Innovation and Entrepreneurship. Fourteen state-level, fifty provincial-level and ten city-level public innovation spaces have been identified. The public innovation blocks have been initially formed in areas with dense universities and developed businesses such as Luxiang and Jiedaokou. In 2016, the plan raised four types of innovative and entrepreneurial clusters, namely "parks, campuses, urban areas and lake areas", proposed 42 innovative valleys including the Happy and Innovation Ecological Peninsula in Jiang'an District, and started the construction of 13 "innovative valleys", including the one in Nantaizi Lake.

226　CHAPTER FOUR • Planning During Regional Integration

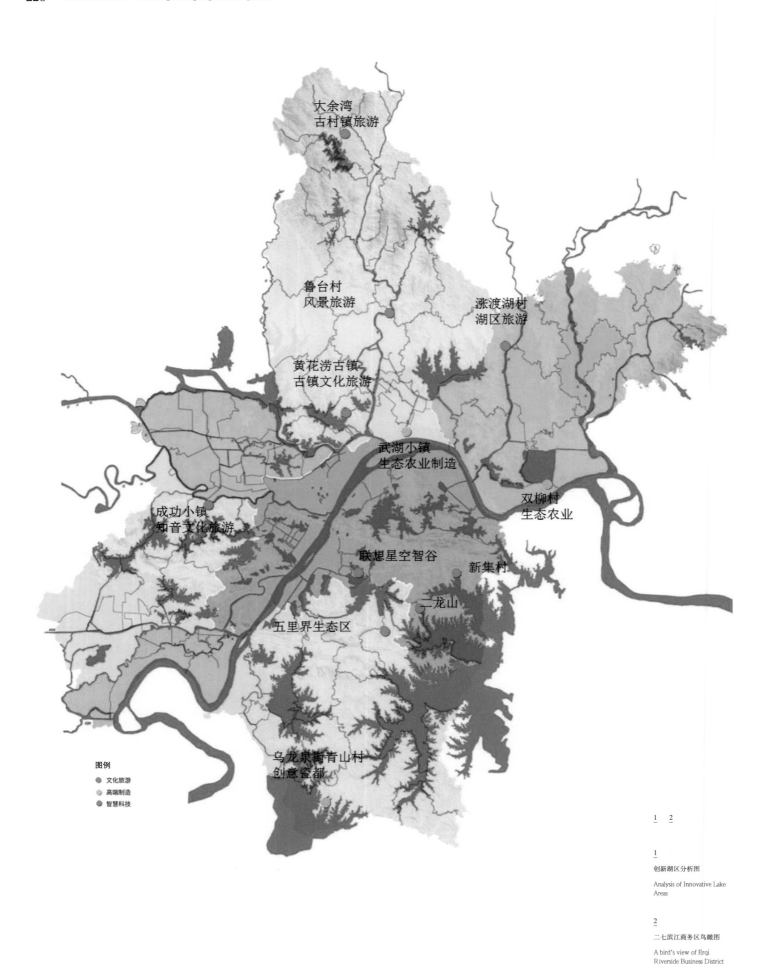

1
创新湖区分析图

Analysis of Innovative Lake Areas

2
二七滨江商务区鸟瞰图

A bird's view of Erqi Riverside Business District

5 - 面向国家现代服务业中心的功能区规划

Functional Area Planning for the National Modern Service Center

伟大的城市必须有伟大的梦想。武汉正致力做足做好"长江"这篇大文章,以长江为城市空间主轴和时间主轴,串起伟大的城市梦想——努力打造"历史之城""当代之城""未来之城",变长江天堑为靓丽画轴,串联伟大城市梦想,复兴城市荣光,打造有世界影响力的亮点城市。为此,武汉市委市政府提出,要以长江文明之心、长江主轴、长江新城(新区)为重点,优化空间结构,重构城市格局,增强城市未来发展竞争力,让武汉成为世界亮点城市,以"三城"建设勾勒出武汉未来发展蓝图。

与此同时,在建设国家中心城市和"三化"大武汉的进程中,还需要若干功能板块拱卫和支撑,形成引领大武汉复兴的"亮点区块"和"重点功能区"空间载体组合。如果说,新城区及开发区是工业倍增的"主战场",中心城区则转向发展现代服务业。因此,在 2010 版总体规划获批后,武汉将工作重心转向规划实施,拟通过七大重点功能区规划实施,探索更加适应市场主体的规划编制模式,为建设国家中心城市提供空间支撑。

一 重点功能区规划

● 二七沿江商务区规划(2009 年)

为加快推进汉口沿江地区发展,以江岸车辆厂等企业搬迁为契机,以打造国际滨江商务区为目标,2009 年 6 月开展了《江岸沿江商务区规划》编制工作。规划延续江岸沿江商务区功能定位,以现代服务业为核心,强化该区域商贸、娱乐及居住功能,将该地区打造成为集商贸商务、文化娱乐、现代生活于一体的城市综合发展区,成为 21 世纪"两型"社会引导区。

● 武昌滨江商务区规划(2009 年)

为引导武昌滨江地区由生产型向现代服务型的转变,促进产业结构调整和转型,实现长江南岸滨江景观的重塑,将该地区建设成为武汉主城区新的现代服务业集聚区和具有滨江特色、标志性、多元化功能区。《武昌滨江商务区规划》(2009 年)提出了城市设计与市场开发的协同运作机制,对可开发地块进行了调查研究,建立了项目库,并对车辆厂、月亮湾片、裕大华片、积玉桥片等重点建设项目进行了规划意向方案设计,提出规划控制要求及分期建设措施。

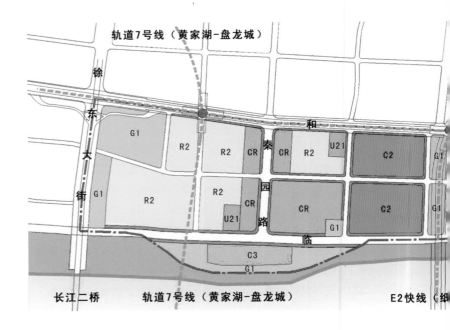

Great cities must have great dreams. Wuhan was making great efforts to expand along the Yangtze River. With the Yangtze River as the main axis of urban space and time, Wuhan was linking up great urban dreams — striving to build the "historical city", "modern city" and "future city" and turning the natural moat of the Yangtze River into a beautiful scroll painting. Wuhan connected the great urban dreams, revived the urban glory and built a bright city with global influence. For this reason, Wuhan Municipal Party Committee and Municipal Government proposed to optimize the spatial structure and reconstruct the urban pattern by focusing on the civilization heart, main axis and new urban area (new district) of the Yangtze River, thus enhancing the competitiveness of the city's future development, making Wuhan a bright spot in the world, and outlining the blueprint for Wuhan's future development with the "three cities" construction.

Meanwhile, in the construction of a national central city and the "three modernizations" of Wuhan, several functional blocks were needed for arch protection and support, which would form a space carrier combination of "bright spots" and "key functional areas" leading the revival of Wuhan. If the new urban area and the development zone were the main battlefields for industrial multiplication, the central urban area would turn to develop modern service industries. Therefore, after the approval of the 2010 Master Plan, Wuhan shifted its focus to the plan implementation, and planned to explore a more market-oriented planning mode through the implementation of the plan for seven key functional areas to provide spatial support for the construction of a national central city.

— Planning for Key Functional Areas

• *Plan for Erqi Riverside Business District (2009)*

In order to speed up the development of Hankou's riverside areas, taking the relocation of Jiang'an Vehicle Plant and other enterprises as an opportunity and

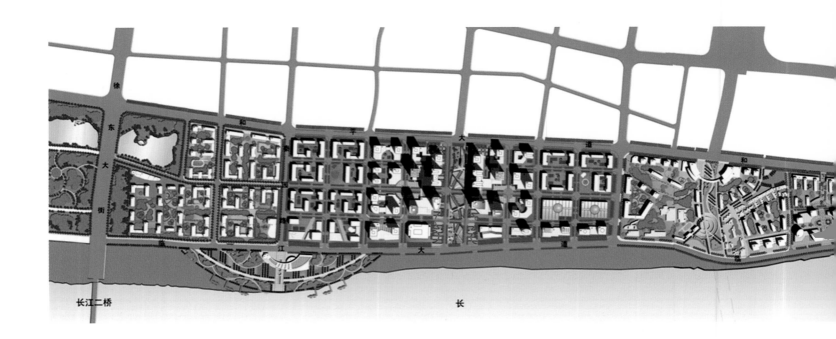

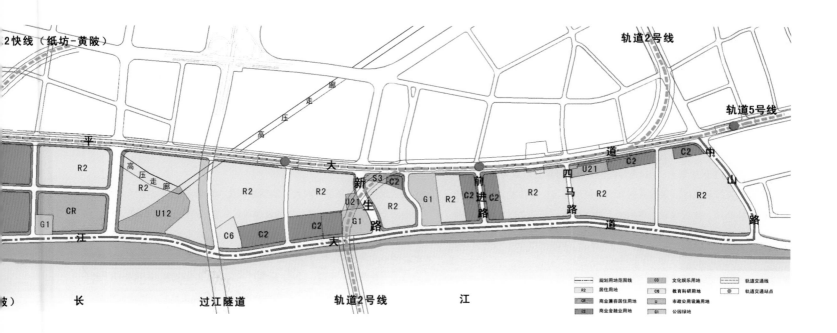

building an international riverside business district as the goal, Wuhan prepared the Plan for Riverside Business District in June 2009. It's planned to continue the functional orientation of the riverside business district, take modern service industry as the core, strengthen the business, entertainment and residential functions of the region and build the region into a comprehensive urban development area integrating business, culture, entertainment and modern life as well as a guiding area of the "two-oriented" society in the 21st century.

- **Plan for Riverside Business District in Wuchang (2009)**

The Riverside area in Wuchang was facing new development opportunities with the "two-oriented" society construction as an opportunity and relying on the space guarantee of industrial retreat. To guide the transformation of the region from production to modern service, promote the adjustment and transformation of industrial structure, and realize the reconstruction of the Binjiang landscape on the southern bank of the Yangtze River, the Binjiang area would be built into a new modern service industry cluster and a landmark and diversified functional area with Riverside characteristics in the main urban areas of Wuhan. Plan for Binjiang Business District in Wuchang (2009) put forward the cooperative operation mechanism of urban design and market development, investigated and studied the developable plots, established the project database and designed the plan intention scheme for key construction projects such as Vehicle Plant, Moon Bay, Yudahua, Jiyuqiao, etc., and proposed the plan control requirements and phased construction measures.

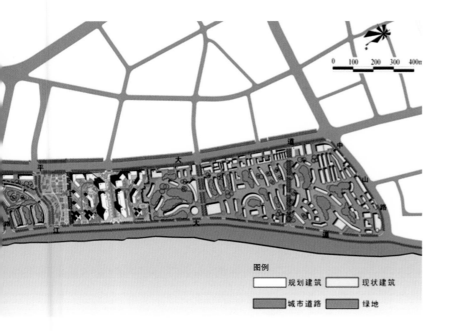

1
武昌滨江商务区用地规划图（2017年）
Land Use Plan of Binjiang Business District in Wuchang (2017)

2
武昌滨江商务区总平面图（2017年）
General Plan of Binjiang Business District in Wuchang (2017)

● 杨春湖高铁商务区规划（2014 年）

作为武汉城市总体规划确定的六大城市副中心之一，杨春湖商务区定位为"国家高铁之心及武汉重要门户展示窗口"。2014 年 5 月，杨春湖商务区实施性规划获武汉市规委会审议通过。由于北洋桥垃圾填埋场生态处置方式发生重大调整，后续开展了北洋桥垃圾填埋场周边城市设计优化及公园概念设计工作。

《杨春湖高铁商务区规划》（2014 年）将杨春湖商务区打造成为功能完善、宜居宜业的先行示范区。该商务区开展垃圾填埋场生态修复，强调城市空间的立体利用，体现绿意盎然的生态景观特色，开展"以人为本，步行友好"的街道环境设计，树立人居环境与可持续发展典范。

● 中法武汉生态示范城总体规划（2016 年）

"中法武汉生态示范城"位于武汉市蔡甸区，是在中法两国元首的见证下由政府直接推动设立的可持续发展示范合作项目，中法双方组建联合技术团队共同开展总体规划编制工作。《中法武汉生态示范城总体规划》成果在 2016 年底经住房与城乡建设部审查同意后，于 2017 年 2 月经湖北省人民政府正式批复。生态城规划用地面积 39 平方公里，建设用地面积 17 平方公里，总人口 20 万。

中法武汉生态城是最新时代（新型城镇化）背景下的建城项目、最新理念的示范项目、最新经济形态的发展项目。空间结构上强调绿色、生态、低碳，构建足量生态空间为基底的空间结构；产业经济上强调转型、互补、融合，重点发展创新型第三产业，培育多样化科技创新空间；规划方法上强调混合、和谐、交融，强调产城融合、文化交融，实现职住平衡，探索各类用地功能高度混合的用地规划模式，提出了社会和谐指标体系。

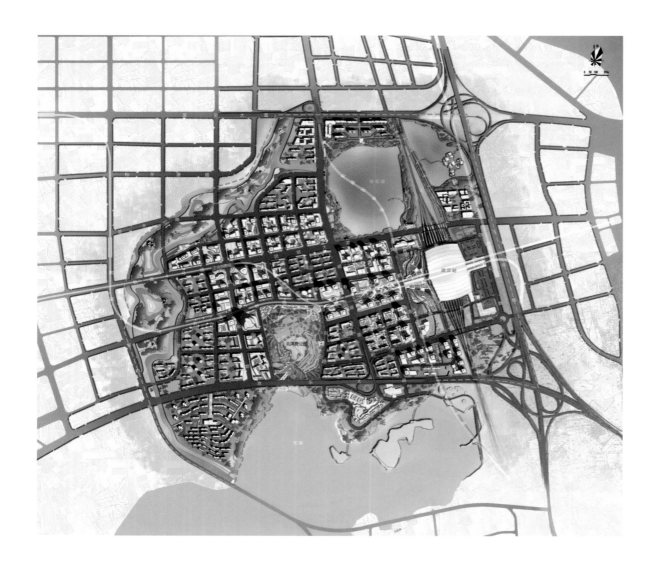

第肆章 • 区域一体化时期的规划　231

1	2
3	

1
杨春湖商务区城市设计总平面图

General Plan of Urban Design for Yangchunhu Business District

2
杨春湖商务区北洋桥景观大道效果图

Effect Drawing of Beiyangqiao Landscape Avenue in Yangchunhu Business District

3
杨春湖商务区总体鸟瞰图

An overall bird's view of Yangchunhu Business District

CHAPTER FOUR • Planning During Regional Integration

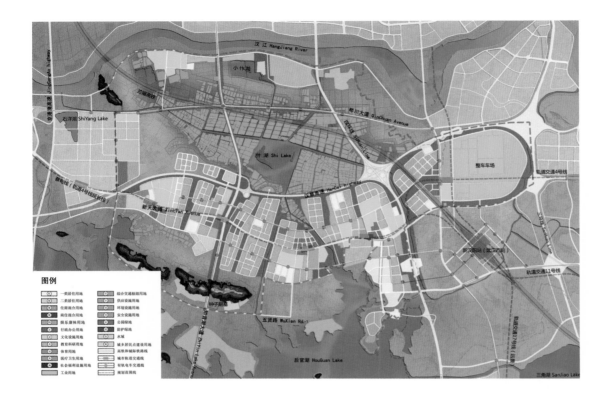

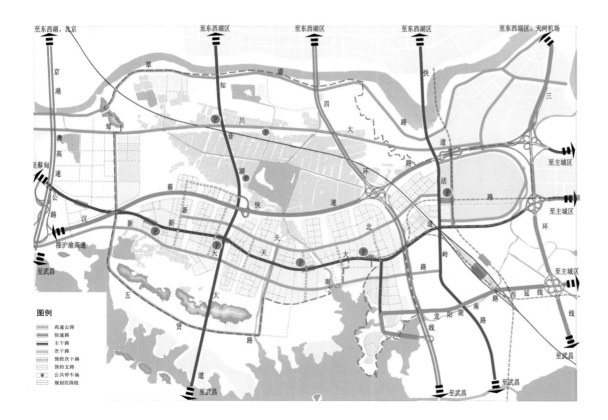

1

中法武汉生态城用地规划图

Land Use Plan of Wuhan Sino–French Ecological Demonstration City

2

中法武汉生态城道路交通系统规划图

Road Traffic System Plan of Wuhan Sino–French Ecological Demonstration City

3

中法武汉生态城意向效果鸟瞰图

An bird's view of the expected effect of Wuhan Sino–French Ecological Demonstration City

- *Plan for Yangchunhu High-speed Railway Business District (2014)*

As one of the six sub-centers identified in Wuhan's urban master plan, Yangchunhu Business District was positioned as "the heart of the national high-speed railway and the exhibition window of Wuhan's important portal". In May 2014, the implementation plan of Yangchunhu Business District was approved by the Municipal Planning Commission. As the ecological disposal mode of Beiyangqiao landfill site had been greatly adjusted, the urban design optimization and park concept design around Beiyangqiao landfill site had been carried out subsequently.

Plan for Yangchunhu High-speed Railway Business District (2014) aimed to build Yangchunhu Business District into a well-functioning, livable and industrial pilot demonstration area. For this reason, Wuhan carried out ecological restoration of landfill sites to emphasize the three-dimensional utilization of urban space and reflect the green ecological landscape features, and also designed the "people-oriented, pedestrian-friendly" street environment to set up a model of living environment and sustainable development.

- *Master Plan for Wuhan Sino-French Ecological Demonstration City (2016)*

Located in Caidian District of Wuhan, the Sino-French Ecological Demonstration City in Wuhan is a demonstration cooperation project on sustainable development directly promoted by the government in the presence of the heads of state of China and France. China and France set up a joint technical team to work together on the overall plan. The results of Master Plan for Wuhan Sino-French Ecological Demonstration City were officially approved by Hubei Provincial People's Government in February 2017 after being reviewed and approved by the Ministry of Housing and Urban-Rural Development at the end of 2016. The eco-city had a planned land area of 39 square kilometers and a construction land area of 17 square kilometers, with a total population of 200,000.

Sino-French Wuhan Ecological Demonstration City is the highest-level cooperation project, the construction project in the latest era (under the condition of new urbanization), the demonstration project of the latest concept and the development project of the latest economic form. The spatial structure emphasized green, ecological and low carbon, and constructed a spatial structure based on sufficient ecological space. Industrial economy emphasized the transformation, complementarity and integration, focused on the development of innovative tertiary industry and cultivated the diversified space for scientific and technological innovation. The plan method emphasized the mixing, harmony and integration — city-industry integration and cultural integration — to realize the job-housing balance, and explored the land plan mode with highly mixed functions of various lands, and put forward the index system of social harmony.

一 亮点区块规划

● 长江主轴概念规划（2017年）

为集中展示长江文化、生态特色、发展成就和城市文明，武汉市第十三次党代会提出优化建设武汉长江主轴，把长江主轴武汉段作为世界级城市中轴文明景观带来打造，变"长江天堑"为"城市靓丽画轴"，创新大武汉的"中轴结构"。自2017年起，按照"总体规划—专项规划"模式开展工作，形成了一个概念规划和江滩景观提升、左右岸大道、城市阳台、沿江立面整治、码头岸线整治、桥梁美化等6个近期实施专项规划。

《长江主轴概念规划》（2017年）按照建设"交通轴、经济轴、文化轴、生态轴和景观轴"的目标，串联"历史之城、当代之城、未来之城"，提出"五轴一体、五个层次、五个节点"的概念规划框架。其中：功能上按照"五轴一体"思路进一步丰富长江主轴的功能内涵；空间上按照"长江水体—桥梁、码头与滩涂—江滩公园—堤防及道路—沿岸建筑"5个层次进行优化，构建长江主轴的空间布局框架；时序上遵循"一年打基础、两年出样板、三年初见效、五年成规模、十年基本建成"这5个时间节点，统筹长江主轴的整体建设安排。

为配合近期主轴的全面提升，形成6个专项规划。其中，江滩景观提升规划聚焦江滩区域景观环境提升，重点为打通断点，提升设施配置水平；左右岸大道以近期道路建设为重点，初步形成交通轴；城市阳台专项提出启动中华路、月亮湾、汉正、江汉关4个城市阳台建设；沿江立面专项以两岸核心区老旧建筑为主，指导建筑立面改造；码头岸线专项重点为清理、集并生产码头，优化客运和旅游码头，还滨水岸线于民；桥梁美化专项则按照"桥梁博物馆"的目标，美化、亮化、彩化长江沿线桥梁。

1	3
2	4

1–2
中华路阳台图
Zhonghua Road Balcony

3–4
汉正阳台图
Hanzheng Road Balcony

— Planning of Featured Plots

• *Conceptual Plan for of the Yangtze River Main Axis (2017)*

To display the culture, ecological features, development achievements and urban civilization of the Yangtze River, the 13th Party Congress of Wuhan City proposed to plan and optimize the construction of Yangtze River main axis in Wuhan, and build this section into a world-class civilized landscape belt in the urban central axis, transforming the "natural moat of the Yangtze River" into the "beautiful urban scroll painting" to innovate in the "central axis structure" of Wuhan. Based on the "master plan—special plan" model, Wuhan has carried out work since 2017 to form a conceptual plan and six recent special plans concerning river beach landscape upgrading, left and right bank avenues, city balconies, riverside facade renovation, wharf shoreline renovation and bridge beautification.

According to the goal of building "traffic axis, economic axis, cultural axis, ecological axis and landscape axis", Conceptual Plan for the Yangtze River Main Axis (2017) connected the "historical city, modern city and future city" in series, and proposed a conceptual plan framework of "five axes in one, five levels and five nodes". In terms of function, the functional connotation of the Yangtze River main axis was further enriched according to the idea of "five axes in one". In terms of space, the spatial layout framework of the Yangtze River main axis was constructed according to the five levels of "Yangtze River water body — bridges, docks and mudflats — Jiangtan Park — dikes and roads — coastal buildings". In terms of schedule, the overall construction arrangement of Yangtze River main axis was coordinated in accordance with the five nodes of "laying the foundation in one year, producing the model in two years, producing the initial result in three years, achieving the scale in five years and completing the construction in ten years".

Six special plans have been formed to coordinate with the recent overall upgrading of the main axis. The Jiangtan landscape improvement plan focused on the improvement of the landscape environment in this region, emphasizing the getting through breakpoints and improving facility allocation level. The left and right bank avenues focused on the recent road construction and initially formed the traffic axis. The special urban balcony project proposed to start construction of four urban balconies: Zhonghua Road, Moon Bay, Hanzheng and Jianghan Pass. The special riverside elevation project focused on the old buildings in the core areas on both banks to guide the elevation renovation of buildings. The special wharf coastline project focused on cleaning up, collecting and producing the wharf to optimize the passenger and tourist wharf and return the waterfront shoreline to the people. The special bridge beautification project focused on beautifying, brightening and colorizing bridges over the Yangtze River according to the goal of the "Bridge Museum".

1	3
2	4

1
月亮湾阳台图
Moon Bay Balcony

2
江汉关阳台图
Jianghan Pass Balcony

3–4
长江文明之心效果图
Effect Drawing of Civilization Heart of the Yangtze River

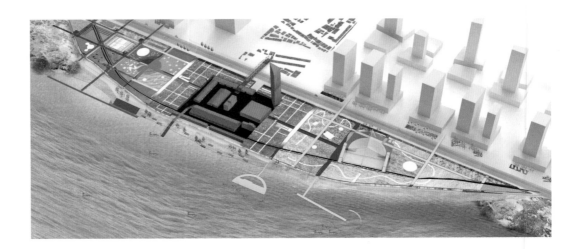

● 长江文明之心概念规划（2018年）

市委十三届四次全会提出"以长江文明之心为重点，提升建设历史之城"，市政协十三届二次会议将其列为1号建议案。市国土资源和规划局通过多轮调研座谈，先后组织开展了武汉历史之城——暨长江文明之心国际征集和概念规划研究。

长江文明之心是历史之城的核心区域，是以两江交汇的南岸嘴为圆心，以长江武汉段为蓝轴、龟蛇绵延山系为绿轴，以3.5公里为半径，覆盖了两江四岸的核心区域。《长江文明之心概念规划》（2018年）打造世界最负盛名的大河文明对话区、长江生态文化最集聚的展示区、武汉历史文化精髓的浓缩区，使长江文明之心形成"一园两轴、三镇六片"的整体空间结构。同时，概念规划针对文化资源缺乏整合提升、区域风貌不协调、产业发展阻滞等问题，提出"文化提升、空间弥合、功能统筹、政策设计"四大策略，共同打造面向未来的历史之城。

238 CHAPTER FOUR • Planning During Regional Integration

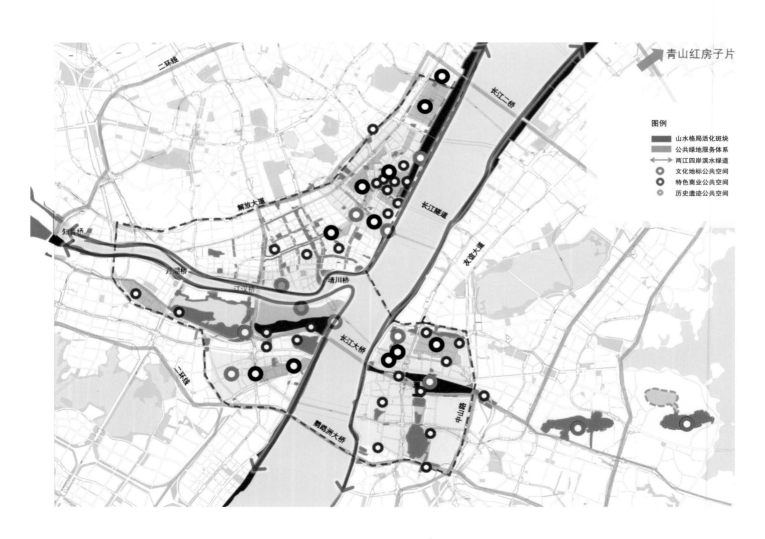

1
长江文明之心交通体系图
Traffic System Plan of
Civilization Heart of the
Yangtze River

2
长江文明之心景观体系图
Landscape System Plan of
Civilization Heart of the
Yangtze River

3
长江新城规划结构图
Structure Plan of the Yangtze
River New City

● 长江新城总体规划（2018 年）

为贯彻落实市十三次党代会"规划建设长江新城"的战略决策，按照"开门规划、扩大影响、规范高效、精选机构"的总体要求，高标准开展规划编制工作。

发展定位上，《长江新城总体规划》（2018 年）对标世界先进城市和雄安新区，积极承担国家使命，立足区域功能发展，明确"创新名城、生态绿城、现代智城、国际友城、创富民城"的目标定位，努力打造长江经济带高质量发展示范区。

空间布局上，该规划突出以水定城、弹性发展。构筑长效水安全格局，依托武湖、草湖核心水体构建城市水库，实施上、下湖隔堤工程；营造共生绿楔，重视大别山到木兰山到长江的"山湖林田河"的整体涵养与保护，构筑两级通风廊道；传承山水文脉，采用"台架建城、高地建城、特地特用"方式，构建组团化开发模式。

基础设施上，该规划围绕"未来"与"智慧"两大核心主题，构筑高效、韧性和超前的支撑系统，形成空中、地面、地下一体化交通结构，建设集约高效的供水系统和循环再生的污水处理系统，建立低碳清洁的综合能源体系，提升综合防灾能力和环境保护能力。

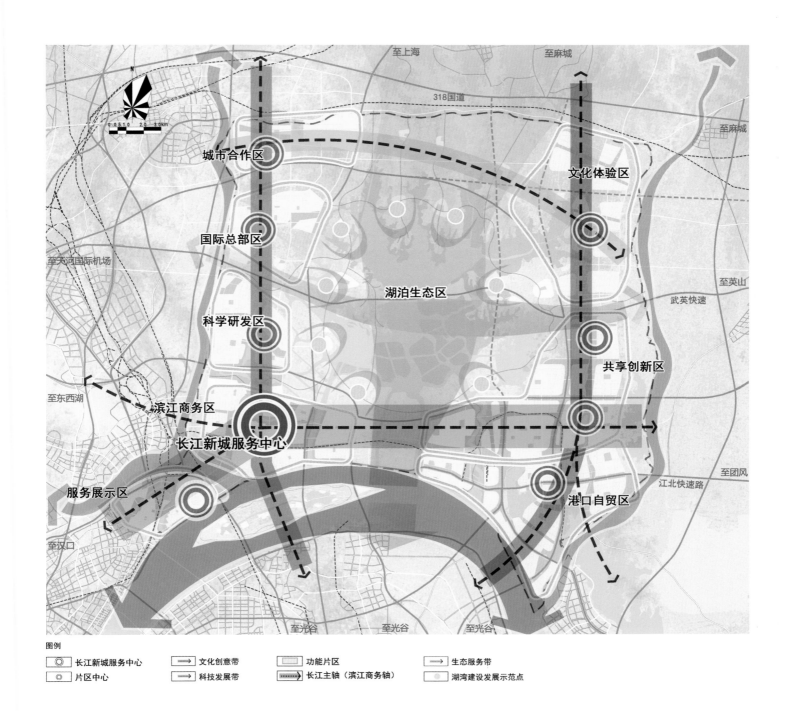

- *Conceptual Plan for the Civilization Heart of the Yangtze River (2018)*

The Fourth Plenary Session of the 13th CPC Municipal Committee put forward the proposal of "focusing on the civilization heart of the Yangtze River and promoting the construction of the historical city". The Second Session of the 13th CPC Municipal Committee listed it as Proposal No.1. After several rounds of research and discussion, the Municipal Land and Resources Planning Bureau successively organized and launched the international solicitation and conceptual plan research of Wuhan historical city and the civilization heart of the Yangtze River.

The civilization heart of the Yangtze River is the core area of the historical city. With a radius of 3.5 kilometers, it covers the core area of four banks of two rivers, with the southern bank mouth where the two rivers meet as its center, the Wuhan section of the Yangtze River as its blue axis, Turtle Mountain and Snake Mountain as its green axis. Conceptual Plan for the Civilization Heart of the Yangtze River (2018) aimed to build the world's most famous civilization dialogue area for big rivers, the display area where the ecological culture of the Yangtze River is most concentrated, and the concentration area of Wuhan's historical and cultural essence, so as to form the overall spatial structure of "one park, two axes, three towns and six plots" at the civilization heart of the Yangtze River. Meanwhile, the conceptual plan raised the four strategies of "cultural promotion, spatial bridging, functional coordination and policy design" to jointly build a future-oriented historical city, so as to solve these problems of lack of integration and promotion of cultural resources, uncoordinated regional features and blocked industrial development, etc.

- *Master Plan for the Yangtze River New City (2018)*

To implement the strategic decision of the 13th Municipal Congress of Party Representatives to "plan and build the Yangtze River New City", the plan was prepared in accordance with the overall requirements of "open planning, expanding influence, standardizing efficiency, and selecting institutions".

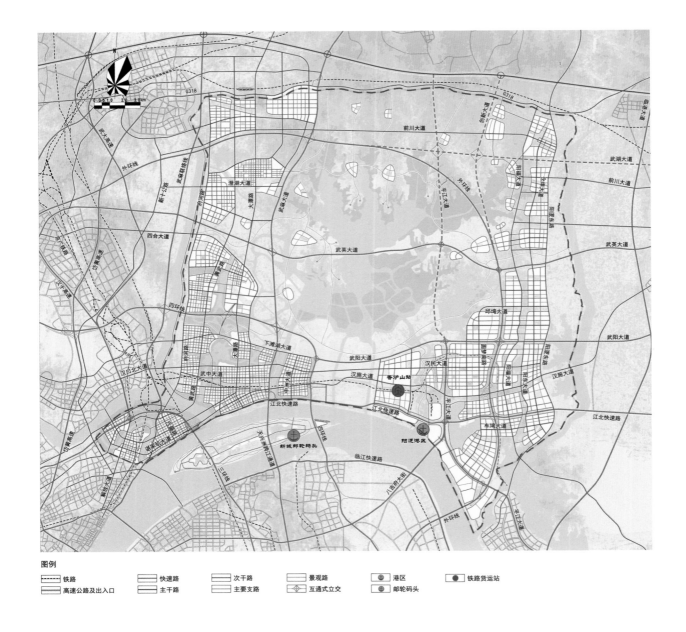

In terms of development orientation, Master Plan for the Yangtze River New City (2018) aimed to emulate the world's advanced cities and Xiong'an New Area, actively undertake the national mission, define the target orientation of "innovative famous city, ecological green city, modern wisdom city, international friendly city, and capital investment city" based on the development of regional functions, and strive to build a high-quality development demonstration area along the Yangtze River Economic Belt.

In terms of spatial layout, the plan highlights that Wuhan is characterized by water and flexible development. The plan aimed to construct a long-term water security pattern, build urban reservoirs relying on the core waters of Wuhu Lake and Grass Lake, and implement the upper and lower lake dike separation projects; build the symbiotic green wedge, and attach importance to the overall conservation and protection of mountains, lakes, forests, fields and rivers from Dabie Mountain to Mulan Mountain to the Yangtze River to build a two-stage ventilation corridor; inherit the context of mountains and rivers and construct the cluster development mode based on the mode of "building cities on platforms, building cities on uplands and using special land for special purposes".

In terms of infrastructure, the plan focused on the two core themes of "future" and "wisdom", and built an efficient, flexible and advanced support system to form an integrated air, ground and underground transportation structure. It also built an intensive and efficient water supply system and a recycled sewage treatment system, established a low-carbon and clean comprehensive energy system, and improved Wuhan's comprehensive ability to prevent disasters and protect the environment.

1 2

1
长江新城道路系统规划图
Road System Plan of the Yangtze River New City

2
长江新城用地规划图
Land Use Plan of the Yangtze River New City

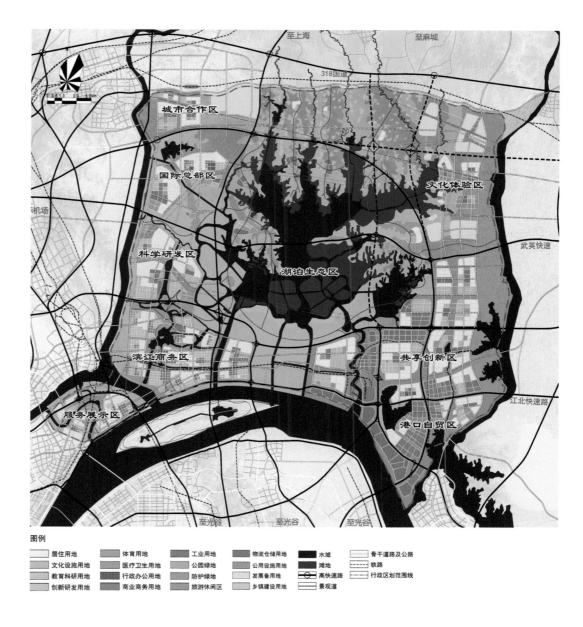

6 - 面向国家综合交通中心的枢纽规划
Hub Planning for a National Integrated Transport Center

20世纪80年代，通过实施"两通起飞"和"两个开通"，把武汉建设成为内联华中、外通海洋的经济中心、商业中心、交通运输中心的发展战略，武汉市交通基础设施获得了长足的进步。但是，也要清醒地看到，这一时期也是武汉社会经济排名大幅下滑的时期，天河机场、汉口新火车站虽然提出时间较早，但建成时间基本都在1990年代中后期，滞后于国内同类型城市基础设施建成时间。因此，在进入21世纪的前几年，武汉的交通市政基础设施存在较多历史欠账和短板。

在此背景下，从2008年开始，武汉的城市建设重点迅速聚焦到"大交通"和"大城建"上来，对外强化城市枢纽联络和辐射能力，对内补齐超大城市交通市政基础设施短板，提出了打造国家级综合交通枢纽城市以及城建"攻坚、跨越和品质"三个计划。

重新提出打造国家级综合交通枢纽城市的目标，实际是对武汉区位优势的再思考。武汉的地理区位比较稳定，一直处于"胡焕庸线"以南区域的几何中心，得中独优。这里说的区位更多是指交通区位，从车船到"铁、水"、再到铁路时代，武汉的区位优势从相当重要、到唯一垄断、再到区域节点，腹地范围从湖湘、到陕甘豫赣、再回缩至本省，经历了倒U形的曲线滑落。因此，在海运和航空时代，武汉市不仅要继续补足铁路短板，特别是重视沿长江高铁的建设，更需要在"铁、水、公、空"各个领域做到中部第一，在高铁网络规模、航空国际国内航线、"一带一路"国际通达能力等方面走在全国前列，全力打造全国综合交通枢纽。

与此同时，武汉市还必须坚持大规模推进城市建设，应对"机动化"时代的到来和克服"分散化、碎片化"地理格局的先天缺陷。通过城建攻坚计划，补足交通短板，重点实施城市过江交通、环线和放射线快速路，实现主城区"30分钟"畅通工程；通过城建跨越计划，实现功能升级，重点实施轨道交通一期至四期计划，实现轨道交通从线到网的蝶变；通过城建品质计划，转变城建模式，重点实施中山大道综合改造、东湖绿道等项目，从"大拆大建"到"让城市静下来"，从"缩减交通空间"到"增加交往场所"。城建思路实现了从强功能、到提升承载力、再到注重城市品质环境的转型。

In the 1980s, Wuhan made great progress in transportation infrastructure by implementing the strategy of making Wuhan an economic center, a commercial center and a transportation center connecting central China with overseas markets through "supporting transportation and the circulation of commodities" and "opening to the mainland and abroad". However, we are also aware that in this period, Wuhan's social and economic ranking has declined sharply. Although the construction of Tianhe Airport and Hankou New Railway Station were proposed very early, they were completed basically in the mid and late 1990s, lagging behind the construction of infrastructure in domestic cities of the same type. Therefore, in the first few years of the 21st century, there are many shortcomings in Wuhan's transportation and municipal infrastructure.

Under this background, starting from 2008, the focus of urban construction in Wuhan has been quickly shifted to "mass transportation" and "big city construction". To strengthen the communication and radiation capacity of the urban hub, and to make up for the shortcomings of transportation and municipal infrastructure as a megacity, the government proposed to build a national comprehensive transportation hub city and to "overcome the difficulties, improve the existing infrastructure and value quality" in urban construction.

The goal of building a national comprehensive transportation hub city is actually a rethinking of Wuhan's regional advantages. Wuhan's geographical location is relatively stable, and it has been at the geometric center of the area south of the Heihe-Tengchong Line. As a center, Wuhan has a lot of advantages. The location here refers more to transportation location. From boats and carriages to railways combined with boats and finally to railways, the location of Wuhan has been quite important, and then Wuhan enjoyed a monopoly position and became the regional node. The hinterland covered Hunan, Shaanxi, Gansu, Henan and Jiangxi, and then reduced to Wuhan, forming an inverted "U" curve. Therefore, in the era of shipping and aviation, Wuhan should continue to make up for the shortage of railways, especially the construction of the high-speed rail along the Yangtze River. Wuhan should come on top of central China in rail, water, road, and air transportation. Wuhan should take the lead in the high-speed rail network, the international and domestic routes of aviation, and the international access capability of the "one belt and one road". Wuhan should make every effort to build a comprehensive national transportation hub.

At the same time, Wuhan must adhere to large-scale urban construction, cope with the era of "motorization" and overcome the congenital defects of the "decentralized and fragmented" geographical pattern. Wuhan should make up for the shortcomings of traffic through the urban construction plan, focusing on the implementation of urban river-crossing traffic, ring and radiation expressways, and realize the "30 minutes" smooth traffic project in the downtown. Through the urban construction improvement plan, Wuhan should realize the function upgrading. The plan focuses on the implementation of the first to the fourth phases of the rail transit plan, completing the transformation of rail transit from line to network. Through the quality plan of urban construction, Wuhan should change the urban construction model, and focus on the comprehensive transformation of Zhongshan Avenue and Donghu Greenway, from the "demolition and construction" pattern to "building a tranquil city", and from "reducing traffic space" to "increasing communication places". The idea of urban construction has transformed from function improvement to capacity enhancement, and then to improvement of urban environment quality.

第肆章 · 区域一体化时期的规划 243

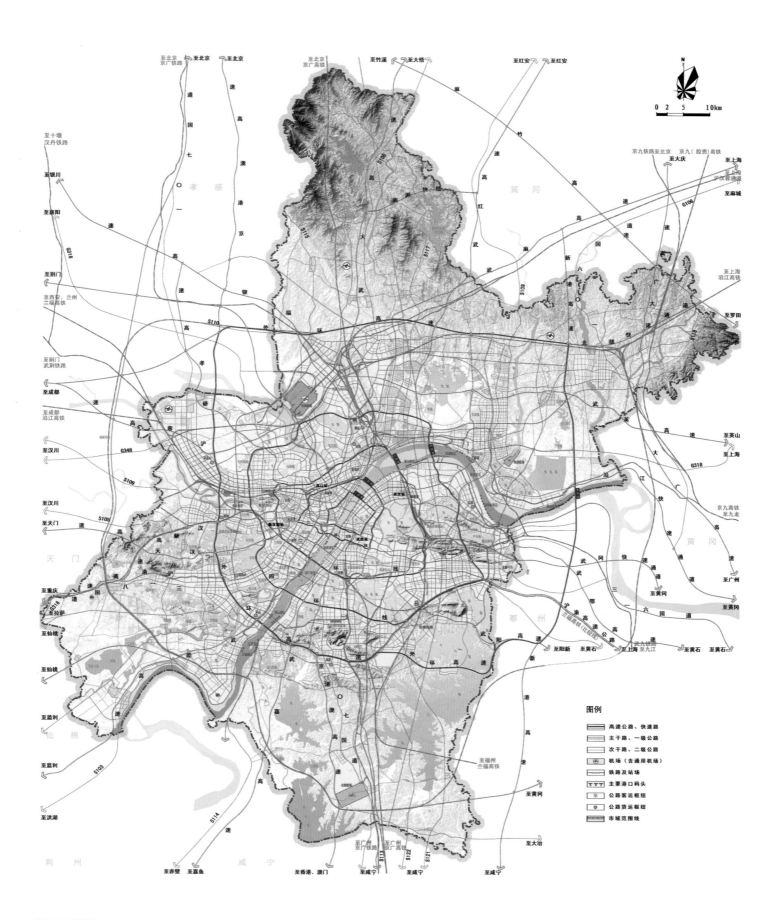

市域综合交通规划图

Urban Comprehensive Traffic Plan

一 城市综合交通规划

为配合新一轮城市总体规划的编制工作，武汉市国土资源和规划局于2016年初开展了规划编制工作，突出了交通规划在"三规"修编中的先行作用，实现了对城市总体规划、土地利用总体规划的支撑作用。

围绕《武汉2049远景发展战略规划》（2016年），高效整合武汉交通枢纽资源，以"空铁换乘"和"铁水联运"为抓手，构筑高度一体化的枢纽设施和服务体系，全面提升空港、水港、陆港的对外辐射能力，打造以国际客运枢纽和国际物流中心为核心的枢纽城市，成为中部地区联系世界的门户，构建"绿色、人本、高效、智慧、一体化"的综合交通体系，打造以"轨道+慢行"为主导的交通出行结构。

枢纽方面，规划重点围绕"天河空铁枢纽"和"阳逻陆港枢纽"两大枢纽体系建设。其中，轨道交通方面，打造"轨道上的大武汉"，构建紧凑集约的城市空间结构，打造1小时"主城区门到门、大都市区点到点、城市圈站到站"的交通圈；路网方面，优化道路骨架网络，加强复合交通廊道建设，完善微循环支路系统，形成"环射成网、循环连通"的路网布局。慢行交通方面，贯彻绿色发展理念，营造高品质的街道环境，实现"慢行复兴"；交通政策方面，积极调控机动车增长和使用，实施差别化的停车设施供给和管理措施，加强交通精细化管理，加快"互联网+智慧交通"发展，实现城市交通畅通有序发展。

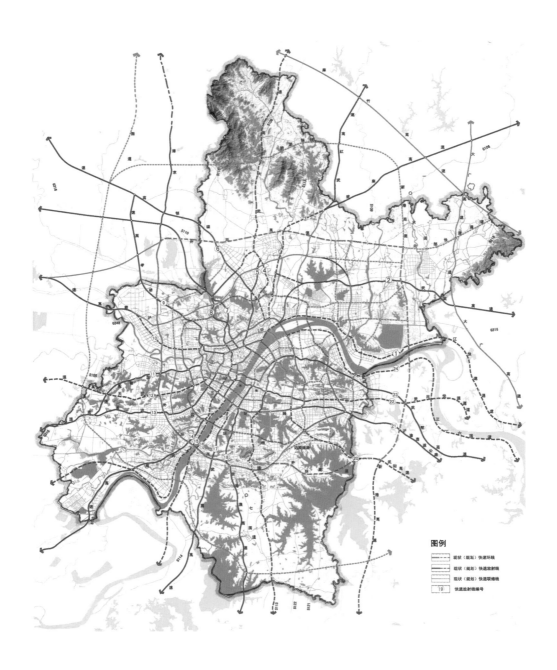

— Urban Comprehensive Transportation Planning

To coordinate with the compilation of the new overall urban plan, the Municipal Ministry of Natural Resources and Planning Bureau launched the planning preparation in early 2016, highlighting the leading role of traffic planning in the revision of the "Three Plans". It has realized the supporting role of the overall urban plan and the land use plan.

To make Wuhan a communication hub, it should efficiently integrate the transportation resources in accordance with *Wuhan 2049 Long-Term Development Strategy Study*. Wuhan should build a highly integrated hub facility and service system through "air-rail transfer" and "iron-water intermodal transport". Wuhan should comprehensively enhance the external radiation capacity of airports, water ports and inland ports, building a hub city of international passenger transport hub and international logistics center, and becoming a gateway for the central China to connect with the world. Wuhan will build a "green, people-oriented, efficient, intelligent and integrated" comprehensive transportation system, and a "rail + non-motorized" traffic structure.

The focus of hub construction is on the two systems, "Tianhe Sky Train Hub" and "Yangluo Land Port Hub". In terms of rail transit, the government should build a "Greater Wuhan on Track", construct compact and intensive urban spatial structure, and a one-hour traffic circle of "door to door in the downtown, point to point in the metropolitan area and station to station in the city circle". In terms of road network, the government should improve the road skeleton network, strengthening the construction of composite traffic corridors and integrating the microcirculation branch system. It should form a road network of "linking the radial traffic line to from a net, and improving the communication". For slow-moving traffic, Wuhan should implement the concept of green development, create a high-quality street environment, and realize the "revival of non-motorized traffic". For transport policies, Wuhan should actively regulate and control the growth and use of motor vehicles, implementing differentiated supply and management measures for parking facilities. It should also strengthen the meticulous management of transportation, speeding up the development of "Internet + Intelligent Transportation" to achieve smooth and orderly development of urban traffic.

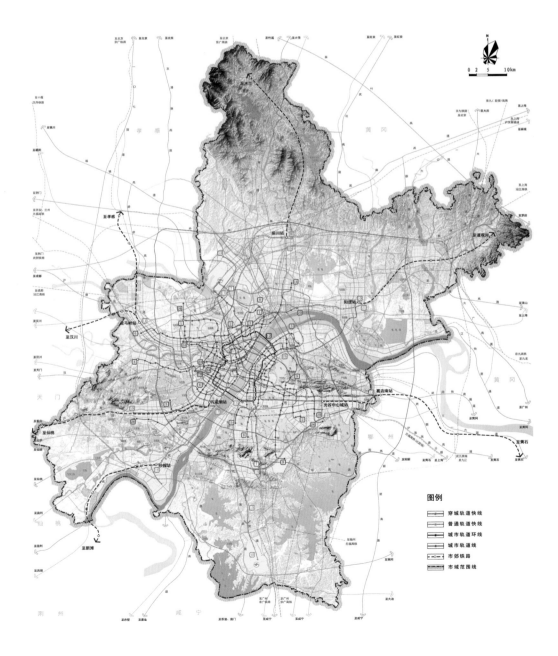

1 | 2

1
"六环二十四射"高快速路系统规划与建设图

Planning and Construction of "Six Rings and Twenty-Four Radial Lines" High Expressway System

2
市域轨道交通规划图

Urban Track Traffic Plan

— 全国高铁之心，京广和沪汉蓉建成奠定了高铁之心的地位

根据《国家中长期铁路网规划（2016~2025年）》，武汉铁路枢纽定位为全国铁路四大枢纽之一、六大路网性客运中心之一、六大客车机车检修基地之一。目前，已形成"三站鼎立"（武汉站、汉口站、武昌站）的客运格局，武汉火车站主要承担京广高铁、武黄城际以及普速铁路客流，汉口火车站主要承担沪汉蓉高铁、汉孝城际以及普速铁路客流，武昌火车站主要承担普速铁路通过与到发，三大火车站通过轨道交通串连。武汉北铁路编组站位于滠口地区，建成后将成为亚洲最大的铁路编组站。

至2035年，为进一步夯实国家级铁路枢纽体系，在目前"三主一辅"的基础上扩充为"五主两辅"的铁路车站格局，分别为武汉站、汉口站、武昌站、新汉阳站（新增）、光谷站（在建）、天河站（新增）、长江新城站（新增）。

为构建武汉与京津冀、长三角、珠三角、成渝及关中海西等主要城市群的高铁辐射圈，需加密武汉铁路枢纽对外联系网络，提升铁路通道能力。《武汉市城市总体规划（2017~2035年）》规划新增西武福高铁（2向）、京九高铁（2向）、沿江高铁（2向）、武贵高铁（1向）、武杭高铁（1向）等骨干高铁线路和京广高铁（2向）、沪汉蓉高铁（2向）4个方向的既有高铁线路，形成以武汉为中心的"两纵两横两连十二方向"高速铁路网。

— Heart of National High-speed Railway Built Upon the Beijing-Guangzhou and Shanghai-Wuhan-Chengdu High-speed Railways

According to the National Medium- and Long-Term Railway Network Planning, Wuhan Railway Hub is positioned as one of the four major railway hubs, one of the six network passenger transport centers and one of the six locomotive maintenance bases. At present, the passenger transport pattern of "tripartite station (Wuhan Station, Hankou Station and Wuchang Station)" has been formed. Wuhan Railway Station mainly undertakes the passenger flow of Beijing-Guangzhou High-speed Railway, Wuhan-Huangshi Intercity Railway and normal-speed railways. Hankou Railway Station mainly undertakes the passenger flow of Shanghai-Wuhan-Chengdu High-speed Railway, Hankou-Xiaogan Intercity Railway and normal-speed railways. Wuchang Railway Station mainly undertakes the passing and arrival of normal-speed railways. The three stations are connected by rail transit. Wuhan North Railway Marshalling Station is located in the Shekou area. Once completed, it will become the largest railway marshalling station in Asia. By 2035, to further consolidate the national railway hub system, the railway station pattern of "five main stations and two auxiliary stations" has been extended

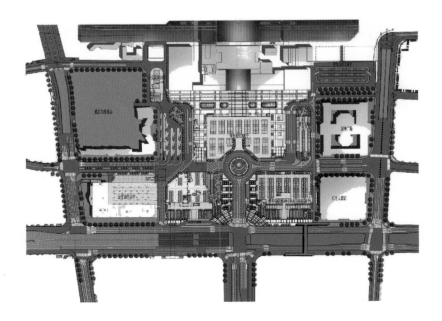

1

大都市区铁路系统布局图

Layout of Railway System in the Metropolitan Area

2

汉口火车站实景

Effect Drawing of Hankou Railway Station

3

汉口火车站总平面规划图

General Plan of Hankou Railway Station

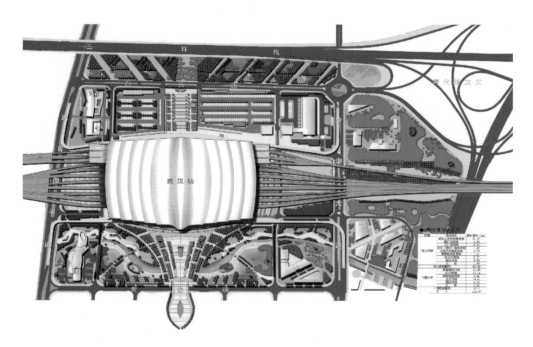

on the basis of "three main stations and one auxiliary station", namely, Wuhan Station, Hankou Station, Wuchang Station, New Hanyang Station (newly added), Guanggu Station (under construction), Tianhe Station (newly added) and Yangtze River New Town Station (newly added).

To build the high-speed railway circle connecting Wuhan and major city clusters such as Beijing, Tianjin, Hebei, Yangtze River Delta, Pearl River Delta, Chengdu, Chongqing and the central Shaanxi plain, it is necessary to bring the communication of Wuhan Railway Hub more closely to other regions and enhance the railway channel capacity. The *Wuhan City Master Plan* (2017—2035) proposes to add new high-speed railway lines, such as Xi'an-Wuhan-Fuzhou High-speed Railway (2-way), Beijing-Kowloon High-speed Railway (2-way), Yangtze-Riverside High-speed Railway (2-way), Wuhan-Guizhou High-speed Railway (1-way), and Wuhan-Hangzhou High-speed Railway (1-way) to link with existing high-speed railway lines such as the Beijing-Guangzhou High-speed Railway (2-way) and Shanghai-Wuhan-Chengdu High-speed Railway, forming a high-speed railway network of "two vertical lines, two horizontal lines, two connections and twelve directions" with Wuhan as the center.

第肆章 • 区域一体化时期的规划　249

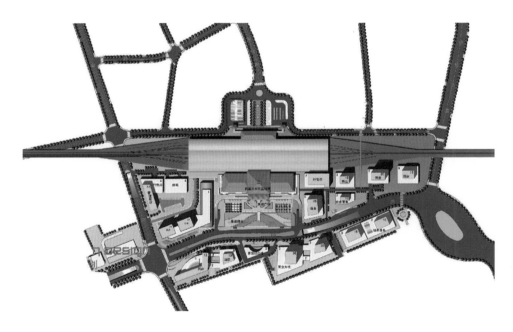

1	3
2	4

1

武汉火车站实景

General Plan of Wuchang Railway Station

2

武汉火车站总平面规划图

General Plan of Wuhan Railway Station

3

武昌火车站实景

Effect Drawing of Wuchang Railway Station

4

武昌火车站总平面规划图

General Plan of of Wuhan Railway Station

一 门户枢纽

2004年,机场二期工程开工。T2航站楼历时3年建设,于2007年底竣工,2008年4月正式投入使用。2009年3月国家民航局和湖北省政府批复了武汉天河机场总体规划,认为武汉天河国际机场是全国民用机场布局规划确定的区域干线机场,在我国民用机场布局中处于重要的位置。2012年7月机场三期工程正式开工,包括新建T3航站楼和第二跑道等。T3航站楼建设工程于2010年顺利通过国家发改委批准立项,2017年8月31日正式启用(T1、T2航站楼停用),航站楼综合体面积是T1、T2航站楼总面积的3倍,实现国内航班与国际航班间的转乘,转乘时间由"T2时代"的最长1小时缩短至20分钟。机场第二跑道于2016年8月正式启用。

《武汉市城市总体规划(2017~2035年)》提出了打造以国际客运枢纽和国际物流中心为核心的枢纽城市的目标,致力于提升武汉天河枢纽的地位和能级、积极参与全球分工合作。为此,武汉市加快推进了天河空港基础设施建设和提档升级,提出了将多条高铁线路直接引入机场,形成全国最大的"空铁"换乘枢纽。规划布置5条跑道,形成年客运吞吐量1.2亿人次的规模,建成国家门户机场枢纽。同时,加强与鄂州顺丰国际货运机场的合作,形成"客货双枢纽"的航空枢纽体系格局。

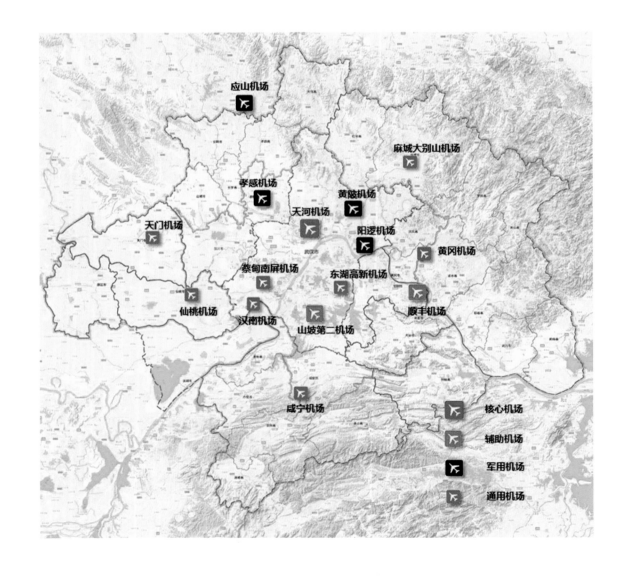

1
天河机场T3航站楼
Terminal T3 of Tianhe Airport

2
武汉城市圈机场体系布局图
Layout of Airport System in Wuhan City Circle

3
天河机场总体规划平面图
Overall Planning of Tianhe Airport

4
天河机场T3航站楼及航站区规划平面图
Planning of Terminal T3 and Terminal Area of Tianhe Airport

— Gateways & Hubs

In 2004, the second phase of the airport project started. It took three years to construct Terminal T2. Terminal T2 was completed at the end of 2007 and was put into operation in April 2008. In March 2009, the Civil Aviation Administration of China and the Hubei Provincial Government approved the overall plan of Tianhe Airport in Wuhan. Tianhe Airport is the regional air route airport determined by the layout plan of the national civil airport, which plays an important role in the layout of civil airports in China. In July 2012, the third phase of the airport project was officially initiated, including building the new Terminal T3 and the second air landing strip. The Terminal T3 construction project was approved by the National Development and Reform Commission in 2010, and officially put to use on August 31, 2017 (The T1 and T2 terminal buildings have been out of service since then). The terminal complex is three times of the total area of T1 and T2 terminal buildings. The transfer time between domestic flights and international flights is shortened from up to one hour of "T2 era" to 20 minutes. The second air landing strip of the airport was officially opened in August 2016.

The Wuhan City Master Plan (2017—2035) points out the goal of building a hub city which is the international passenger transport hub and international logistics center. It is committed to improving the status and capacity of Wuhan Tianhe hub and actively participating in global division of labor and cooperation. Therefore, Wuhan has accelerated the infrastructure construction and upgrading of Tianhe Airport. The government proposed to introduce several high-speed railway lines directly into the airport to form the largest air-rail transfer and hub in the country. Five air landing strips are planned to form an annual passenger throughput of 120 million people, building Tianhe Airport a national portal Airport hub. The government will strengthen cooperation with Ezhou SF-Express International Airport, forming an airline hub system of "hub for both passenger and freight transport".

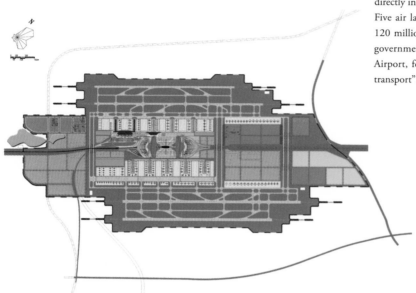

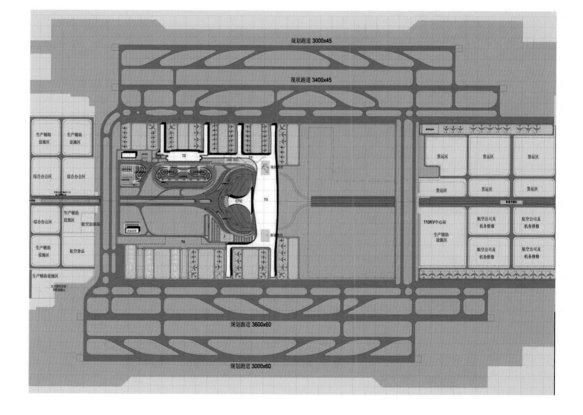

— 长江中游航运中心

2008年初，湖北省委省政府提出建设"武汉新港"的重大战略构想，作为促进湖北中部地区崛起和推动武汉城市圈建设"两型"社会综合配套改革试验区的重要突破口。2009年2月国家交通运输部和湖北省人民政府联合批复《武汉新港总体规划》及三个专项规划（空间、产业、集疏运规划）。规划范围为：西以武汉市域与咸宁市域为界，北临武汉绕城高速—武英高速公路，东接大广高速—沪蓉高速—京港澳高速公路，南至咸宁市域界限，总面积约7925平方公里。

2009年4月，武汉新港阳逻港"江海直达航线"开始正式运营。同年12月，武汉新港江北快速路工程开工。2010年4月，台湾台中港直达武汉阳逻港，汉台快航正式启运。同年7月，武汉新港阳逻集装箱二港区正式开港。2011年5月，湖北省对武汉新港的范围进行了拓展，将咸宁、鄂州、黄冈三市所辖长江岸线全部纳入武汉新港，岸线长度由420公里增加至784.3公里。2012年，交通运输部与湖北省人民政府进行了多次沟通和协调后，对范围拓展后的《武汉新港总体规划》达成共识。2014年，武汉新港成为长江中上游港口中第一个突破"百万标箱"的内河港口，正式迈入世界内河集装箱港口第一方阵，全面实现了《武汉新港总体规划》确定的"亿吨大港、百万标箱"目标。

— Maritime center of the middle reaches of the Yangtze River

At the beginning of 2008, Hubei Provincial Party Committee and Provincial Government put forward a major strategic concept of building "Wuhan New Port" as an important breakthrough to promote the rise of central Hubei and speed up the construction of Wuhan City Circle as a "two-oriented" comprehensive reform pilot area. In February 2009, the Ministry of Transport of the People's Republic of China and Hubei Provincial Government jointly approved the General Plan for Wuhan New Port and three special plans (space, industry, collection and distribution). The planning scope is as follows: the boundary in the west is set between Wuhan and Xianning, in the north by Wuhan Belt Expressway and Wuhan-Yingshan Expressway, in the east by Daqing-Guangzhou Expressway — Shanghai–Chengdu Expressway — Beijing-Hong Kong-Macao Expressway, and in the south by Xianning City. The total area is about 7,925 square kilometers.

"River-Sea Direct Line" in Yangluo Port of Wuhan New Port began to run officially in April 2009. The construction of Jiangbei Expressway in Xingang, Wuhan, started in December of the same year. In April 2010, Taichung Port of Taiwan can connect directly to Yangluo Port of Wuhan, and the Wuhan-Taiwan Express was officially put to use. In May 2011, Hubei expanded Wuhan New Port by bringing the coastline under the jurisdiction of the Yangtze River, Xianning, Ezhou and Huanggang into the New Port. The length of the coastline increased from 420kilometers to 784.3kilometers. In 2012, a consensus on the expanded areas in General Plan for Wuhan New Port was reached after many rounds of communication and coordination between the Ministry of Transport and Hubei Provincial Government. In 2014, Wuhan New Port became the first inland port in the middle and upper reaches of the Yangtze River with the capacity of dealing with millions of standard containers, and formally ranked itself among the leading list of inland container ports worldwide. The target of "building a port with the capacity of dealing with hundreds of millions of tons of goods and millions of standard containers" set out in the General Plan for Wuhan New Port has been fully realized.

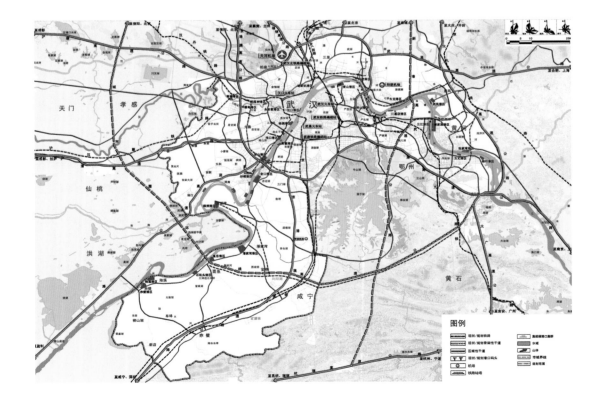

1

武汉新港港区布局与区域交通系统规划图

Layout of Wuhan New Port Area and Planning of Regional Traffic System

2

武汉新港用地规划图

Land Use Plan of Wuhan New Port

3

武汉新港空间结构图

Space Structure of Wuhan New Port

第肆章 · 区域一体化时期的规划

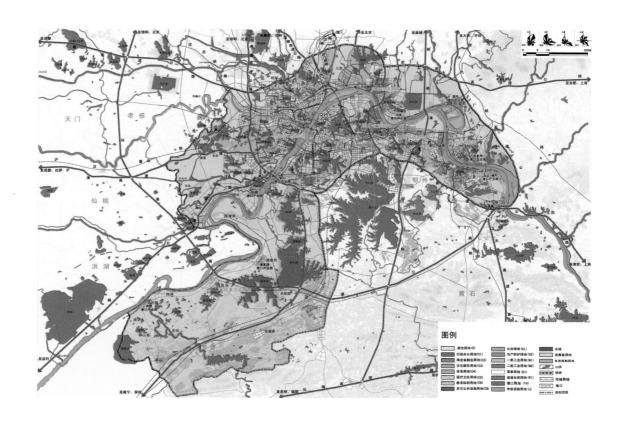

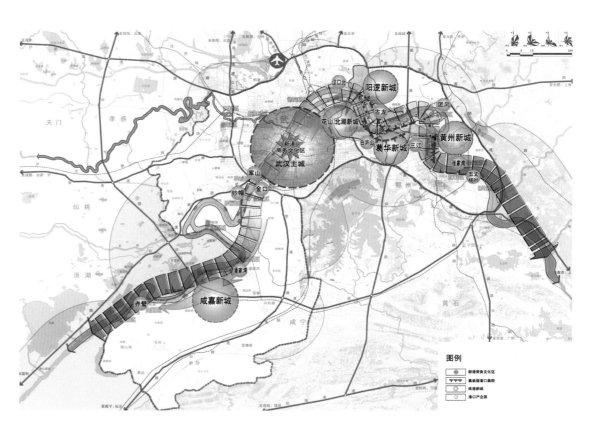

7 – 面向国际滨水文化名城的品质规划
Quality Planning for an International Well-known Waterfront Cultural City

一 重大事件行动规划

● 世界军人运动会城市环境综合整治提升规划（2018年）

2019年10月第七届世界军人运动会在武汉举行，在市委市政府统一部署下，武汉市全面开展了迎军运会环境综合整治提升工作。按照"办赛事"和"建城市"相统一的总体要求，武汉市国土资源和规划局成立了工作专班，组织开展了城市环境综合整治提升各项规划编制工作，运用"城市双修"理念，全面指导军运会环境整治提升工作。

《世界军人运动会城市环境综合整治提升规划》（2018年）在参考杭州、厦门、青岛等城市环境整治标准基础上，结合本地实际情况，针对道路路面、建筑立面、生态绿化、城市家具、户外广告和夜景亮化等六大专项整治导则，提出武汉版环境综合整治提升极致标准，指导优化各专项规划导则，统筹高标准实施方案的编制。通过细致调研，采用"点、线、面"结合的形式，确定对全市25条重点线路、196条基础保障线路等线路两侧，84处场馆和酒店、重要交通枢纽等点位周边进行环境整治，并结合各自空间特点，提出精细化整治要求。

为全面统筹环境提升工作，专班在梳理各区、各部门工作情况基础上，完成了调研线路图绘制、全市"五边五化"作战图、全市拆迁一张图绘制、市政府督查室现场巡查等工作，并形成定期维护更新机制，全面推进军运会工作。

1
武汉市迎军运会城市环境
综合整治提升工作作战图
Action Plan of
Comprehensive Renovation
and Improvement
of Wuhan's Urban
Environment for the World
Military Games

2
武汉市"三旧"改造规划
总图
Wuhan "Three Old"
Renovation Plan — General
Plan

一 "三旧"改造系列规划

● "三旧"改造系列规划（2013至今）

为贯彻落实国家关于开展城镇低效用地再开发试点以及加快推进棚户区改造等有关要求，按照武汉市委市政府确定的2020年前基本完成三环线内"三旧"改造总体目标，2013年市国土资源和规划局组织开展了统筹全市"三旧"改造规划工作纲领性文件的编制工作。2013年9月，武汉市政府正式批复了该规划。

《武汉市"三旧"改造规划（纲要）》（2013年）确定了武汉市"三旧"改造范围为三环线内及相邻重点区域，改造对象为旧城区（棚户区、危旧房）、旧厂房、旧村庄（城中村）。明确了按照成片改造，修缮整治和拆除重建相结合的改造模式进行"三旧"改造。在空间布局方面，要求重点推进两江四岸地区"三旧"改造，集中发展金融证券、跨国总部、商务贸易、国际会展、专业服务、文化旅游等高端服务业；加快推进二环线内"三旧"改造，积极发展现代服务业；有序推进二环线和三环线之间"三旧"改造，大力发展战略性新兴产业，并布局生态宜居社区。同时将"三旧"改造与城市基础设施和轨道交通建设相结合，加强城市功能区之间的联系。在人口和产业安置方面，为充分满足征收人需求，提出就地安置、就近安置、疏解安置、回购房安置和货币安置等多方式综合的居住还建安置方案，并要求优先在居住安置区完善公共服务配套设施和交通设施，切实提升还建居民生活环境。结合都市产业园对城中村、旧厂房的旧工业进行改造，集中迁并，提档升级。为使"三旧"改造能有序推进，市政府出台了"三旧"改造计划，保证2020年基本完成三环线内"三旧"改造总体目标。以此为基础，市国土资源和规划局陆续开展了江岸区百步亭丹水片、江汉区省电影制片厂、青山区34、35街坊等次级功能区片的规划。

— Action Planning for Major Events

• *Comprehensive Renovation and Promotion Planning of Urban Environment for World Military Games (2018)*

Wuhan will hold the 7th World Military Games in October 2019. Under the joint arrangement of the municipal Party committee and the municipal government, Wuhan has carried out comprehensive improvement and upgrading of urban environment for the event. In accordance with the overall requirements of "organizing competitions while building the city", the Municipal Ministry of Natural Resources and Planning Bureau has set up a special working group to organize and carry out the comprehensive renovation and improvement of urban environment. The group applies the concept of "Ecological Restoration and Urban Renovation" as the general guidance in renovation and improvement.

Based on the renovation standards of urban environment in Hangzhou, Xiamen and Qingdao, the group has fully considered the local actual situation and proposed to aim at six special rectifications, namely, road pavement, building façade, ecological greening, urban furniture, outdoor advertisement and night scene brightening. The group put forward Wuhan's ultimate standard of comprehensive improvement and upgrading of urban environment. The group has optimized the guidelines of the six rectifications. The group coordinates the preparation of high-standard implementation plans. Through thorough investigations and researches, the group decides to carry out environmental renovation in the form of "the combination of dots, routes and regions" on both sides of 25 key traffic routes, 196 basic traffic routes, 84 gymnasiums and stadiums, hotels, and important transportation hubs in the city. The group also requires the renovation to consider features of each place and the management to be more refined.

To coordinate the work of improvement in all aspects, after going through the work progress of each district and department, the special group has completed the mapping of investigation route and the renovation map of "Station-side, Road-side, Construction Site-side, Railway-side, River-and Lake-side, Road Cleaning, Facade Beautifying, Landscape Brightening, Water Purifying and Ecological Greening" of the whole city. The group has also finished drawing the demolition and relocation areas in one map. The group has completed the on-site inspection led by Municipal Government Inspection Office, etc. The group also formed a regular maintenance and updated mechanism to speed up the preparation for the World Military Games in all aspects.

— "Three Old" Transformation Planning

• *"Three Old" Transformation Planning (2013 to Present)*

To implement the requirements of the State on the pilot project of the redevelopment of urban land with low utilization and accelerating the reconstruction of shacks, Wuhan Ministry of Natural Resources and Planning Bureau has organized and complied the programmatic outline of the overall planning of "Three Old" transformation in the whole city in 2013 in accordance with the general goal of finishing "Three Old" reconstruction in the Third Ring Road by 2020 set by the municipal Party committee and the municipal government. In September 2013, the municipal government formally approved the plan.

The Outline defines that the transformation area of the "Three Old" in Wuhan should be area within the Third Ring Road and its surroundings. The transformation should target on old town (squatter settlements and dangerous dated buildings), old plants and old villages (villages in the city). The Outline has made clear that the "Three Old" transformation should be carried out in accordance with the transformation mode of regional transformation, combining repair and renovation with demolition and reconstruction. For spatial distribution, the Outline requires to focus on the transformation of the "Three Old" areas along the two rivers and four riverbanks. The transformation should give priority to high-end services such as financial securities, transnational headquarters, business and trade, international exhibitions, professional services and cultural tourism. Wuhan should accelerate the transformation of the "Three Old" within the Second Ring Line, developing modern service industries. Wuhan should promote the transformation of the "Three Old" area between the Second Ring Line and the Third Ring Line step by step. Wuhan should make every effort to support strategic emerging industries and lay out ecological livable communities. Wuhan should combine the "Three Old" transformation with urban infrastructure and rail transit construction, bringing the links urban function areas closer to each other. For population and industrial resettlement, Wuhan has put forward a comprehensive scheme for residential rehabilitation and resettlement including resettling in the original land and in near areas, population distributing to other settlements, relocated settlement and monetary compensation. Wuhan requires that priority should be given to improving public service facilities and transportation facilities in resettlements. The living environment of residents should be guaranteed. The transformation of villages in the city and the old factory building should be in line with the urban industrial park. The relocation should be carried out in a massive scale. The industry should go through refinement and upgrading. To push on the transformation of the "Three Old" in an orderly manner, the municipal government has issued the "Three Old" transformation plan to ensure that the overall goal of the transformation of the "Three Old" within the Third Ring Road will be basically completed by 2020. On this basis, Wuhan Ministry of Natural Resources and Planning Bureau has successively carried out the planning of Baibuting Danshui Film in Jiang'an District, Jianghan Provincial Film Studio, Block 34 and Block 35 in Qingshan District and other sub-functional areas.

- 后湖片、南湖片、四新片城市品质提升和功能再造策略研究（2016年）

为系统评估2010版《武汉市城市总体规划（2010—2020年）》批复以来主城区若干功能片区的实施状况，解决交通拥堵、民生乏力、发展滞后、环境品质不佳等"城市病"相关问题，武汉市国土资源和规划局组织开展了四新、后湖、南湖等3个不同类型片区的评估工作，并提出一系列改造提升建议。这3个项目评估，实质上也是对规划编制、管理调控、土地开发、公共服务配套、环境改善、市政基础设施建设等方面的一次反思。自此，规划的时间维度也引起了管理者重视，规划系统内部也逐步建立起"凡居住项目必评估"的工作机制。

○ 四新片城市品质提升和功能再造策略研究（2016年）

在三个片区中，四新片是最后开展规划建设的综合性片区，其问题也是所有新区开发初期都会遇到的，即人口流入较慢、设施配套不足等。《四新片城市品质提升和功能再造策略研究》（2016年）梳理了四新地区成长历程和规划体系，以详实的现状调研为基础，针对四新地区较为突出的建设滞后问题，提出了补充性的建设工程和建设时序要求；针对国际博览中心相关产业链不完善等问题，规划在功能提升、产业结构和实施机制等方面提出了可行的策略；结合全面的人口摸底，明确实际的规划人口，并提出完善公共设施、道路交通和市政配套设施的策略。

○ 后湖片城市品质提升和功能再造策略研究（2016年）

后湖是城市总体规划确定的以居住功能为主的综合组团之一，也是汉口地区较早开展建设的新区。在多年的发展过程中，后湖的居住、交通、市政配套等有了较大幅度的提升，但依然面临着产业用地较少、内部道路微循环不畅、教育和医疗配套不足等问题。《后湖片城市品质提升和功能再造策略研究》（2016年）充分运用GIS技术建立了多数据分析平台进行空间识别，明确未来空间优化的重点及改造方式，制定了片区和零星用地整治图则。规划还就增补创新产业功能、加大服务设施均好覆盖、畅通路网等提出了相关策略建议。

○ 南湖片城市品质提升和功能再造策略研究（2016年）

南湖是1996版、2010版城市总体规划确定的以居住功能为主的综合组团之一，也是武昌地区最早开展建设的新区。《南湖片城市品质提升和功能再造策略研究》（2016年）以提升城市品质、城市升级为宗旨，一方面"谋远"提出系统优化建议，另一方面"应急"提出近期重点优化项目。规划提出将南湖地区单一居住职能调整为科创、居住、休闲等多元复合功能。依托周边高校集聚的科教资源，打造依河、畔湖、夹山的泛南湖大学之城示范区，导入农业科研、文化创意等产业资源，打造生活便捷、环境优美、活力繁荣的宜居宜业社区，构建良好的生产、生活、生态空间体系，形成"学区+园区+社区"的"三创"实验区。在此基础上，规划还提出片区人口规模、公共服务设施及交通市政设施配套要求。

南湖片交通设施控制图

Coverage of Traffic Facilities in Nanhu District

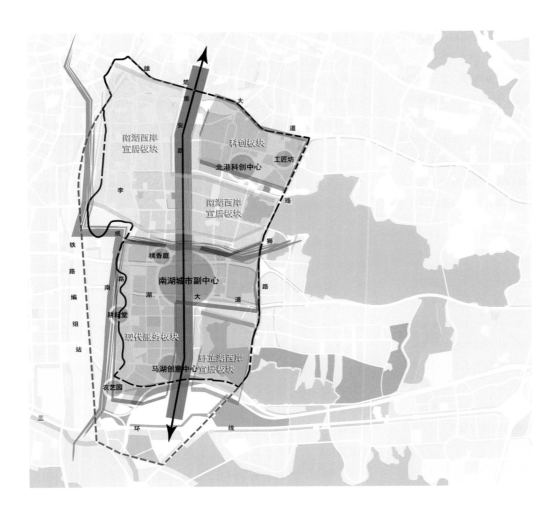

1
南湖片功能结构图
Functional Structural of Nanhu District

2
南湖片公服设施控制图
Coverage of Public Service Facilities in Nanhu District

• **Research on quality improvement and functional reengineering strategies of Houhu District, Nanhu District and Sixin District (2016)**

To systematically evaluate the implementation of several functional areas in the main urban area since the approval of the 2010 urban master plan, and to solve the problems related to "metropolitan diseases" such as traffic congestion, people's livelihood difficulties, lagging development and poor environment quality, Wuhan Ministry of Natural Resources and Planning Bureau has organized evaluation on three different types of districts, namely, Sixin, Houhu and Nanhu, and brought up a series of suggestions for improvement and upgrading. The evaluation of these three projects is in essence a reflection on planning, management and regulation, land development, public service support, environment improvement and municipal infrastructure construction. Since then, the time consideration of planning has also attracted managers' attention. And the working mechanism of "Every residential project must first beevaluated" has been gradually introduced within the planning system.

○ Research on quality improvement and functional reengineering strategies of Sixin District (2016)

Compared with the other three districts, Sixin is the last to be under planning and construction. It also encountered problems occurred in the early stage of development like all new districts, namely slow population inflow, inadequate facilities and so on. The Research went through the growth process and planning system of Sixin District. The Research puts forward supplementary construction projects aiming at the prominent lagging construction problem and the time and order requirements for the construction on the basis of detailed investigations and researches on the current situation. The research put forward the feasible strategies in terms of functional promotion, industrial structure and implementation mechanism etc. aiming at the problems such as the imperfect industrial chains related to the International Expo Center, specified the actual planned population by combining with the comprehensive censure mapping, and put forward the strategies for perfecting public utilities, road transportation and municipal supporting facilities.

○ Research on quality improvement and functional reengineering strategies of Houhu District (2016)

Houhu District is one of the comprehensive groups with residential function as the main part determined by the urban master plan. It is also a new district constructed in the early years in the Hankou region. After years' development,

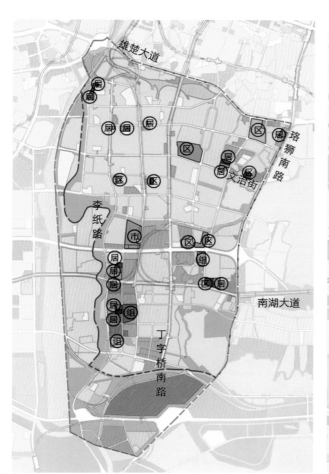 

the condition of residence, transportation and municipal supporting facilities in Houhu have been greatly improved. However, it still faces such problems as inadequate industrial land, poor microcirculation of roads inside the district, insufficient education and medical support. The Research establishes a multi-data analysis platform for spatial identification and clarifies the key points of future spatial optimization and transformation methods through full use of GIS technology, formulating renovation plans for the district and scattered land. The planning also puts forward relevant strategies and suggestions on supplementing the functions of innovative industries, increasing the coverage of service facilities and reducing the stress on traffic.

○ Research on quality improvement and functional reengineering strategies of Nanhu District (2016)

Nanhu is one of the comprehensive groups with residential function as the main part determined by the 1996 and 2010 editions of the urban master plan. It is also the earliest new district to be constructed in the Wuchang region. The purpose of the planning is to improve the quality of the city and upgrading. It puts forward systematic "long-term development" and recently key optimization projects as "emergency responses". The Research proposes to extend the original residential function of Nanhu to multiple functions, such as scientific research and innovation, residence and leisure. Nanhu should build a demonstration zone of Pan-Nanhu University City beside the lake, surrounded by rivers and mountains by introducing the scientific and educational resources provided by the surrounding universities. Nanhu should build livable and industrially viable communities with convenient life, beautiful environment and prosperity through industrial resources such as scientific research in agriculture and cultural and creative industry. Nanhu should establish a good spatial system of production, life and ecology, forming an experimental zone with "three-innovation" of "School District + Park + Community". On such basis, Nanhu puts forward that public service facilities and transportation and municipal facilities should meet the needs of its population.

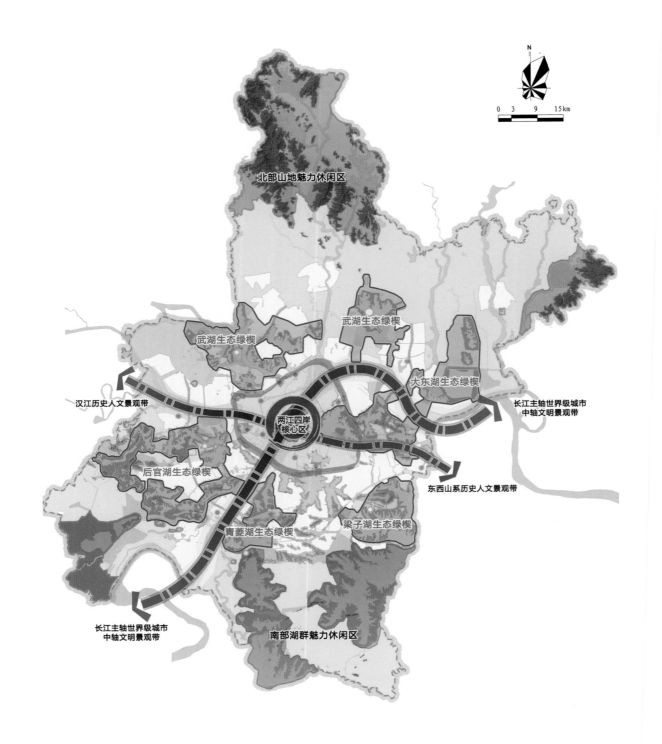

一 蓝绿融合系列规划

● **概念设想：滨水生态绿城规划（2017年）**

武汉市第十三次党代会提出，水是武汉决胜未来的核心竞争力，要做足做好水文章，大力推进"四水共治"，努力打造国内外知名的滨水生态绿城。为落实市委市政府的决策部署，《武汉市建设滨水生态绿城综合规划》（2017年）提出充分发挥武汉全国独有、世界少有的资源禀赋优势，突出"两江交汇、三镇鼎立"的独特城市格局，保护"河湖密布"的自然生态特质，营织"蓝绿融城"的空间网络，凸显滨江、滨湖城市特色，建成国际知名的滨水生态绿城。

该规划明确了主要任务，可概括为"1343"体系：在战略层面，将"水"作为城市赶超发展最重要的战略资源，提出"一核两轴、六楔百湖、蓝绿融城"的空间框架；在系统层面，提出"以水定城、以水润城、以水优城"三大策略，构建高效的城市水安全保障、开放连续的蓝绿网络和功能多元的滨湖功能区等三大体系；在实施层面，以"四水共治"为切入点，空间上聚焦"黑臭水体、低洼易渍、大湖周边"等重点区块，时间上衔接"四水共治、城建计划、绿水青山"等行动计划，围绕"护蓝、优绿、理岸"形成四大计划和若干工程；在机制层面，建立"以水定城"的城市建设协调预警机制，建立完善共建推进机制以及滨水区城市设计刚性管控机制。

- **以水定城：水系规划及大东湖水网联通工程规划**

2006年，武汉市成为国家"两型"社会建设试点和水生态修复与保护示范城市，同年《城市蓝线管理办法》和《武汉市湖泊保护条例》发布实施，水系问题成为当时的研究重点和热点。恰逢当时新一轮城市总体规划和国家城市水系规划规范基本编制完成，为破解此前因城市加速扩张而引发的城市水系的自然结构形态受到破坏等一系列水系问题，武汉市国土资源和规划局、水务局、园林局组织联合编制了《武汉市水系规划》（2006年），将生态优先、环境先导的城市发展基本原则落到实处。

该规划研究范围为武汉市市域范围，重点为都市发展区范围的水系。充分依托武汉历史水系脉络和"两江分三镇、九水通百湖"的河渠网络现状，在对水系及相关要素进行分类、分级评价基础上，结合城市空间拓展，与府河、长江和汉江一起，构建都市发展区"两江—河四区"的水系基本结构体系。四个区以湖泊为核心、港渠为纽带，构建不同特色的四片生态水网，建立联系城市内外的生态廊道和城市风道，促进六大生态绿楔的形成，同时串连主要湖泊，编织结构合理、生态良好、水流畅通、环境宜人的覆盖全市的环城水网。

该规划采取"分类划定、功能控制"的方式，划定了市域范围105个湖泊和126个明渠蓝线，针对湖泊区位、规模、功能和周边建设情况的不同，确定了湖泊、渠道功能和岸线控制，构建了污染控制、生态修复、水网建设三大工程体系，并进一步确定了湖泊空间控制的"三线一路"指引，明确了水面、滨水绿化区、滨水建设区和环湖路的控制范围线划定标准和控制要求，从根本上锁定湖泊岸线、固定湖泊形态。

在此基础上，武汉市随后又陆续开展编制了《"大东湖"水网明渠用地范围界定规划》（2006年）《水质提升总体规划》（2006年）等一系列规划，武汉市水系的规划管控已由原来的都市发展区进一步拓展到武汉市全域范围，由最初的湖泊、明渠管控扩展到了河、湖、港、渠等重点水体。

1　2

1
滨水生态绿城空间景观框架图

Spatial Landscape Framework of Ecological Green Waterfront City

2
"大东湖"生态水网系统规划图（2006年）

Planning of "Greater Donghu Lake" Ecological River Network System (2006)

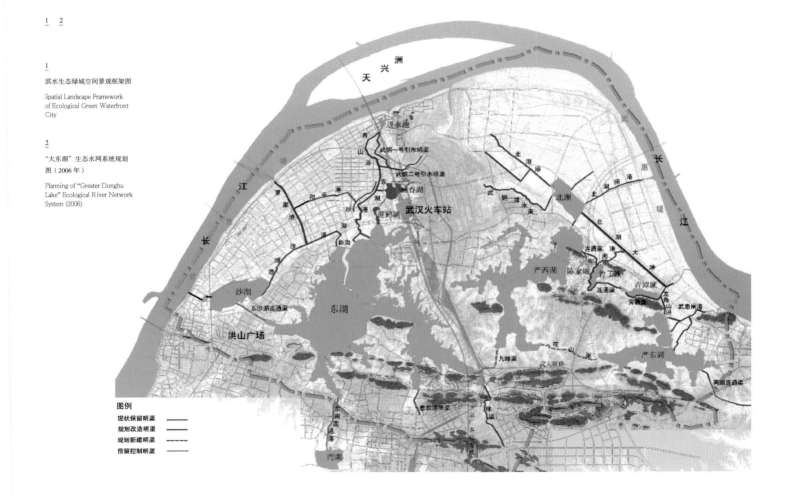

— Water and Green land Integration Planning

● **Conceptual envision: waterfront eco-green city planning**

The 13th Party Congress of Wuhan put forward that rivers and lakes are the core competitiveness of Wuhan in the future. Wuhan should lay emphasis on rivers and lakes, vigorously "managing the four rivers at the same time". Wuhan should strive to build a well-known waterfront ecological green city at home and abroad. To carry out the decision made by the municipal Party committee and the municipal government, the Plan proposes to give full play to Wuhan's unique and rare resource advantages. The Plan should highlight the unique urban layout of "the convergence of two rivers with three towns standing side by side" and protect the natural feature of "densely covered with rivers and lakes". The Plan should establish the space configuration of water and green land integration, and highlight the city feature of locating by rivers and lakes, building an internationally renowned ecological green waterfront city.

The Plan defines the main tasks, which can be summarized as "1343" system. On the strategic level, the plan regards rivers and lakes as the most important strategic resources for urban development, proposing to from a spatial framework of "one core, two axes, hundred lakes in six directions, and water and green land inside the city". For the system, the Plan puts forward three strategies of "building the city around water, nourishing the city with water, and making the city better with water", establishing three systems: efficient urban water security system, open and connected Blue-Green Network system and multi-functional lakeside area system. In terms of practice, the Plan focuses on key aspects like "black and odorous water body, low-lying and easily waterlogged areas, and areas around the large lakes" with "managing the four rivers at the same time" as the breakthrough point. As for the time, the plan is a link to action plans of "managing the four rivers at the same time, urban construction plan, clear water and green hills". The planning forms four plans and several projects around "protecting rivers, increasing green land and managing river banks". For the mechanism, the plan establishes a coordinated early warning mechanism for urban construction, a mechanism for promoting co-construction and a rigid management and control mechanism for urban design in a waterfront city.

● **Building the city around water: river system planning and greater Donghu Lake water network interconnection project planning**

In 2006, Wuhan became a national pilot "two-oriented" city and a demonstration city of water ecological restoration and protection. In the same year, the Urban Blue Line Management Measures and Wuhan Lake Protection Regulations were issued and implemented. The river system became the hotspot and focus of researches at that time. Meanwhile, a new round of urban master plan and

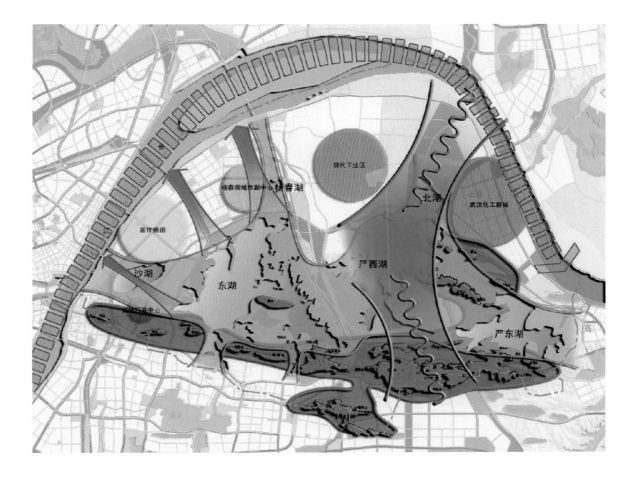

1
"大东湖"水上游线示意图
（2006年）

Plan of "Greater Donghu Lake" Upstream Line (2006)

2
"大东湖"生态结构图
（2006年）

Ecological Outcome of "Greater Donghu Lake" (2006)

national urban river system planning norms were basically compiled. In order to solve a series of river system problems, such as the destruction of natural structure of urban river system caused by accelerated urban expansion, Wuhan Ministry of Natural Resources and Planning Bureau, Water Affairs Authority and Landscape Bureau jointly compiled Wuhan River System Planning to put the basic principles of eco-priority and environment-oriented urban development into practice.

The Planning mainly covers city area, with emphasis on the river system within the urban development district. The Planning has considered the historical river system of Wuhan and the current rivers and canals network of "three towns divided by two rivers and nine rivers connected with hundreds of lakes". Based on the classifying and grading evaluation of the river system, the Planning has combined urban space expansion to bring up the idea of constructing the basic structure of river system of "relying on the Yangtze River and the Hanjiang River, together with the Fuhe River, to link Four Districts". The four districts, with lakes as the focus and harbors and canals as the link, should construct four ecological river networks with different characteristics. The four districts should establish ecological corridors and urban wind channels linking the inside and outside of city, speeding up the formation of six ecological green wedges. The four districts should also connect the main lakes to each other, building a river network around the city with a reasonable structure, good ecological environment, unobstructed river flow and pleasant environment.

By "classification and functional control", the planning delineated 105 lakes and 126 open channel blue lines. Considering the differences of the location, scale, functions and surrounding construction of lakes, it makes clear the functions of lakes, channels and coastline control. The Planning should construct three major engineering systems of pollution control, ecological restoration and river network construction. The Planning further sets up the "three-line and one-road" guidelines for the spatial control of lakes. The Planning defines the demarcation criteria and control requirements for the control limits of the surface of the water, waterfront greening area, waterfront construction area and lake-encircling road. The Planning fundamentally determines the lake shoreline and the lake shape.

Based on the Planning, Wuhan subsequently complied a series of plans, such as the "Greater Donghu" Lake River Network Open Channel Land Plan, and the Master Plan for Water Quality Improvement. The planning, management and control of river system in Wuhan has been further extended from the original urban development district to the whole city. The planning and management of Wuhan's river system has expanded from the original lake and open channel management to the key areas such as rivers, lakes, harbors and canals.

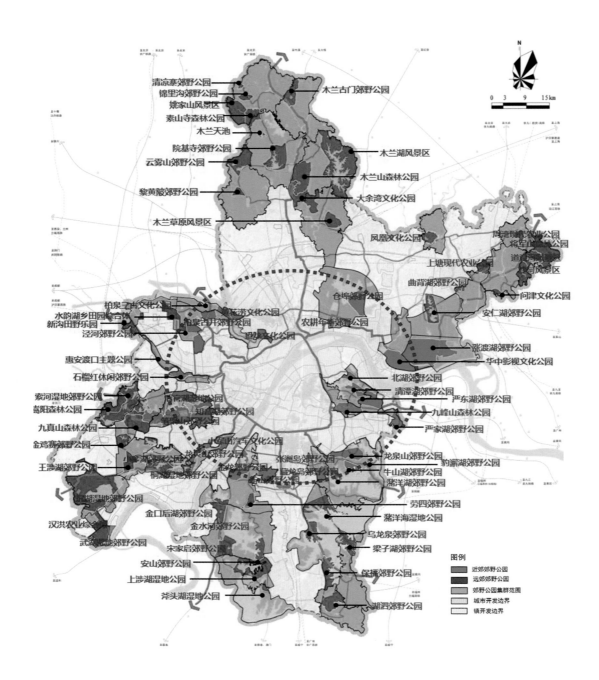

- **以水润城：郊野公园战略实施规划（2018年）（含谦森岛、劳四村）**

为构建开放连续的蓝绿网络，近年来武汉市突出了以湖群为基础的郊野公园集群规划和建设，并以此作为城市总体规划的实施规划。《郊野公园战略实施规划》（2018年）拟在非集建区策划形成郊野公园集群，以建促保，主动实施，实现生态资源的合理利用，引领武汉六大生态绿楔功能升级。

该规划以问题为导向，对国内外案例进行研究，研判游憩、休闲、农业产业发展等方面的发展趋势，以风景名胜区、自然保护区、森林公园、湿地公园等生态功能区为基础，整合农田林网、河湖水系，营造"蓝绿"生态网络，明确郊野公园可承载的"涵养保育、自然观光、田园体验、主题度假、文化创意"等五大基本功能，提出11种郊野公园的类型及功能业态指引，形成"一环两翼"的郊野公园群空间格局，结合土地整治相关政策，从高标准基本农田建设、耕地占补平衡、城乡建设用地增减挂钩等方面，提出配套政策建议。2016年以来推动了黄陂区祁家湾谦森岛郊野公园、江夏区郑店劳四村郊野公园等的建设。

同时，为全面落实"让城市安静下来"的发展理念，在山水边构建安全、连续、便捷、舒适、魅力的慢行空间网络系统，构筑了"三边四岸、一街一带"的武汉市慢行系统及绿道体系，引导形成"轨道公交+慢行"的交通出行模式。以路权规划为核心，融合"完整街道"理念，拟定标准，落实规划控制和管理；以目标为指引，以实施为导向，形成全过程"规划+项目清单"。

• *Nourishing the city with water: Strategic Implementation Plan for Country Parks (2018) (including Qiansen Island and Laosi Village)*

To build an open and continuous rivers and mountains network, in recent years, Wuhan has highlighted the planning and construction of country park clusters based on lake clusters which were the implementation plan of urban master plan. It's planned to form a country park cluster in non-construction areas, promoting construction through protection. Wuhan should actively implement, realize the rational use of ecological resources, and lead the upgrading of the six ecological green wedges.

The Plan is problem-oriented. It studies domestic and foreign cases of the development trend of recreation, leisure and agricultural industry. Based on the ecological function areas such as scenic spots, nature reserves, forest parks and wetland parks, the Plan integrates farmland forest network, river and lake water system to build a rivers and mountains ecological network. The Plan defines five basic functions of country parks: conservation and protection, natural sightseeing, pastoral experience, theme vacation, cultural creativity. The Plan proposes 11 types of country parks and functional format guidelines, forming a spatial pattern of "one ring and two wings" of the country park group. In accordance with the relevant policies of "land consolidation", the Plan puts forward suggestions for the policy on high standard basic farmland construction, balance of arable land occupation and compensation and balance of the increase and decrease of urban and rural construction land. The Plan has recently promoted the construction of Qijiawan Qiansen Island Country Park in Huangpi District and Zhengdian Laosi Village Country Park in Jiangxia District.

At the same time, to fully implement the development concept of "building a tranquil city", the planning has constructed a safe, continuous, convenient, comfortable and charming slow-moving space network system around mountains and rivers. The Plan has also constructed a slow-moving system of "three sides, four banks, one street and one belt" and a green road system in Wuhan, guiding the formation of travel model of "rail + non-motorized traffic". The Plan takes road right planning as the core, integrating the concept of "complete street". On this basis, the Plan formulates standards and puts planning control and management into practice. The Plan is guided by objectives and implementation-oriented, forming of "planning + project list" for the whole process.

1 2

1
武汉市郊野公园空间布局图
Spatial Layout of Country Parks in Wuhan

2
武汉市郊野公园总体空间结构图
Overall Spatial Structure of Country Parks in Wuhan

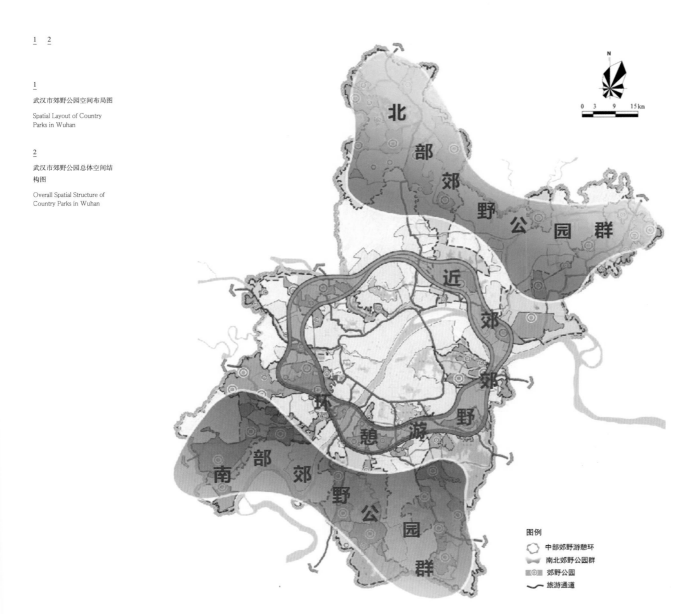

● 以水优城："大湖+"功能区实施性规划（2018年）

《武汉市"大湖+"功能区实施性规划》（2018年）重在把水资源禀赋转化为功能优势，彰显"百湖之市"城市特色，提升城市功能品质，促进城市与湖泊融合发展，以湖泊水体、周边绿地和功能区为对象，探索"大湖+"保护与利用新模式。在坚持生态优先、突出对湖泊水体"蓝线"和周边绿地"绿线"保护的基础上，该规划根据湖泊区位和功能特点，提出城市型、郊野型、生态型3种类型湖泊功能区，强调绿色发展，突出对康体休闲、创新创意、科创研发等多种公共功能的植入，增加新功能、营造新空间、培育新动力，打造一批不同主题、各具特色、多元活力的湖泊功能区。

● 行动规划："拥抱绿水青山"五年行动规划（2016年）

为进一步加强武汉生态文明建设有关要求落地，抓紧落实武汉市政府工作报告中提出的"编制实施拥抱绿水青山"行动方案，按照"突出重点，以点带面"的总体要求，《"拥抱绿水青山"五年行动规划》（2016年）从结构、品质、总量、人均指标四个方面提出都市发展区内"拥抱绿水青山"行动计划的建设目标。在生态框架结构方面，实现2010版总规"两轴两环、六楔入城"生态框架的基本贯通。在生态绿化总量方面，贴近2010版总体规划确定的目标，建成区绿化覆盖率达到41%；建成区绿地率达到35.8%；全市人均公园绿地面积达到12平方米。在生态品质特色方面，建成一批独具特色的滨水公园集群，打通一批通江、达湖、成网的高品质"绿脉"。在生态服务均好方面，完善"市级—区级—社区级"公园绿地系统，落地一批贴近市民生活需要的小公园、小绿地、小森林——"三小"绿地。

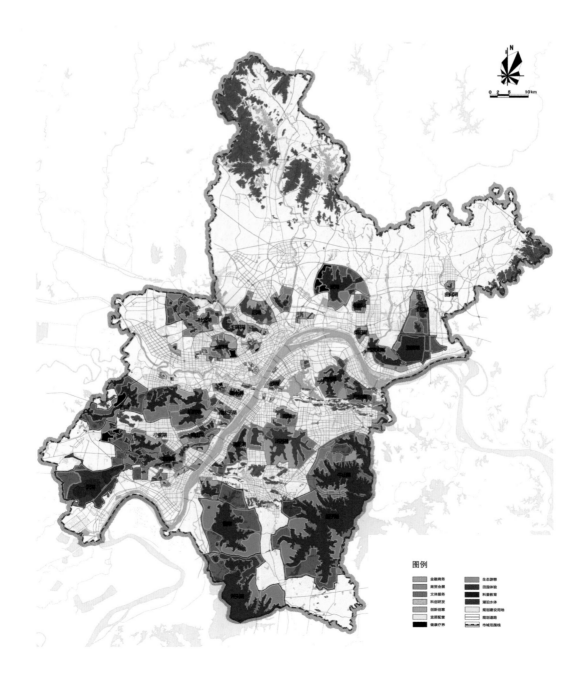

● *Making the city better with water: Implementation Plan for "Great Lakes+" Functional Areas (2018)*

The Plan focuses on transforming water resources into functional advantages, highlighting the urban feature of "City of Hundreds of Lakes". The Plan aims at improving the quality of urban functions, and promoting the integrated development of cities and lakes. This Plan takes lakes, surrounding green spaces and functional areas as the object, exploring a new mode of protection and utilization of "Great Lakes+". On the basis of giving priority to ecological benefits and highlighting the protection of "Blue Line" of lakes and "Green Line" of surrounding green spaces, the Plan puts forward three types of Lake functional zones: urban, rural and ecological in accordance with the location and functional characteristics of lakes. The Plan emphasizes green development, highlighting the implantation of various public functions such as recreation, innovation, scientific research and development. The Plan should add new functions, new spaces and new driving forces to lake functional areas, building a number of lake functional areas with different themes, different characteristics and diversified vitality.

● *Action plan: Five-year Action Plan for "Embracing Clear Waters and Green Hills" (2016)*

To further implement the construction of ecological civilization in Wuhan and speed up the organization of "embracing clear waters and green hills" put forward in the government work report, the action plan proposes four requirements for "embracing clear waters and green hills" in the urban development area in the four aspects of structure, quality, total amount and per capita indicators, in accordance with the general requirement of "highlighting key points and taking points and areas as a whole". In terms of the ecological framework, the ecological framework of "two axes, two rings and six wedges in the city" in the 2010 Master Plan has been basically integrated. For the total amount of ecological greening, it is close to the target set by the 2010 Master Plan, with the green coverage rate of the built-up area reaching 41%, the green coverage rate of the built-up area reaching 35.8%, and the per capita park green area reaching 12 square meters in the whole city. For the ecological quality, a number of unique waterfront park clusters have been built, opening up a number of high-quality "green veins" that connect rivers and lakes into a network. For ecological services, the green space system of "city-district-community" parks has been improved and a number of small parks, green spaces and small forests have been built to meet the needs of citizens.

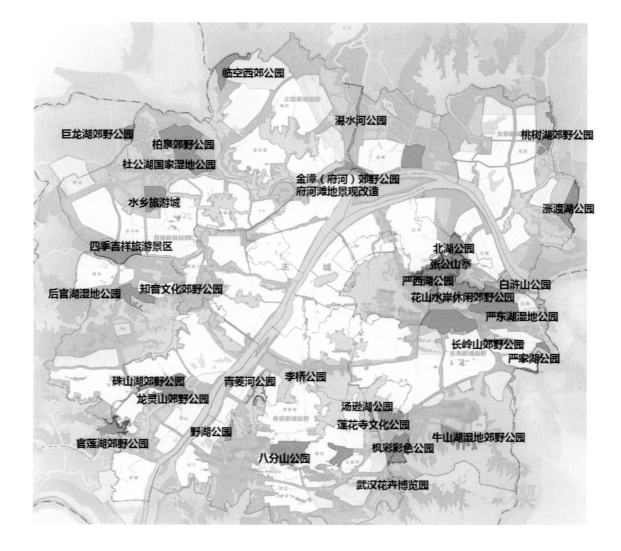

1
"大湖+"主题功能区规划布局图
Layout of "Great Lakes +" Theme Functional Area Plan

2
"拥抱绿水青山"五年行动规划平面图
Five-year Action Plan for "Embracing Clear Waters and Green Hills"

1
谢家院子平面布局规划图
Plane Layout Planning Map of Xie's Courtyard

2
谢家院子规划效果图
Planning Rendering of Xie's Courtyard

图例
1、天井围屋
2、洋务学堂
3、旅游服务中心
4、村委会
5、幼儿园
6、藏风塘
7、得水塘
8、虎山亭
9、西塘
10、东塘
11、梯田广场
12、景观步道
13、谢家陵园
14、天井新居

一 历史文化名城保护规划

● 空间格局：从散点保护拓展到全域保护

○ 历史文化名城专题研究及专项规划

武汉市非常重视文化名城和历史文化街区的保护工作，分别于1984年、1988年、1996年、2006年先后开展了4轮历史文化名城保护规划的编制工作，其中"2006版保护规划"于2010年与《武汉市城市总体规划（2010～2020）》一起获国务院审批，确定了汉正街传统商贸风貌区、汉口原租界风貌区、汉阳旧城风貌区、武昌旧城风貌区4片旧城风貌区，保护面积为14.1平方公里。江汉路及中山大道片、青岛路片、"八七"会址片、一元路片、昙华林片、首义片、农讲所片、洪山片、珞珈山片、青山"红房子"片10个区域确定为历史地段，划定保护范围和控制地带总面积为415.5公顷。

根据武汉市新一轮城市总体规划编制要求，2017年开展的《武汉历史文化名城专题研究》中，提出了历史文化遗产保护的全面遗产观以及"三层多类"的保护框架体系，即构建"区域—市域—主城"三个保护圈层，补充历史文化名镇（村）、工业遗产等多种保护类型。

近年来，武汉市国土资源和规划局依据2010版《武汉市城市总体规划（2010～2020年）》，编制了20余项旧城风貌区和历史地段的保护规划。

○ 市域历史文化名镇名村保护规划（以谢家院子为例）

在对武汉市新城区内所有具有历史文化物质和非物质遗产的67个镇、2807个行政村、15580个自然村湾普查基础上，2012年武汉市国土资源和规划局组织编制了《武汉市历史镇村保护名录规划》，推荐了51个历史文化名镇名村，包括2个省级历史文化名镇、4个国家级、20个省级、25个市级历史文化名村。武汉市国土资源和规划局在此基础上陆续开展了历史镇村保护规划的编制工作。

2016年，武汉市国土资源和规划局组织开展了谢家院子资源调查和保护规划工作。通过多种技术手段，《市域名镇名村保护规划》（2016年）勾勒出谢家院"九间半"建筑群的全貌及历史变迁，逐步明确9处优秀历史建筑、6处传统建筑、3条历史街巷、8处历史环境要素及5类传统文化和非物质文化遗产。2017年9月谢家院子被武汉市人民政府列为第一批"武汉市历史文化名村"。

该规划秉承"历史村湾的保护即为建筑原真与生活真实"的理念，在保护、发展与实施三方面，对村民的生活可持续进行深入思考，利用"众规武汉"平台开展"开门做规划"，探讨保护与利用之间的平衡关系，合理确定保护范围及要素；从"区域—行政村—村湾"层层剖析，重点对村湾住房空间选择、村民住房需求、历史建筑三个方面利用GIS分析手段，提出有机更新的方案；从村民的经济动力和社会动力的成因入手，建立发挥村民能动性机制，同时合理界定政府、企业和村委三大主体的权责，最后形成适应合作性的规划分期引导，并通过百度词条名词入库、众规武汉讨论宣传等，进一步推进历史村庄宣传工作。

— Conservation Program of Famous Historical and Cultural Cities

• *Spatial pattern: expansion from conservation for scattered points to conservation of the whole region*

○ Monographic study and specialized plan of famous historical and cultural cities

Wuhan has paid much close attention to the conservation work of famous cultural cities and historical and cultural blocks. It successively carried out 4 rounds of compiling work of conservation programs for famous historical and cultural cities in 1984, 1988, 1996 and 2006 respectively. "Conservation Plan in 2006" was approved together with *Urban Master Planning of Wuhan City (2010—2020)* by the State Council in 2010 to confirm 4 old city view areas, including Hanzheng Street Traditional Business View Area, Hankou Former Foreign Settlement View Area, Hanyang Old City View Area and Wuchang Old City View Area, with the conservation area of 14.1 square kilometers. 10 sections including Jianghan Road and Zhongshan Avenue Section, Qingdao Road Section, "August 7th Meeting" Conference Site Section, Yiyuan Road Section, Tanhualin Section, Shouyi Section, Peasant Movement Institute Section, Hongshan Section, Mountain Luojia Section and Qingshan "Red House" Section were confirmed to be historical sites and were designated with conservation scope and control regions with the area of 415.5 hectares.

According to the compilation requirements for a new round of urban master planning in Wuhan City, a comprehensive heritage view of historical and cultural heritage conservation and a "three-tier and multi-category" conservation framework system was put forward, namely building three conservation circle layers "region-urban area-major city" and supplementing famous historical and cultural towns (villages), industrial heritages and other various conservation types.

In recent years, according to the Urban Master Planning in 2010, over 20 conservation plans were compiled for old-city view sections and historical sites.

○ Conservation program of famous historical and cultural towns and villages in the administrative region of the city (taking Xie's Courtyard as an example)

In 2012, based on the general investigation for all the 67 towns, 2807 administrative villages and 15580 natural villages with historical and cultural material and intangible heritages in new urban district of Wuhan City, Wuhan National Land Resources and Planning Bureau organized and compiled *Conservation Directory Planning of Historical Towns and Villages in Wuhan City* and recommended 51 famous historical and cultural towns and villages, including 2 provincial famous historical and cultural towns as well as 4 national, 20 provincial and 25 municipal famous historical and cultural villages. On this basis, the compilation of conservation planning in historical towns and villages has been carried out.

In 2016, Wuhan National Land Resources and Planning Bureau organized to carry out survey of resources and conservation and planning work in Xie's Courtyard. By various technological means, the plan drew the outline of the complete picture and historical changes of building group with "9.5 rooms" in Xie's Courtyard and gradually made clear 9 outstanding historical buildings, 6 traditional buildings, 3 historical streets and alleys, 8 historical environment elements and 5 categories of traditional cultures and intangible cultural heritages. In September 2017, Xie's Courtyard was listed into "famous historical and cultural villages in Wuhan City" in the first batch by Wuhan Municipal People's Government.

Adhering to the concept that "conservation of historical villages means restoring the authenticity of architecture and life", the Plan conducted a deep reflection for villagers' life sustainability from three aspects including conservation, development and implementation and utilized *Zhonggui Wuhan* Platform to "open the gate to make plans" to discuss about the balance relationship between conservation and utilization and to reasonably confirm conservation scope and elements; an organic updated scheme was put forward through an analysis from layer to layer in "region—administrative village—village" with GIS analysis means in three aspects including housing space selection of villages, villagers' housing demand, and historical buildings as a key point; to start from the cause of villagers' economic stimuli and social dynamics, it established a mechanism to give play to villagers' initiative, meanwhile reasonably defined the rights and liabilities of the government, enterprises and village committees, finally formed planning phased guidance to adapt to cooperativeness and further promoted the publicity work of historical villages through noun entrance of Baidu entry, discussion and publicity of Zhonggui Wuhan and so on.

● 时间维度：从新中国成立前拓展到新中国成立后

○ 武汉市1949~1978年历史建筑保护名录规划（2014年）

为充分发掘新中国成立后武汉市的历史建筑资源，保护和传承城市历史文脉，加强历史建筑保护工作的动态性和延续性，2014~2016年武汉市国土资源和规划局联合市文化局等分别开展了武昌、江岸、江汉、硚口、青山、洪山、汉阳7个中心城区1949~1978年历史建筑的资源调查工作，初步摸清了各区在这一时段历史建筑的"家底"。在此基础上，2018年武汉市国土资源和规划局组织开展了规划编制工作，对7个分区调查成果进行整合和优化。结合武汉自身情况，《历史建筑保护名录规划》（2014年）构建了登录与评判相结合的评价体系，并确定了新中国成立后历史建筑的分级保护名录，共计历史建筑256处。其中，一级建筑32处、二级建筑93处、三级建筑131处。该名录作为优秀历史建筑、文物保护单位等的资源后备库，为下一步申报以及后续的紫线划定工作奠定基础。

● 类型演化：从区片规划拓展到线路展示

○ 武昌古城、汉阳旧城、汉口原租界

为举办辛亥百年庆典，2008年编制了《武昌古城保护与复兴规划》。该规划通过对武昌旧城风貌区整体协调、分级保护、功能提升、板块推进、改善空间环境、激活产业经济等规划手段，带动武昌古城的整体复兴。

3500年的历史沉淀，为汉阳留下了独具武汉传统的街巷文化，一种新的城市建设模式在汉阳旧城风貌区完美地演绎了"在再生中保护，在发展中再生"的规划理念，保护并延续了老汉阳独特的历史文脉。

汉口原租界风貌区集"英、法、德、俄、日"等多国租界建筑遗址于一体，集中体现了汉口自开埠以来的城市历史发展进程。作为武汉市重要的历史风貌区，汉口原租界风貌区一直以来受到高度的关注，相继完成了多轮规划研究，2003年编制完成的《汉口一元片区保护规划》采取了分级保护的模式，切实保护了区域风貌。

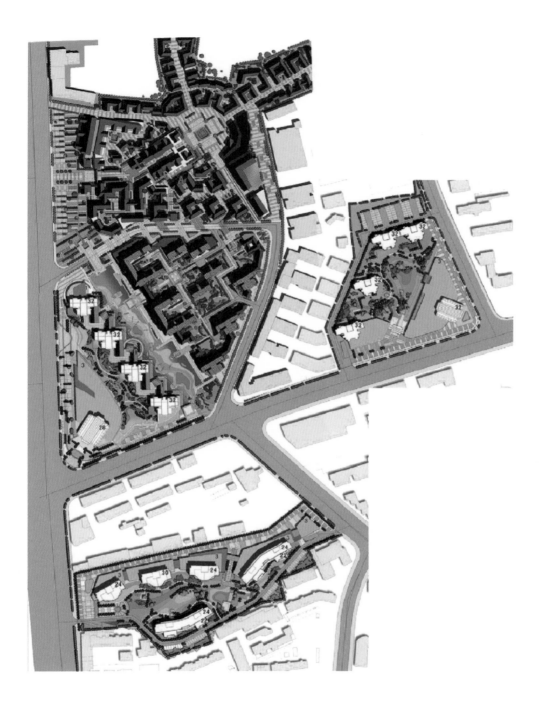

1
汉阳旧城风貌区鸟瞰图
Bird's eye view of Hanyang Old City View Area

2
青山区建国后优秀历史建筑资源调查和保护规划图
Survey and Protection Plan of Excellent Historical Building Resources in Qingshan District after the Founding of the People's Republic of China

3
汉阳旧城风貌区总平面规划图
Master Planning Map of Hanyang Old City View Area

- *Time dimension: expansion from before the founding of the P.R.C. to after the founding of the P.R.C.*

○ Conservation Directory Planning of Historical Buildings (2014) after the Founding of the P.R.C. (1949—1978) in Wuhan City

During 2014 to 2016, to fully explore historical building resources in Wuhan City after the founding of the P.R.C., to protect and inherit urban historical vein, and to strengthen the dynamic nature and continuity of conservation of historical buildings, Wuhan National Land Resources and Planning Bureau, together with Municipal Cultural Bureau and other departments, carried out resource investigation work of historical buildings respectively in seven central urban areas including Wuchang, Jiang'an, Hanjiang, Qiaokou, Qingshan, Hongshan and Hanyang after the founding of the P.R.C. (during 1949 to 1978) and preliminarily figured out resources of historical buildings in all the urban areas after the founding of the P.R.C. On this basis, planning and formulation work was organized and carried out in 2018 to integrate and optimize investigation achievements of seven urban areas. Combined with Wuhan itself, the Planning built the evaluation system combined with login and judgment and confirmed the layered conservation directory of historical buildings after the founding of the P.R.C.: there are a total of 256 historical buildings after the founding of the P.R.C. Among, there are 32 first-class buildings, 93 second-class buildings and 131 third-class buildings. As a resource reserve for excellent historical buildings and culture relic conservation sites, the directory lays a foundation for declaration in the next step and subsequent purple line demarcation.

- *Type evolution: expansion from section planning to routes to show*
  ○ Wuchang Ancient City, Hanyang Old City and Hankou Former Foreign Settlement

To hold for the centennial celebration for the Revolution of 1911, Wuhan Ancient City Conservation and Rehabilitation Plan was compiled in 2008. Through overall coordination, layered conservation, function promotion and plate propulsion of old city view areas in Wuchang, improving space environment, activating industrial economy and other planning manners, the Plan could be used to drive the entire rehabilitation of Wuchang Ancient City.

Historical precipitation for 3500 years has left street culture unique with traditions in Wuhan for Hanyang. A new urban construction mode has perfectly demonstrated the planning concept "conservation in regeneration, regeneration in development" in Old City View Areas of Hanyang, which has protected and continued unique historical context of old Hanyang.

Integrated with architecture sites in many countries such as "UK, France, Germany, Russia and Japan", Hankou Former Foreign Settlement View Area intensively reflects the urban history development process of Hankou since opening as a commercial port. As an important historical view area in Wuhan, Hankou Former Foreign Settlement View Area has always drawn much close attention and many rounds of planning studies have been successively completed. The pattern of layered conservation was adopted in the *Conservation Planning of Yiyuan District, Hankou* compiled in 2003, which practically protected the view in the district.

1
汉口原租界风貌区保护控制图

Conservation and Control Map of Hankou Former Foreign Settlement View Area

2
武昌古城规划结构图
（2008年）

Structure Map of Wuchang Ancient City Planning (2008)

3
武昌古城用地规划图
（2008年）

Map of Wuchang Ancient City Land Use Planning (2008)

4
一元片旧城风貌保护规划功能结构图

Function Structure Map of Old City View Conservation Planning in Yiyuan District

第肆章 • 区域一体化时期的规划 273

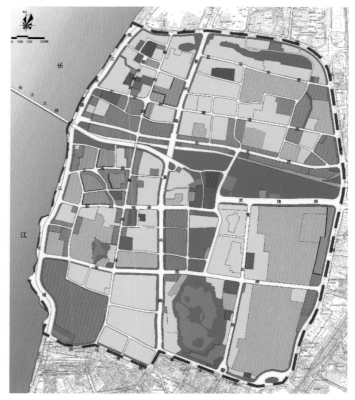

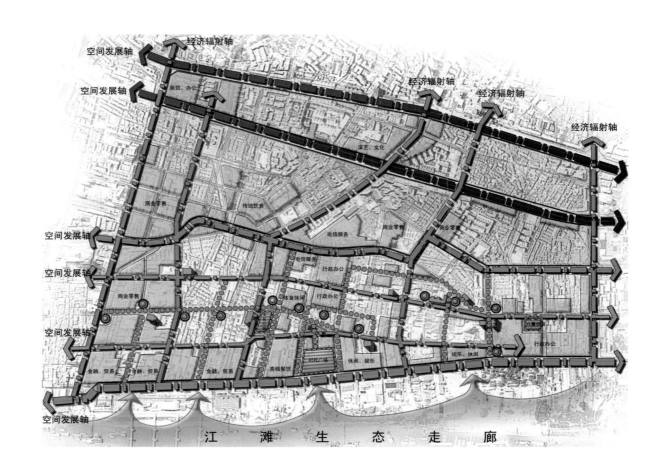

CHAPTER FOUR • Planning During Regional Integration

图例

01 江汉关监督署公署旧址
02 俄国领事馆旧址
03 三北轮船公司旧址
04 李凡洛夫公馆旧址
05 华俄道胜银行旧址
06 源泰洋行遗址
07 汉口俄国志愿者航运公司遗址
08 邦可花园遗址
09 新泰大楼旧址
10 顺丰砖茶厂遗址
11 惠罗公司旧址
12 俄国巡捕房工部局遗址
13 顺丰茶栈旧址
14 巴公房子旧址
15 邦可餐厅旧址
16 万国医院遗址
17 新泰砖茶厂遗址
18 新泰洋行仓库旧址
19 俄国总会旧址
20 汉口东正教堂
21 阜昌茶厂遗址
22 江汉关大楼
23 大智门火车站
24 平和打包厂

○ 武汉万里茶道遗产展示区综合规划

万里茶道是17~20世纪中叶，中国与俄国之间以茶叶为大宗贸易商品的长距离商贸路线，也是继古代丝绸之路衰落后在欧亚大陆兴起的又一条重要的中国农业文明西输及欧洲工业文明东输的国际性经济、文化商道。武汉是万里茶道上最大的贸易集散中心，在茶叶贸易繁盛时期，从武汉出口的茶叶一度达到全国80%以上，被誉为"东方茶港"。

2015年3月，"八省一市""万里茶道文化遗产保护工作推进会"在武汉举行，正式明确湖北省为万里茶道申遗的牵头省份，武汉市为牵头城市，开展《武汉万里茶道申遗展示区综合规划》的编制研究工作，具体分为研究万里茶道、规划特色茶道、复兴万里茶道3个部分。在研究部分，划定申遗展示的研究区和核心区，梳理范围内历史特色和现状问题。在规划部分，提出"打造汇聚茶文化体验、街头博物馆、名人故居等功能为主导的独具俄式风情的万里茶道展示区"的目标，构建"一条茶路、三种风情"的整体空间结构，并依托黎黄陂路串联万里茶道历史遗存，形成茶文化体验及博览核心。在复兴部分，制定建筑整治、交通市政、绿化提升三大线性工程，以及华俄道胜银行地块、邦可花园地块、顺丰茶栈地块等十大重点改造项目，通过分期分批实施实现展示区的复兴。

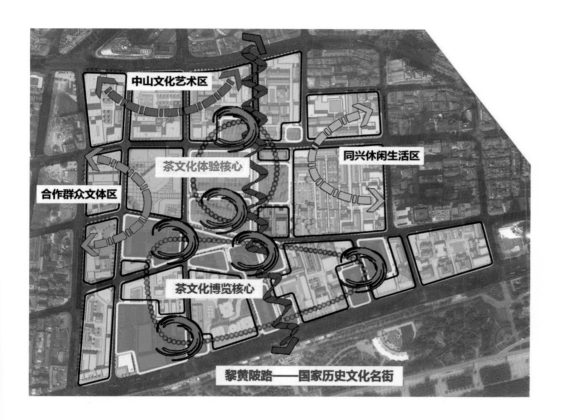

1

万里茶道文化遗存分布图

Distribution Diagram of Cultural Heritages along Wanli Tea Route

2

万里茶道核心展示区结构图

Structure Chart of Wanli Tea Route Core Demonstration Area

○ Comprehensive Plan of Wuhan Wanli Tea Route Heritage Demonstration Area

Wanli Tea Route is a long-distance commerce and trade route between China and Russia with tea as goods in bulk trade from the 17th Century to the middle of the 20th Century and also another important international economic and cultural trade route for Chinese agriculture civilization to be exported to the West and European industrial civilization to be exported to the East rising in Eurasia after the Ancient Silk Road declined. Wuhan is the largest trade distribution center along Wanli Tea Route. In the thriving period of tea trade, tea exported from Wuhan once reached above 80% of that nationwide. It was honored as "Tea Port in the East".

In March, 2015, "Promotion Meeting of Conservation for Cultural Heritage of Wanli Tea Route" in "Eight Provinces and One City" was held in Wuhan to officially confirm Hubei Province as a leading province in the application for Wanli Tea Route as a world heritage and Wuhan City as a leading city and to carry out the compilation and research work in Comprehensive Planning of Demonstration Area to Apply Wanli Tea Route in Wuhan as World Heritage, specifically including three parts, namely researching Wanli Tea Route, planning featured tea route and revitalizing Wanli Tea Route. In the research part, the research area and the core area were designated for demonstration while applying as the world heritage to systemize historic features and current situation within scope. In the planning part, the target "to forge Wanli Tea Route Demonstration Area unique with Russian style led by gathering tea culture experience, street museum, former residences of celebrities and other functions" was put forward to build an overall space structure with "one tea route and three styles", to rely on Lihuangpi Road to connect historical relics along Wanli Tea Route in series and to form tea culture experience and expo core. In revitalization part, three linear projects including regulating structure, traffic and municipal administration and greening and promotion, as well as ten key transformation projects such as Russo-Chinese Bank Plot, Bangke Garden Plot and Shunfeng Tea Station Plot were formulated to realize revitalization of demonstration area through implementation in phases and in batches.

● 保护方式：从静态保护拓展到活化利用

○ 中山大道街区设计改造

中山大道是武汉市中心城区传统风貌保留最为完整、承载城市历史变迁最为集中的地区。随着岁月变迁，其人文特色、商业氛围和街道环境难以彰显这条百年老街的独特魅力。2014年，以地铁6号线建设为契机，武汉市委市政府提出对中山大道最核心地段———元路至武胜路4.7公里地区进行综合改造。《中山大道综合整治规划》（2014年）确定的历史建筑共34处(包括国家级文物保护单位1处，省、市级文物保护单位10处，市优秀历史建筑23处)，一般建筑95处。通过建筑立面、园林景观和市政工程等专项整治，着力将中山大道打造成宜游宜行、感悟历史、体验生活的文化旅游大道。

该规划打破传统道路改造一味拓宽车行道的做法，变车行优先为步行优先，变交通干道为文化旅游街道，主打"慢行交通"，让城市生活"慢下来"。改造后的中山大道（一元路至前进一路段）的路权重新分配，设置为双向两车道，划定公交车专用车道确保公共交通路权，并结合国际先进设计理念对街道铺装、交叉口处理、人行过街、道路绿化、地铁出入口等细节进行了精细化设计。

该规划结合现状道路改造打造了美术馆文化广场、水塔及艺术市场、民众乐园3个特色开放空间节点，让全长4.7公里的街道空间丰富多变，趣味盎然。

中山大道串联原"法、俄、英"等租界，沿线共有各类保护性老建筑及特色里分民居约150余处，历史底蕴深厚。规划结合现状建筑调研、江岸区政府发展需求及市场意愿，提出具有可操作性的功能升级建议，通过历史建筑功能复兴重新焕发中山大道百年老街商业活力。

1 2

1–2
中山大道实施实景图

Realistic Picture of Implementation on Zhongshan Avenue

- *Conservation mode: expansion from static conservation to activation and utilization*

  ○ Design Transformation of Zhongshan Avenue Block

Zhongshan Avenue is the area to retain the most complete traditional scene of central urban area in Wuhan City and to carry the most concentrated urban historical changes. Along with time changes, it is difficult for its human features, commercial atmosphere and street environment to manifest the unique charm of this one hundred years old street. In 2014, to take the opportunity to build Metro Line 6, Municipal Party Committee and Municipal Government put forward comprehensive transformation of the core location of Zhongshan Avenue, location 4.7kilometers from Yiyuan Road to Wusheng Road. There are a total of 34 historic buildings (including 1 national relic protection unit, 10 provincial and municipal relic protection units and 23 municipal outstanding historical buildings) and 95 ordinary buildings confirmed in the Comprehensive Improvement Plan of Zhongshan Avenue. Through special rectification such as building elevation, landscape architecture and municipal works, it strived to build Zhongshan Avenue into a cultural tourism avenue suitable for travel and walk, feeling history and experiencing life.

The Plan broke the practice of traditional road transformation to blindly expand roadway, turned taking the priority to drive a car into taking the priority to take a walk and changed traffic artery into cultural tourism street featured with "slow-moving traffic" to "slow down" urban life. The relocation of road right of Zhongshan Avenue (from Yiyuan Road to Qianjin Road Section I) after transformation was set as a bi-direction two-lane road. The bus lane was designated to ensure right of way for public traffic and the international advanced design concept was combined to conduct refined design for street pavement, intersection treatment, pedestrian crossing street, road greening, subway entrance and other details.

The Plan combined current road transformation to forge three featured open space nodes including Art Gallery Culture Square, Water Tower & Art Market and Public Park to make the street space 4.7kilometers in overall length abundant, changeable and more interesting.

Zhongshan Avenue connects former foreign settlements of France, Russia, UK and other countries in series and there are more than 150 various protective old buildings and featured Lifen dwellings with profound historical background. Combined with the investigation and survey of current buildings, the development demand of district government and the will of the market, the Plan put forward function upgrading suggestions with operability and shone commercial vitality of Zhongshan Avenue, the one hundred years old street again through revitalizing functions of historic buildings.

○ 黎黄陂路片（一期）整治与保护工程规划方案

2014年1月，中共中央办公厅正式批准建立武汉中共中央机关旧址纪念馆。同年1月4日，习近平总书记作出"修旧如旧，保留原貌，防止建设性破坏"的批示。《黎黄陂路片（一期）整治与保护工程规划》（2014年）围绕红色文化、中俄茶叶贸易线路主题，营造体验老汉口多元文化及多样生活的城市特色空间，打造汇聚红色教育场所、街头博物馆、名人故居等功能为主导的独具俄式风情的国家历史文化街区、全国爱国主义教育示范基地、原租界区博物馆聚集区。江岸区依托中山大道，按照一年一条垂江路的做法，开展了黎黄陂路、合作路、兰陵路的实施改造。

改造前

改造后

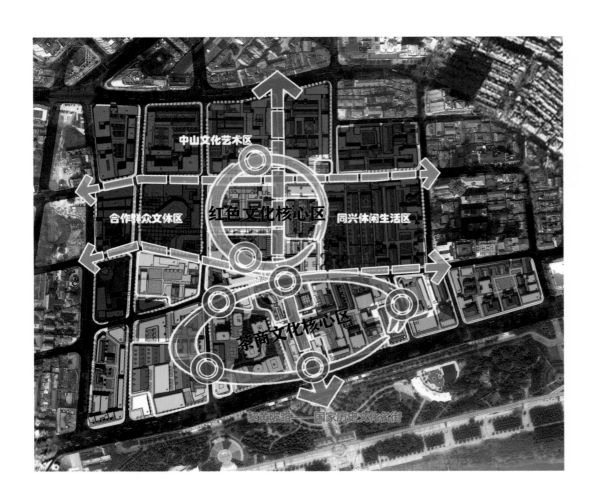

1–2 黎黄陂路对比图
Comparison Chart of Lihuangpi Road

3 黎黄陂路片结构图
Structure Chart of Lihuangpi Road Section

○ Planning Scheme for Renovation and Conservation Project of Lihuangpi Road Section (Phase I)

In January, 2014, General Office of the CPC Central Committee officially approved to establish Wuhan Memorial Hall for Former Site of the CPC Central Committee. On January 4, 2014, General Secretary Xi Jinping gave written instructions "restoring the old as the old, retaining the original appearance and preventing constructive damage". Surrounding themes of red culture and Sino-Russian tea trade route, it was planned to create a featured urban space experiencing multi-culture and diversified life in old Hankou and forge a national historical and cultural block, a national patriotism education demonstration base, and an accumulation area of former foreign settlement museums unique with Russian style led by gathering red education places, street museums, former residences of celebrities and other functions. Relying on Zhongshan Avenue, Jiang'an District carried out implementation and transformation of Lihuangpi Road, Hezuo Road and Lanling Road based on the practice of one riverside road annually.

○ 工业遗产保护与利用规划（2013年、2015年）

武汉是中国重要的工业城市和中国近代工业的发祥地之一，保留有数量众多的工业遗产资源。2013年武汉市政府批复《武汉市工业遗产保护与利用规划》，公布了首批27处工业遗产名单。为进一步做好武汉工业遗产保护和再利用工作，市国土资源和规划局于2015年组织编制了《武汉市第二批工业遗产保护与利用规划》，是继2013年工业遗产专项规划的后续和深化研究。

该规划确定了武汉市工业遗产的评价标准，拟定第二批工业遗产名单，遴选出38处工业遗存推荐纳入武汉市第二批工业遗产名单，依照工业遗产分类分级保护，制订工业遗产管理办法。

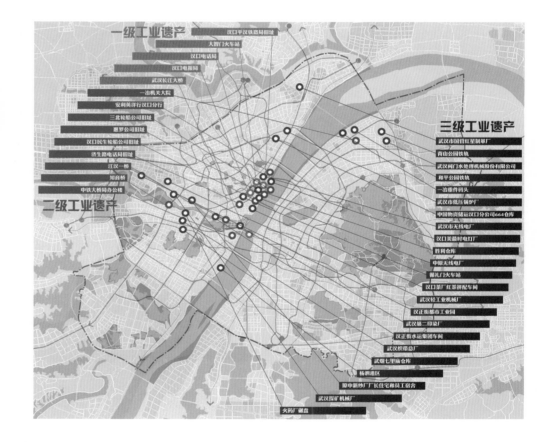

1
2
―
3

1–2
一级工业遗产——大智门火车站候车厅旧址
Class I Industrial Heritage: Former Address of Waiting Hall in Dazhimen Railway Station

3
武汉市第二批工业遗产空间分布图
Spatial Distribution Map of Industrial Heritages for the Second Batch in Wuhan

○ Conservation and Utilization Plan of Industrial Heritages (2013 & 2015)

As an important industrial city in China and one of birthplaces of industry in modern China, Wuhan remains numerous industrial heritage resources. In 2013, Wuhan Municipal Government approved the Conservation and Utilization Plan of Industrial Heritages in Wuhan City and published a list of 27 industrial heritages in the first batch. In 2015, to further complete the conservation and reutilization work of industrial heritages in Wuhan, Municipal National Land Resources and Planning Bureau organized to compile the Conservation and Utilization Plan of Industrial Heritages in Second Batch in Wuhan, which was the follow-up and in-depth study after specialized planning of industrial heritages in 2013.

The plan confirmed evaluation standards for industrial heritages in Wuhan City, formulated a list of industrial heritages in the second batch, selected 38 industrial remains to be recommended to enter into the list of industrial heritages in the second batch in Wuhan and formulated administrative measures for industrial heritages according to the classified and layered conservation in industrial heritages.

- 保护方式：从封闭保护到开放宣传

○ 盘龙城遗址公园规划

盘龙城遗址是迄今为止长江流域发现的夏商时期规模最大、出土遗存最为丰富的城邑遗址，是长江流域青铜文明中心、武汉城市之根。2005年武汉市编制了《盘龙城遗址保护总体规划》，划定了6.55平方公里保护范围进行重点保护，并在周边划出近10平方公里环境协调区进行控制性保护。控制遗址核心区和一般保护区内的一切建设，严格遗址建设控制地带及周边环境协调区内建设项目的审批。古城形制呈方形，南北宽约290米、东西长约260米，面积约75400平方米。城内东北部分布着"前朝后寝"宫殿建筑遗址，面积近2000平方米。城垣为夯土筑成，城垣中部设有城门，城外有壕沟环绕。外围分布有居住区、手工作坊、贵族平民墓地等遗迹。

2017年12月，盘龙城考古遗址公园入选第三批国家考古遗址公园。未来将按照"遗址保护工程、博物馆工程、商业服务与配套工程"三步走的战略实施，将盘龙城国家考古遗址公园打造成武汉靓丽的名片之一，实现由废墟向国家考古遗址公园的转变。

- *Conservation mode: from closed conservation to open publicity*

○ Plan of Panlongcheng Ruins Park

Panlongcheng Ruins are ruins of cities and towns with the largest scale and the most abundant unearthed remains in the period of the Xia-Shang Dynasty discovered in Yangtze River Basin up to now and is the bronze civilization center in Yangtze River Basin and the root of Wuhan City. In 2005, Wuhan City compiled Overall Plan for Conservation of Panlongcheng Ruins to delimit a conservation scope of 6.55 square kilometers for key conservation and an environmental coordination area of nearly 10 square kilometers in the surrounding for controlled conservation. It controlled all construction in the core area of the site and the general conservation area, and strictly coordinated the approval of construction projects in the site construction control zone and the surrounding environment. The shape of the ancient city is square, about 290 meters wide from the north to the south and about 260 meters long from the east to the west with the area of about 75,400 square meters. Palace building ruins including "the court in front and bedrooms behind" are distributed in the northeast of the city with the area of nearly 2,000 square meters. There is a city wall built with rammed earth, a gate set in the center of the city wall and ditches surrounding outside the city. There are residence zones, manual workshops and mobilities and civilians' cemeteries and other ruins distributed in the surrounding.

In December, 2017, Panlongcheng Archaeological Park was selected into national archaeological parks in the 3rd batch. In the future, it will forge Panlongcheng National Archaeological Park into a beautiful namecard of Wuhan and realize the transformation from ruins to a national archaeological park according to the three-step strategy implementation of "relics protection project, museum project and commercial service and supporting project".

第肆章 · 区域一体化时期的规划　281

1

盘龙城遗址公园效果图

Effect Picture of Panlongcheng Ruins Park

2

盘龙城遗址公园规划平面图

Planning Plan of Panlongcheng Ruins Park

一 基本公共服务均等化和生活圈规划

近年来,随着经济社会不断发展,社会结构正在由"倒丁字型"向"土字型"转变,正在向全面建成小康社会构建的"橄榄型"社会迈进。这一时期城市化进程不断加快,城市人口迅速膨胀,新社会阶层和新社会群体将更多,城区流动人口规模将会更大,城市公共服务设施的建设规模和服务能力越来越难以满足老百姓日益多元的基本需求。造成这种现象的原因是多方面的,缺乏系统性的民生规划、社区规划研究是其中较为重要的因素之一。

在社会管理向社会治理的转型期,国家日益重视社会治理能力在城市发展、社会转型中的作用。在此背景下,武汉相继开展了中小学、医疗、养老、体育等市级层面的公共服务设施专项规划,在解决了市级规划的"大设施、大问题、大空间"后,目光逐步转向到被大众忽略的社区规划的"小设施、细问题、微空间"上来,开展了15分钟生活圈规划的探索和戈甲营等社区微改造实践,逐步建立了"社区规划师"和社会公众参与制度,实现了公共服务从"均等"走向"均好"的目标。

● 关注市区:公共服务设施规划(2013年)

一直以来,武汉市非常重视各类公共服务设施规划的编制工作,为解决居民上学难、看病难、养老床位紧张、健身场地不足等问题,市国土资源和规划局分别联合市教育局、卫生局、民政局、体育局开展了中小学、医疗卫生、养老、体育等公共服务设施专项规划,从优化城市功能、建立公共服务设施分级体系、提升设施建设标准、均等化布局等方面,有力地保障了各类公共服务设施用地空间,为实现基本公共服务均等化奠定了较好基础。

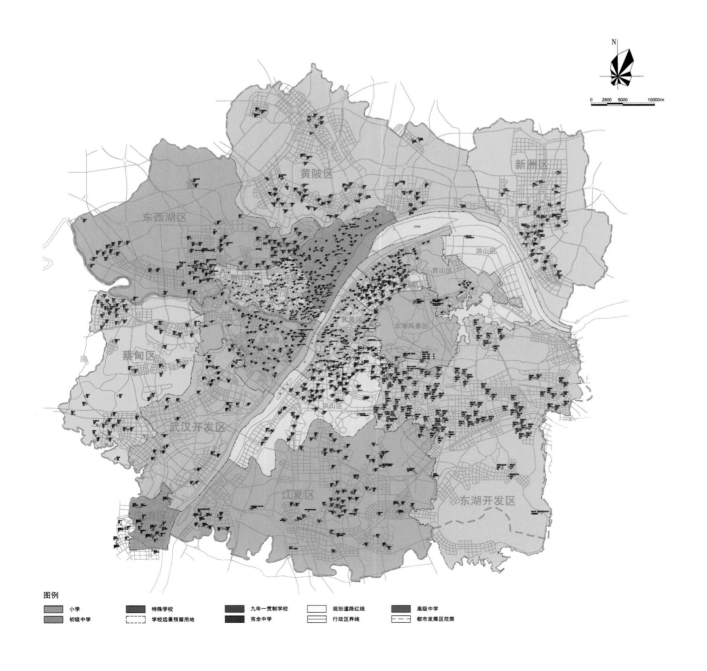

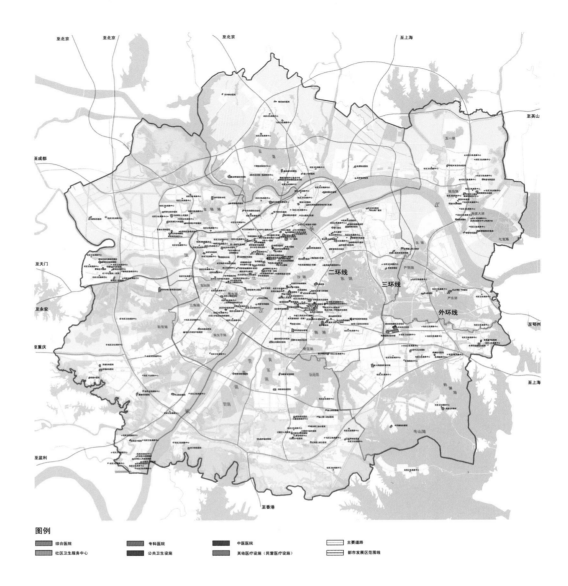

1
中小学布局规划图
Allocation Planning Map of Middle and Primary Schools

2
医疗卫生设施用地规划图
Land Use Planning Map of Medical Treatment and Public Health Facilities

— Equalization of Basic Public Services and Life Circle Planning

In recent years, with continuous economic and social development, the social structure was switching from "inverted T type" to "土 type" and striding forward to the "olive-shaped" society built while building a moderately well-off society in an all-round way. During this period, urbanization process is accelerating, the urban population is expanding rapidly, there will be more new social classes and new social groups, and the floating population in urban areas will be larger. The construction scale and service capacity of urban public service facilities are increasingly difficult to meet the increasingly diversified basic needs of common people. There are various reasons for this phenomenon and the lack of systematic livelihood planning and community planning research is one of the more important factors.

During the transition period from social management to social governance, the state attaches increasing importance to the role of social governance ability in urban development and social transformation. Under this background, Wuhan successively carried out the specialized planning of public service facilities at city level in middle and primary schools, medical treatment, provision for the aged, sports and so on. After solving "big facilities, big problems and big spaces" in city-level planning, it turned its eyes gradually into "small facilities, detailed problems and micro spaces" in community planning ignored by the public, carried out an exploration for planning of life circle within 15 minutes and micro-transformation practice in Gejiaying and other communities, gradually established "community planner" and social public participation system and realized the target of public services from "equalization" to "all good".

● *To focus on urban area: planning of public service facilities (2013)*

Wuhan has always attached great importance to the compilation of planning in all kinds of public service facilities. To solve such problems as difficulties for residents in going to school, difficulties of getting medical service, shortage of nursing beds for the aged, insufficiency of fitness fields and so on, Municipal National Land Resources and Planning Bureau, together with Municipal Education Bureau, Sanitary Bureau, Civil Affairs Bureau, and Sports Bureau, respectively carried out specialized planning for middle and primary schools, medical treatment and public health, provision for the aged, sports and other public service facilities, which powerfully guaranteed land use space of all kinds of public service facilities from optimizing urban functions, establishing layered system of public service facilities, promoting construction standards of facilities, equalizing layout and other aspects, and laid a good foundation to realize equalization of basic public services.

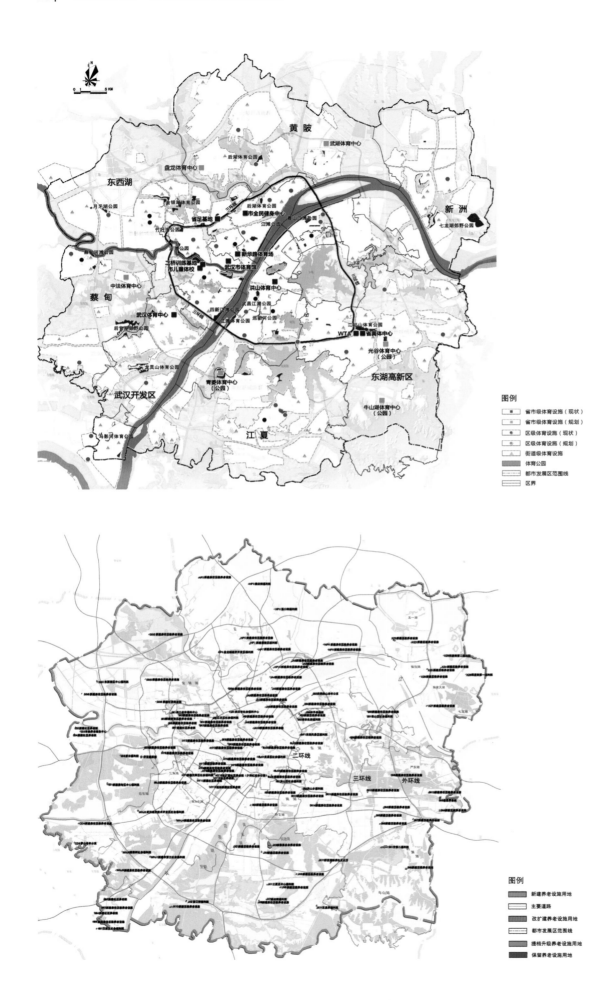

- 关注社区：15分钟生活圈规划

为进一步落实新一轮城市总体规划"15分钟生活圈"战略意图，促进城市公共服务品质和高质量发展，提升基层社区治理和精细化管理的水平，2018年，市国土资源和规划局组织开展了《武汉市15分钟社区生活圈实施行动指引》编制工作，作为新一轮总体规划八大专项战略行动规划之一，为15分钟社区生活圈建设和社区规划提供指导依据。

该规划提出按照"15分钟步行可达、活动半径500~800米空间范围，一般服务常住人口为3万~6万人，规模约1~3平方公里，人口密度约为每平方公里1万~2万人"划定社区生活圈。武汉主城区共划定283个生活圈，划分为修补、提升和新建三大类，针对每一种类型提出精细化的建设策略和模式。

- 关注微改造：戈甲营社区规划、南湖街道社区规划（2017~2018年）

2017年以来，市国土资源和规划局致力于以社区规划推动社会治理的创新实践，以武昌区为先行试点，充分调动街道、社区、居民等多方力量开展了戈甲营社区、华锦社区、都府堤社区、张家湾社区等若干社区规划的试点工作，强调以社区居民全程、广泛参与规划，重点围绕提升空间品质、彰显文化价值、推动创新集聚、营造空间活力等4个领域开展研究，力求在过程中凝聚共识，形成反映居民共同意志的规划、设计、建设与经营管理方案。

南湖街遴选出200多名社区规划师，回收1500分调查问卷，开展了90场联合设计，形成了多方认可的微改造设计方案，项目建成后新增运动设施200套、活动场地5000平方米、停车位500个、扩展绿化面积2000平方米，改造道路5公里，营造优质景观小品80余处。

戈甲营社区是武昌古城北部的一个老旧社区，《戈甲营社区规划》（2017年）旨在激发社区活力，促成社区发展由"政府输血"转向"自我造血"，进而达到社区可持续发展的目标。该规划组建了政府、居民、设计师、专家、艺术家、商家和社会组织等多元主体参与的社区工作坊，以关注居民生活、改善社区环境、发展社区生产为出发点，寻找社区矛盾和问题的源头，通过社区规划重构社会、文化、经济和物质格局，关注弱势群体，化解社会不稳定因素，提高居民满意度、归属感。

1
武汉市体育设施布局规划图
Layout Planning for Sports Facilities in Wuhan

2
养老设施用地规划图
Land Use Planning Map of Facilities for the Aged

3
主城区生活圈分类图
Classification Map of Life Circle in Main Urban Area

- *To focus on community: planning of life circle within 15 minutes*

In 2018, to further implement the strategic intent of a new round of overall urban planning "life circle within 15 minutes", to promote quality of urban public service and high-quality development and to enhance the level of grassroots community governance and refined management, Municipal National Land Resources and Planning Bureau organized to compile the Guidelines of Actions Implemented in Life Circle within 15 Minutes in Wuhan City as one of eight specialized strategic action plans in a new round of overall planning to provide an instruction basis for 15-minute community life circle construction and community planning.

It was put forward in the plan to delimit the community life circle as "space range reachable on foot within 15 minutes with action radius of 500-800 meters, permanent resident population of general services of 30,000 to 60,000, the scale of about 1-3 square kilometers and population density of about 10,000 to 20,000 people per square kilometers". There are a total of 283 life circles being designated in main urban area to be divided into 3 categories including repairing, promotion and rebuilding. Refined construction strategies and models are put forward for each category.

- *To focus on micro-transformation: Gejiaying Community Planning and Nanhu Street Community Planning (2017 — 2018)*

Since 2017, Municipal National Land Resources and Planning Bureau has been devoted to driving innovation practice of social governance with community planning, fully mobilizing street, community, residents and many other strengths to carry out pilot work of some community planning in Gejiaying Community, Huajin Community, Dufudi Community, Zhangjiawan Community and so on with Wuchang District as the first pilot, emphasizing community residents to participate in planning extensively in the whole process, carrying out studies emphatically surrounding four fields including promoting space quality, manifesting culture value, drive innovation gathering and creating space vitality, striving to build consensus in the process and forming a planning, design, construction and operation management scheme to reflect residents' common will.

Nanhu Street selected over 200 community planners, recovered 1,500 questionnaires, carried out 90 joint designs, and formed micro-transformation design schemes approved by all parties. Upon the completion of the project, 200 sports facilities, activity sites for 5,000 square meters and 500 parking spaces will be newly increased, greening area will be expanded for 2,000 square meters, roads will be renovated for 5 kilometers and over 80 high-quality featured landscapes will be created.

Gejiaying Community is an old community in the north of Wuchang Ancient City. *The Plan for Gejiaying Community* is aimed at stimulating the vitality of community and promoting community development from "government subsidy" to "self-development" to reach the goal of sustainable development of the community. The Plan organized community workshops taken part in by multiple subjects such as government, residents, designers, experts, artists, merchants and social organizations to set focusing on residents' life, community environment improvement and community production development as the starting point, to seek for the source of community contradictions and problems, to reconstruct social, cultural, economic and material patterns through community planning, to pay attention to vulnerable groups, to defuse factors of social instability and to promote residents' degree of satisfaction and sense of gain.

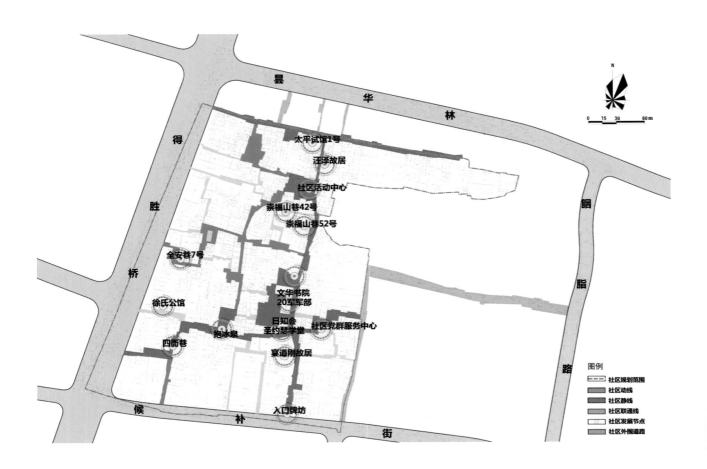

第肆章 • 区域一体化时期的规划 287

1
戈甲营社区规划结构图
Gejiaying Community Planning Structure Chart

2
戈甲营巷效果图
Effect Picture of Gejiaying Alley

3
老年活动中心入口效果图
Effect Picture for Entrance of Senior Citizens Activity Center

4
崇福山巷效果图
Effect Picture of Chongfushan Alley

8 - 乡村振兴战略引领下的村庄规划
Village Planning Under the Guidance of Rural Revitalization Strategy

乡村是具有自然、社会、经济特征的地域综合体，兼具生产、生活、生态、文化等多重功能，与城镇互促互进、共生共存，共同构成人类活动的主要空间。我国人民日益增长的美好生活需要和不平衡不充分的发展之间的矛盾在乡村最为突出。正如业界常说的，中国具有"第一世界的一线城市，第三世界的农村"这种城乡高度二元拼贴的景观，中国城镇化的许多方面都呈现出较突出的对比和冲突，也是中国城镇化高度压缩性、复杂性与矛盾性的体现。这些都是中国在城镇化过程中历史矛盾与现实新矛盾的交织、历时性与共时性矛盾的交织。

近年来，武汉市以统筹城乡发展、加快新型城镇化建设、推进小城镇"四化同步"发展和农村地区全面振兴发展为总体思路，以"家园行动计划"补齐了乡村规划建设短板，塑造了美丽乡村的建设样板，逐步建设形成了一批以武湖街、大集街为代表的发展迅速、魅力独特的特色城镇和以石榴红村、大余湾村、星光村、小朱湾村等为代表产村一体、乡韵乡风的富美村庄。同时，在乡村振兴战略要求下，以田园综合体规划为突破口，探索了农村新产业、新业态的发展路径。

一 家园建设行动规划

2005年，为贯彻落实中央关于建设社会主义新农村的重大战略部署，改变村庄长期缺乏规划指导、农村生产生活条件落后等现实发展问题，武汉市启动了以普惠制为特色的农村地区"家园建设行动计划"，以"全市统筹、试点先行、分期分批"的推进方式，按照每年350个村的进度，用6年时间完成了全市2087个行政村的规划全覆盖。同时，结合"致富门道明晰、基础设施完善、社保体系建立、社会和谐稳定"的"四到家园"政策要求，以村庄规划为指引，从产业发展、基础设施、村湾集并等几个方面，系统性开展了全市新农村建设工作，有效填补了农村地区生产、生活设施的历史欠账，基本补齐了武汉农村基建短板，谋划了"一村一品"特色化的产业发展路径。

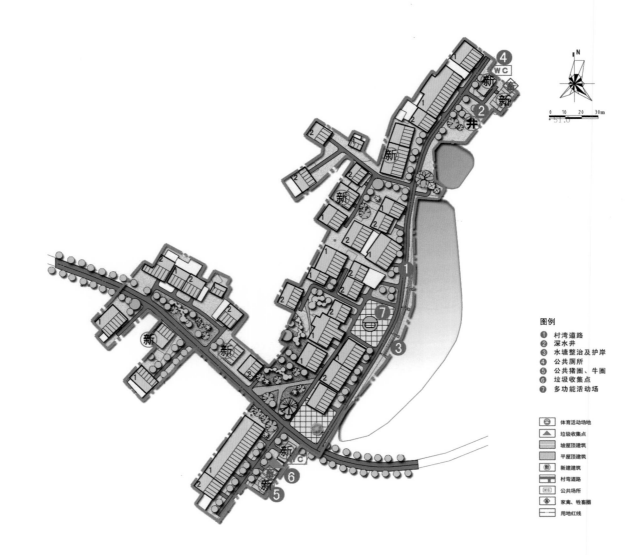

The village is a regional complex with natural, social and economic characteristics. With multiple functions such as production, life, ecology and culture, it promotes and coexists with cities and towns to form the main space for human activities. The contradiction between our people's ever-growing needs for a better life and the unbalanced and inadequate development is most prominent in the countryside. As it's said in the industry that China has the landscape of "first-tier cities in the first world and rural areas in the third world", which is a highly binary collage of urban and rural areas. Many aspects of China's urbanization show sharp contrast and conflict, which is the embodiment of the high degree of compression, complexity and contradiction of China's urbanization. These are all the interweaves of historical and new contradictions in reality, and of diachronic and synchronic contradictions during China's urbanization.

In recent years, Wuhan City has taken the overall idea of coordinating urban and rural development, speeding up the construction of new urbanization, promoting the synchronous development of "industrialization, informatization, urbanization and agricultural modernization" of small towns and the overall revitalization of rural areas. With the "Home Action Plan", it has supplemented the weakness of rural planning and construction, shaped the construction model of beautiful village, and gradually built a number of fast-growing and unique characteristic towns represented by Wuhu Street and Daji Street, as well as rich and beautiful villages represented by Shiliuhong Village, Dayuwan Village, Xingguang Village and Xiaozhuwan Village and others that integrate the production and the countryside and has rural charms and civilization. Meanwhile, under the requirements of the rural revitalization strategy, the development path of new industries and its forms in rural areas is explored with the planning of rural complex as a breakthrough.

— Action Plan for Home Construction

In 2005, in order to implement the central government's major strategic deployment of building a new socialist countryside and change the practical problems of development such as the long-term lack of planning guidance in villages and the backward conditions of production and life in rural areas, Wuhan City launched the "Action Plan for Home Construction" featuring the Generalized System of Preferences. In accordance with the planning method of "Overall Urban Planning and Pilot First by Stages and in Batches", Wuhan City had a complete coverage of the planning of 2087 administrative villages within 6 years according to the progress of 350 villages per year. Also, in combination with the "Four implementing Schemes for Home Construction" policy requirement of "Clear Road to Wealth, Perfect Infrastructure, Establishment of Social Security System and Social Harmony and Stability", and guided by village planning, the City's new rural construction has been systematically carried out from several aspects such as industrial development, infrastructure, and integration of villages, effectively filling up the historical debts of production and living facilities in rural areas, basically filling up the weakness in Wuhan's rural infrastructure, and plotting an industrial development path characteristic of "Leading Product in One Village".

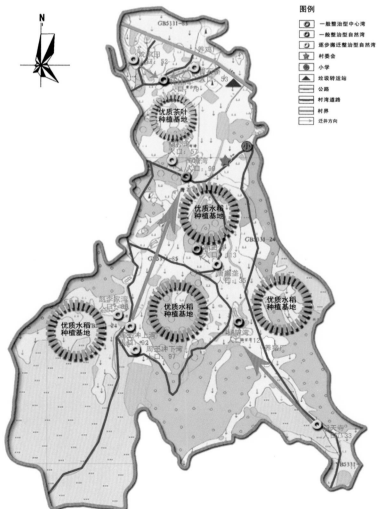

1　2

1
青石桥村村湾建设规划图
Planning Map of Qingshiqiao Village Construction

2
青石桥村村庄体系规划图
Planning Map of Qingshiqiao Village System

— 美丽乡村建设

2011年，为进一步延续和升级新农村建设，增强农村发展活力，武汉市顺应新型工业化、新型城镇化、农业现代化发展趋势，启动了"美丽乡村创建计划"，制订了美丽乡村建设项目申报指南、建设标准、验收办法等文件，创新提出以镇域村庄布局规划为指导，以村庄建设综合规划、村庄整治规划和农房建设规划为落地，以村庄产业发展规划和村湾绿化专项规划为支撑的"1+3+2"美丽乡村规划编制体系，规范了美丽乡村创建的顶层设计。同时，按照"突出重点、打造亮点、整合聚焦"的思路，通过竞评方式在全市遴选了一批中心村（社区）和特色村进行重点建设和发展，并针对不同村庄的资源禀赋和发展类型，开展了美丽乡村规划建设模式的探索，逐步建设出武汉市乡村地区的亮点和样板。

以《黄陂区王家河街胜天村龚家大湾美丽乡村实施规划》（2015年）为例，其目标是建设成一个文化底蕴彰显、建筑特色浓郁、家园生活简洁、区域风貌和谐的武汉市美丽乡村样板，规划提出村湾环境整治、建筑外立面改造以及提升公共设施品质等三项主要内容。该规划在现有建设条件和自然基础下挖掘胜天村龚家大湾历史人文资源，重新构筑地方传统建筑风貌，保留12栋老建筑，改造26栋新建筑，将村湾生活品质提升与区域风景旅游服务拓展相结合，构建"服务带+观光带，乡村生活圈+旅游体验圈"的村湾发展框架。

— 乡村振兴规划实施的探索

2017年"田园综合体"作为乡村新型产业发展的亮点措施被写进中央一号文件，田园综合体至此上升到国家战略高度，成为在城乡一体格局下顺应农村供给侧结构性改革、新型产业发展、农村产权制度改革要求，实现乡村现代化、新型城镇化、社会经济全面发展的一种可持续性模式。在此宏观政策背景下，武汉市在临空港经济技术开发区（东西湖区）、经济技术开发区（汉南区）、新洲区、江夏区规划建设4个大型都市田园综合体，广泛利用社会和市场的力量，推进农业现代化与城乡一体化互促发展，逐步打造出引领武汉市乡村振兴发展的重点极核。针对都市

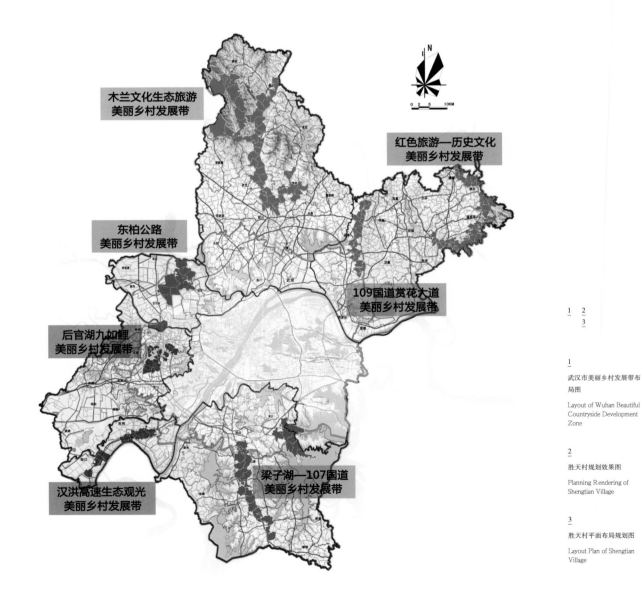

1
2
3

1
武汉市美丽乡村发展带布局图
Layout of Wuhan Beautiful Countryside Development Zone

2
胜天村规划效果图
Planning Rendering of Shengtian Village

3
胜天村平面布局规划图
Layout Plan of Shengtian Village

田园综合体范围较大、需求多样、建设主体多元的发展特征，武汉市从规划体系、功能营造、空间布局、设施配给、管控方式等方面对田园综合体的规划建设模式进行了全面的探索。

以东西湖区田园综合体为例，利用其立足近郊区位条件，打造以绿色渔业、有机水稻、特色果蔬为主导的都市农业；以生态为根本立足点，针对区内水网纵横的生态特色，严格保护蓝线、绿线范围，采取退塘还湖、河渠连通、河道生态化治理、渠道生态化改造等措施，构建生态水网；以基本公共服务均等化为立足点，构建"基本生活圈、拓展生活圈、高级生活圈"三级生活圈体系，按照适度和超前的原则，制定各级生活圈的公共服务设施配置标准；以对接国土空间规划管控为立足点，以田园功能单元为基本对象，制定项目准入、建设强度、设施配套等刚性管控指标体系和产业发展、文化风貌、生态环境等弹性管控指标体系。

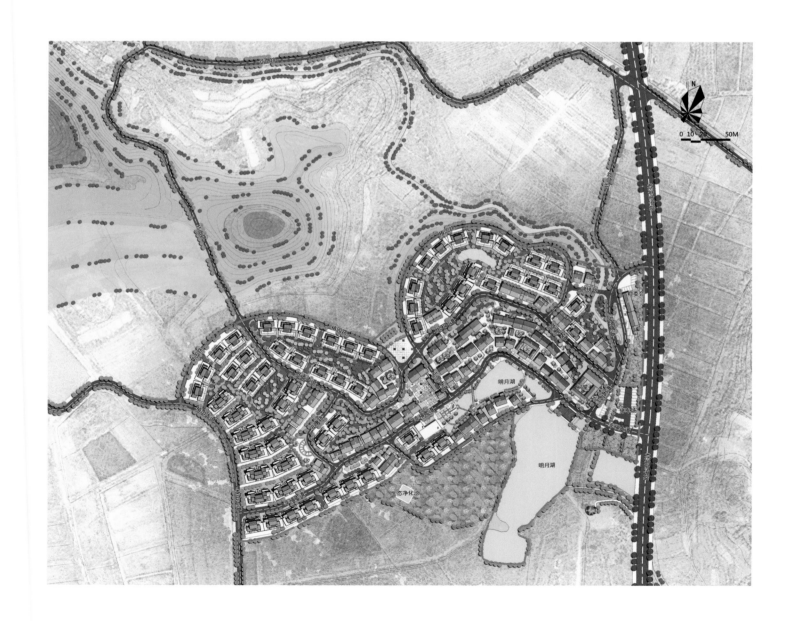

— Beautiful Countryside Construction

In 2011, in order to further continue and upgrade the new rural construction and enhance the vitality of rural development, Wuhan City followed the development trends of new industrialization, new urbanization and agricultural modernization, launched the "Beautiful Countryside Construction Plan", formulated documents such as the application guidelines, construction standards and acceptance methods for Beautiful Countryside Construction Project, innovatively proposed the "1+3+2" beautiful countryside planning formulation system under the guidance of the layout plan of village in the town area, supported by the village industrial development plan and the special village greening plan and implemented with Comprehensive Planning for Village Construction, Village Renovation Planning and Rural Housing Construction Planning, and standardized the top-level design created by beautiful countryside. Meanwhile, according to the idea of "stressing the key points, creating the highlights, integrating the focuses", it has selected a group of central villages (communities) and characteristic villages in the entire city for major construction and development. Besides, based on resource endowment and development type of different villages, it has carried out the exploration in the mode of beautiful countryside planning and construction so as to forge the highlights and models of the rural areas in Wuhan gradually.

For example, the "Beautiful Countryside Implementation Plan of Gong Jia Da Wan, Shengtian Village, Wangjiahe Street, Huangpi District (2015)" is intended to build a beautiful countryside model of Wuhan City with profound cultural deposits, rich architectural features, simple home life and harmonious regional landscape. Three main contents, namely, environmental enhancement of the village and bay, renovation of the building facade and improvement of the quality of public facilities, are proposed in the Plan. It is planned to excavate the historical and cultural resources of Gong Jia Da Wan in Shengtian Village under the existing construction conditions and natural basis, reconstruct the local traditional architectural style, preserve 12 old buildings and renovate 26 new buildings, combine the improvement of the quality of life in the village with the expansion of regional scenic tourism services, and establish the development framework of the village of "Service Zone + Sightseeing Zone, Village Life Sphere + Tourism Experience Circle".

— Exploration on the Implementation of Rural Revitalization Plan

In 2017, the "Rural Complex" was written into the No. 1 Central Document as a novel measure for the development of new rural industries, so it has risen to the height of national strategy and become a sustainable mode to meet the requirements of structural reform on the rural supply side, the development of new industries and the reform of rural property rights system under the pattern of urban and rural integration, and to realize the modernization of villages, new urbanization and all-round social and economic development. Under the background of macroscopic policy, Wuhan City plans to build four metropolitan rural complexes in Airport Economic and Technological Development Zone (Dongxihu), Economic and Technological Development Zone (Hannan), Xinzhou District and Jiangxia District, and take extensive use of social and market forces so as to promote the mutual development of agricultural modernization and urban and rural integration, and gradually forge the key polar nucleus leading the revitalization and development of villages in Wuhan. In view of the development characteristics of metropolitan rural complex, such as large scope, various needs and diverse construction subjects, Wuhan City has made a comprehensive exploration on the planning and construction mode of rural complex from the aspects of planning system, functional construction, spatial layout, facility allocation, and management and control methods and so on.

For instance, the Rural Complex in Dongxihu District built the metropolitan agriculture dominated by green fishery, organic rice and featured fruits and vegetables based on the suburban location conditions. Taking the ecology as the fundamental foothold and aiming at the vertical and horizontal ecological characteristics of the water network in the District, it strictly protected scope of the blue and green line and took measures such as returning ponds to lakes, connecting rivers and canals, ecological treatment of rivers, ecological transformation of channels and so on in order to construct an ecological water network. According to the equalization of basic public services, it set up the system of the three-tiered "basic, expanded and advanced" life sphere and followed the principle of moderation and transcendence to formulate the allocation standards for public service facilities in the life spheres at all levels. In accordance with link between planning and management & control on land and space, it regarded the rural functional units as the basic objects and established the rigid management & control indicator systems comprised of project access, construction intensity, equipment arrangement and so on and flexible ones comprised of industrial development, cultural features and ecological environment and so on.

9 – 韧性城市理念引领下的基础设施规划

Infrastructure Planning Guided by Concept of Resilient City

改革开放后，城镇化进程进入高速发展阶段，但前期粗放式的城镇化发展模式使基础设施滞后、城市灾害适应能力脆弱等问题凸显。步入 21 世纪之后，城市面临的不确定性因素和未知风险也不断增加。武汉市提出建设韧性城市，强调通过工程改进、设施完善等物质层面和公众参与、制度创新等社会层面相结合的系统构建，提高城市系统面对不确定性因素的抵御力、恢复力和适应力，提升城市规划的预见性和引导性，实现城市的可持续高质量发展。

一 建立高韧度的市政基础设施体系

在韧性城市理念的引导下，市政基础设施规划思想从"够不够"向"稳不稳"逐渐过渡。以往的规划以支撑城市发展的规模需求为目的、以资源承载力为前提来确定各类基础设施的体系和规模，而从提升城市韧性，需要统筹考虑基础设施的稳定性和面对风险干扰的脆弱性，增强城市综合承载能力。

在此背景下，武汉市自 2011 年起进行了每年一度的市政基础设施统计分析，重点关注市政系统资源供给与节能、环境保护、综合防灾三大服务职能，并形成《武汉市市政基础设施年度报告》面向公众发布，反映了市政基础设施建设状况、服务水平变化、设施建设与城市规划发展的匹配关系。

2017 年《武汉市市政基础设施承载力评估》应运而生，首次对武汉市各类市政基础设施运转情况进行了深度"体检"，全面核查了市政基础设施服务缺口，并结合城市"三旧"改造契机，提出市政设施建设与"三旧"改造同步实施，为市政基础设施建设落地提供了空间保障。同时，该评估还建立了城市开发与市政建设的过程控制机制，实时跟踪解决城市发展中市政基础设施建设配套严重脱节的问题，为新型城镇化健康发展提供了制度保障。

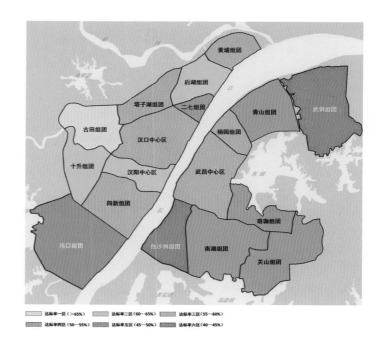

1 东西湖田园综合体用地规划布局图
Land Use Layout Plan of the Rural Complex in Dongxihu District

2 主城区分区市政设施综合水平评价图
Evaluation Map of Comprehensive Level of Municipal Facilities of Subarea of Main City

After the reform and opening up, the urbanization process has entered a stage of rapid development. However, the extensive urbanization mode in the early stage has highlighted the problems of lagging infrastructure and weak urban resilience to disasters. After entering the 21st century, cities are facing increasing uncertainties and unknown risks. Wuhan City proposes to build a resilient city and emphasizes the systematic construction through the combination of project improvement and facilities improvement materially and public participation and system innovation socially so as to improve the resistance, resilience and adaptability of the urban system to uncertainties, enhance the predictability and guidance of urban planning, and realize the sustainable and high-quality urban development.

— Establishment of the Highly Resilient Municipal Infrastructure System

Under the guidance of the concept of a resilient city, the idea of planning municipal infrastructure gradually transits from "sufficiency" to "stability". The previous plan was aimed to support the scale demand of urban development and determine the system and scale of various infrastructure based on the premise of resource carrying capacity. However, in order to strengthen the resilience of the city, it is necessary to fully consider the stability of infrastructure and the vulnerability to risk interference to enhance the comprehensive urban carrying capacity.

Under this background, Wuhan City has conducted an annual statistical analysis on municipal infrastructure since 2011, focused on the three major service functions of municipal system resource supply and energy conservation, environmental protection, and comprehensive disaster prevention, and formulated the Annual Report on *Wuhan Municipal Infrastructure* to release it to the public, which reflects the construction status of municipal infrastructure, changes in service level, and the matching relationship between facilities construction and urban planning & development.

In 2017, "Wuhan Municipal Infrastructure Carrying Capacity Assessment" came into being. It was the first in-depth "physical examination" of the operation of various municipal infrastructure in Wuhan City. It comprehensively checked the service gap of municipal infrastructure. Besides, combined with the opportunity of the urban "Three Old" transformation, it was proposed that the municipal infrastructure construction and the "Three Old" transformation be implemented simultaneously so as to offer space guarantee for the implementation of construction of municipal infrastructure. Meanwhile, it established process control mechanism for urban development and municipal construction in order to track and solve the serious disconnection between construction and accessory of municipal infrastructure in urban development in real time, offering institutional guarantee for the healthy development of new urbanization.

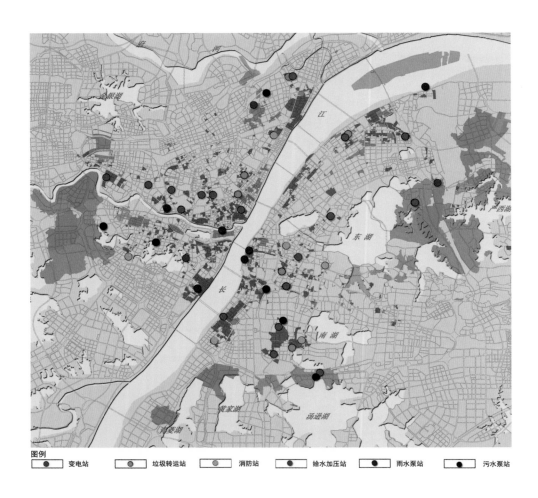

1
General Layout of New Facilities Combined with "Three Old" Planning

2
Overall Structural Drawing of Hankou River Beach Park (2003)

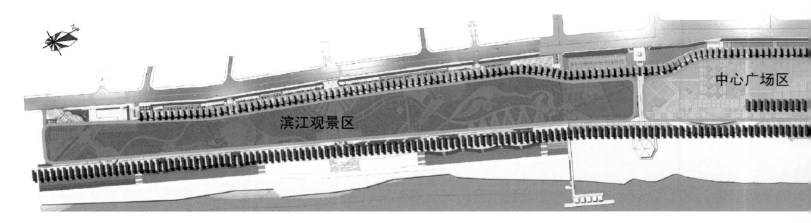

一 牢筑"四水共治"水安全体系

新中国成立以来,武汉治水思想和任务伴随着社会经济的发展不断演变、互相推动,由此前"以单一功能为导向的工程治水"逐渐转向"以人水和谐共生为目标的综合理水"过渡转变。1949~1999年,通过堤防加固、供水、雨水、污水系统完善等一系列工程建设,使得武汉三镇初步形成了自成一体的防洪体系和水设施总体布局。

1998年大水灾发生之后,促使了治水思路的转变和发展战略的转型,开始侧重水资源和水环境的保护,尤其是进入21世纪之后,治水作为生态文明建设的重要组成部分,逐步提升到了国家战略高度。"水十条""海绵城市"相继提出,武汉自此进入了"水生态、水环境、水景观、滨水空间利用与城市特色塑造并重"的综合水治理时代,围绕长江大保护、打造滨水生态绿城的总体目标,大力推进"四水共治"——"防洪水、排涝水、治污水、保供水",初步走出一条具有武汉特色的治水现代化道路。

● 防洪水:汉口江滩防洪及环境综合整治规划

2001年《汉口江滩防洪及环境综合整治规划》出台,首次将城市防洪工程与滨江地区的空间建设相结合起来,把汉口江滩打造成为体现武汉滨江城市特色的重要标志性景区,并作为城市核心区内的滨江休闲绿化带,同时满足景观游憩、绿化生态和休闲娱乐功能。汉口江滩的建设实现了防御水患与生态修复、滨水景观与环境提升的融合,对汉口主城区人居环境的优化、城市功能的提升和城市形象的塑造起到了画龙点睛的重要作用。

— Consolidate the Water Safety System of "Flood Control, Drainage of Stagnant Water, Sewage Treatment and Guaranteed Water Supply".

Since the founding of People's Republic of China, Wuhan's water control thoughts and tasks have evolved and pushed forward with the development of social economy. From then on, the former "project water control oriented by single function" has gradually shifted to the "comprehensive water management with the goal of harmonious coexistence of human beings and water". From 1949 to 1999, through a series of engineering construction such as dike reinforcement, water supply, rainwater and sewage system improvement, the Wuchan, Hankou and Hanyang in Wuhan initially formed a self-contained flood control system and overall layout of water facilities.

After the Great Flood in 1998, it prompted the transformation of water management concept and development strategy, and began to focus on the protection of water resources and environment. Especially after the 21st century, water control, as an important part of the construction of ecological civilization, has gradually risen to the height of national strategy. "Action Plan for Water Pollution Prevention" and "Sponge City" have been put forward one after another. Since then, Wuhan has entered an era of comprehensive water management that "pays equal attention to water ecology, water environment, water landscape, waterfront space utilization and urban characterization". Centering on the overall goal of the Yangtze River Protection and construction of a green city of waterfront water ecology, Wuhan has vigorously promoted the Water Safety System of "Flood Control, Drainage of Stagnant Water, Sewage Treatment and Guaranteed Water Supply". It has initially created a modern road of water management with Wuhan characteristics.

● *Flood Control: The Hankou River Beach Park's Comprehensive Flood Control and Environmental Improvement Plan*

In 2001, the Hankou River Beach Park's Comprehensive Flood Control and Environmental Improvement Plan was issued and for the first time combined the urban flood control project with the space construction in the riverside area, making Hankou River Beach Park an important landmark scenic spot embodying the characteristics of Wuhan riverside city and the riverside recreational green belt in the core area of the city so as to satisfy the functions of landscape recreation, greening ecology and leisure & entertainment. The construction of Hankou River Beach Park has realized the integration of flood prevention and ecological restoration, waterfront landscape and environmental improvement and played an important role in the optimization of human settlement environment, the improvement of urban function and the creation of urban image in the main urban area of Hankou.

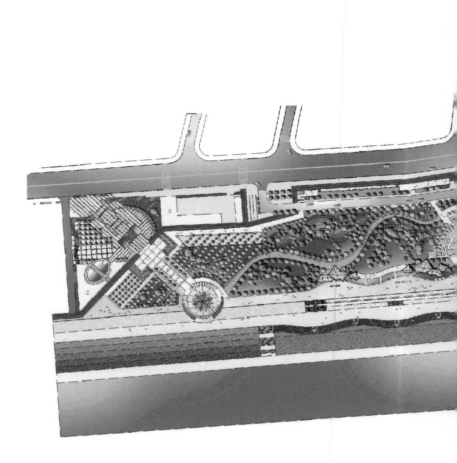

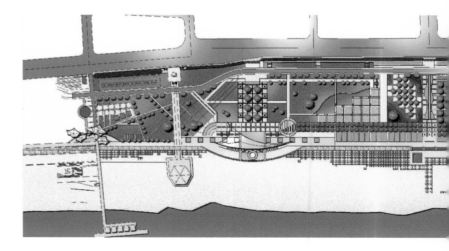

1

汉口江滩一期总平面规划图
（2003年）

Hankou Jiangtan Phase I
Master Plan (2003)

2

汉口江滩二期总平面规划图
（2003年）

Hankou Jiangtan Phase II
Master Plan (2003)

3

汉口江滩三期总平面规划图
（2003年）

Hankou Jiangtan Phase III
Master Plan (2003)

第肆章 · 区域一体化时期的规划

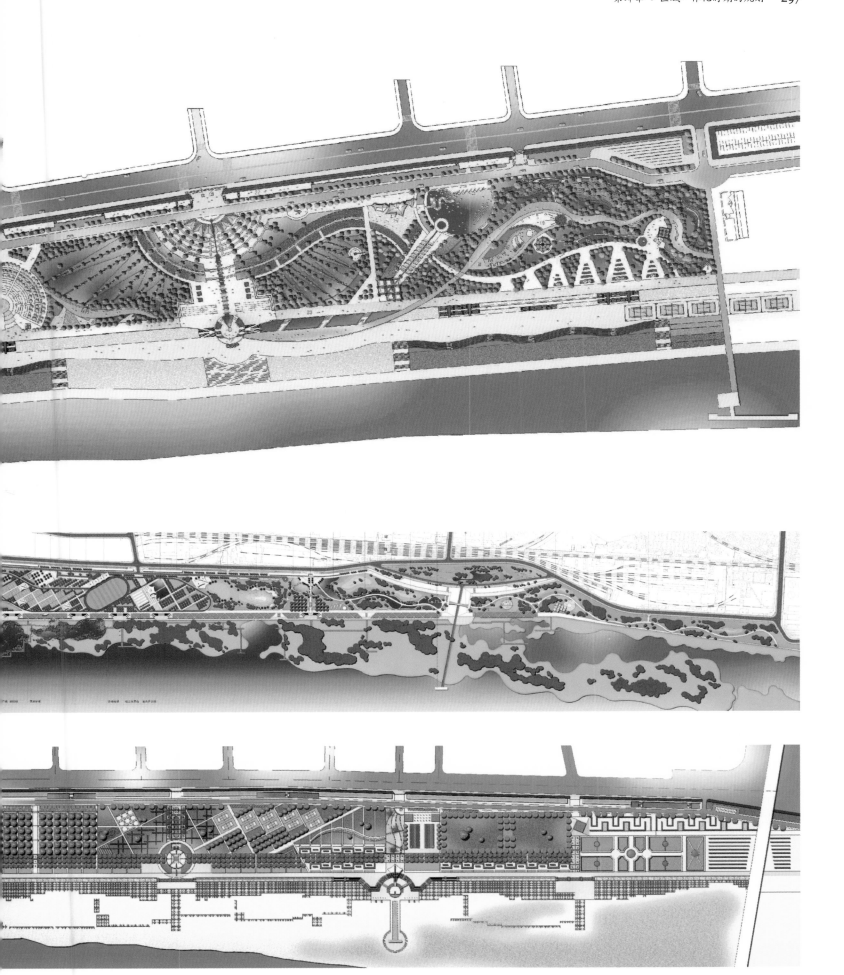

● 排涝水：武汉市排水防涝规划

2011 年 6 月武汉市连续遭遇 5 轮暴雨袭击，内涝影响较大。其后，北京、成都、南京、广州等特大城市相继出现严重的城市内涝问题。城市的排水问题已成为中国城市集体面对的现代性难题。

为破解内涝忧患，《武汉市排水防涝专项规划》于 2012 年出台。规划实现了从单纯排水、蓄排结合向综合防涝，从人工设计向模型仿真的两大转变。首次将传统排水扩展为排水防涝，确定了武汉市的防涝目标，提出多层次、多渠道加强城市建设与内涝防治的紧密融合，构建了具有武汉特色的排水防涝结构体系和与之对应的标准体系，并以地方标准的形式加以推广，为有序推进防涝设施建设、引导城市科学规避内涝风险提供了重要依据。

● 治污水：武汉市主城区污水收集与处理专项规划

为有效控制主要污染物排放量，明显改善生态环境质量，加快推进污水收集、处理系统的建设，2009 年《武汉市主城区污水收集与处理专项规划》出台。该规划开启了武汉市污水治理由"污水控制"向"污水管理"，"污水单一达标排放"向"水的健康循环利用"的思想转变。在主城区共布局了 13 座污水处理厂，第一次系统构建了主城区污水收集和处理的骨架系统，将武汉市主城区划分为 13 片生活污水集中处理区和 2 片分散污水处理区，整合提出与武汉水环境相适应的污水系统布局及分合流排水体制分区，有效指导了其后的污水系统规划和建设。

武昌大东湖地区是武昌北部围绕东湖、沙湖、严西湖、严东湖等大型湖泊的区域。随着该区城市建设发展，坐落在该区的沙湖、二郎庙、落步咀污水厂周边逐步形成了被居住和公建等建筑包围的态势，导致污水厂扩建需求与周边环境品质提升要求之间的矛盾逐步凸显。

2015 年，为保护和提升大东湖地区水环境品质，以沙湖和二郎庙污水厂搬迁为契机，编制出台了《武昌大东湖地区污水厂集并及深隧系统规划》。首次应用深隧的污水转输功能实现"四厂集并、排水下移"——沙湖、二郎庙、落步咀和白玉山污水系统范围内污水通过深隧集并到新北湖污水厂，统筹解决该区域存在的污水收集与处理、径流污染控制、内涝防治等排水问题。

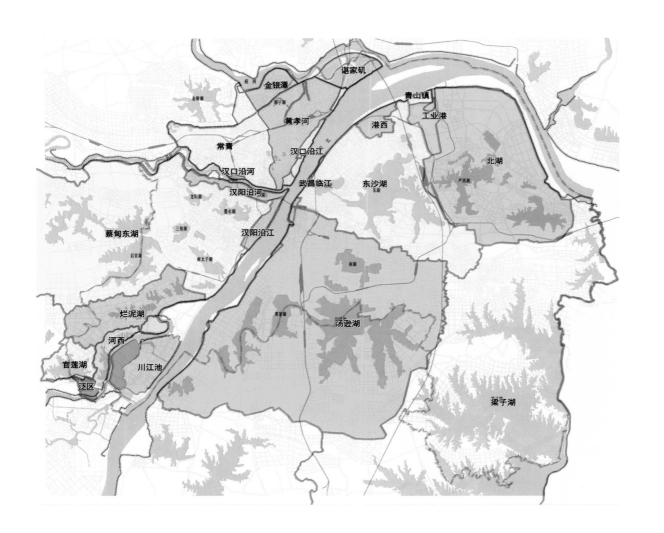

1

规划排水分区图

Map of Subarea with Planned Drainage

2

重大设施布局图

Layout of Major Facilities

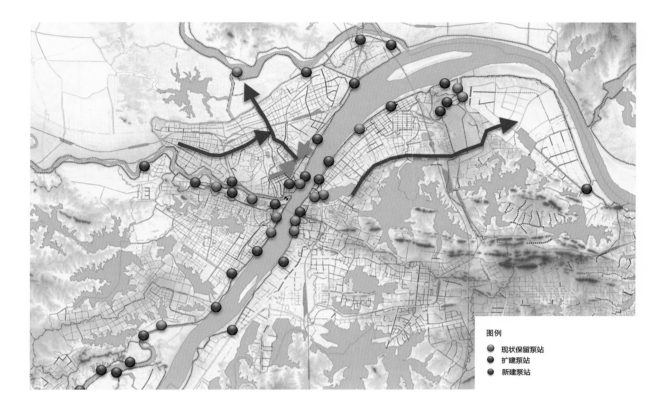

- *Drainage of Stagnant Water: Drainage and Waterlogging Prevention Planning for Wuhan City*

In June 2011, Wuhan City was hit by 5 rounds of torrential rain in a row and greatly affected by waterlogging. Since then, Beijing, Chengdu, Nanjing, Guangzhou and other mega-cities have experienced serious urban waterlogging. The urban drainage has become a modern problem faced by Chinese cities.

In order to solve the waterlogging problem, the Special Drainage and Waterlogging Prevention Plan for Wuhan was issued in 2012. The Plan has realized two major changes from simple drainage and combination of storage and drainage to comprehensive waterlogging prevention, and from manual design to model simulation. For the first time, it expands the traditional drainage into drainage and waterlogging prevention, determines the waterlogging prevention goal of Wuhan City, proposes to strengthen the close integration of urban construction and waterlogging prevention through multiple levels and channels, constructs a drainage and waterlogging prevention structural system with Wuhan characteristics and corresponding standard system, and promotes it in the form of local standards, which offers an important basis for orderly promoting of the construction of waterlogging prevention facilities and guiding the city to scientifically avoid waterlogging risks.

- *Sewage Treatment: Special Plan for Sewage Collection and Treatment in Main Urban Area of Wuhan City*

In 2009, the Special Plan for Sewage Collection and Treatment in Main Urban Area of Wuhan City was issued in order to effectively control the discharge of major pollutants, significantly improve the quality of ecological environment, and accelerate the construction of sewage collection and treatment system. The Plan started the ideological transformation of sewage treatment in Wuhan from "sewage control" to "sewage management", "sewage discharge with a single standard" to "healthy water cyclic utilization". A total of 13 sewage treatment plants were laid out in the main urban area. For the first time, the framework system for sewage collection and treatment in the main urban area was systematically constructed. The main urban area in Wuhan was divided into 13 centralized domestic sewage treatment areas and 2 decentralized sewage treatment areas. A sewage system layout suitable for Wuhan's water environment and a distributed and combined drainage system zone were put forward in an integrated manner, which effectively guided the planning and construction of subsequent sewage systems.

The Wuhan East Lake Eco-tourism Scenic Area in Wuchang is an area in northern Wuchang surrounding large lakes such as Donghu Lake, Shahu Lake, Yanxi Lake and Yandong Lake. With the development of urban construction in this area, the peripheral areas of Shahu Lake, Erlang Temple and Luobuzui Wastewater Treatment Plant located in this area are gradually surrounded by residential and public buildings, resulting in the gradual significant contradiction between the expansion demands of sewage plants and the requirements for quality improvement of the surrounding environment.

In 2015, Planning for the Integration of Sewage Plants and Deep Tunneling System in the Wuhan East Lake Eco-Tourism Scenic Area in Wuchang was compiled and issued in order to protect the water environment in the Wuhan East Lake Eco-Tourism Scenic Area and improve the quality of it by taking the opportunity of relocation of Shahu Lake and Erlang Temple Wastewater Treatment Plant. For the first time, the sewage transfer function with application of deep tunneling makes it possible to integrate the waste water within the sewage systems comprised of Shahu Lake, Erlang Temple, Luobuzui and Baiyushan Treatment Plant into the New Beihu Lake Wastewater Treatment Plant via deep tunnels so as to comprehensively solve the drainage problems such as sewage collection and treatment, runoff pollution control, waterlogging prevention and control and so on in the region.

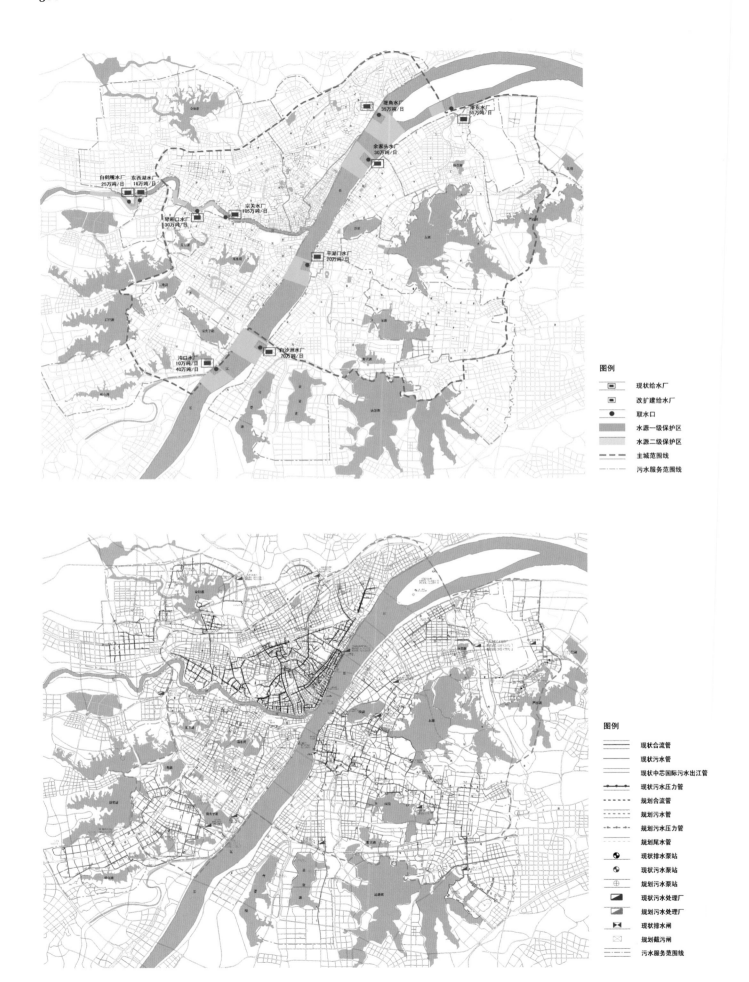

第肆章 · 区域一体化时期的规划　301

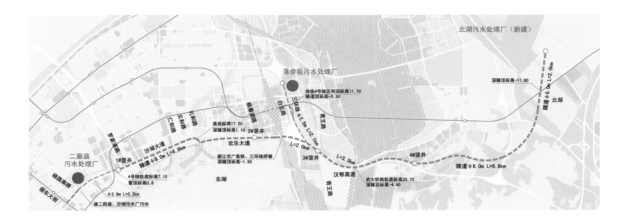

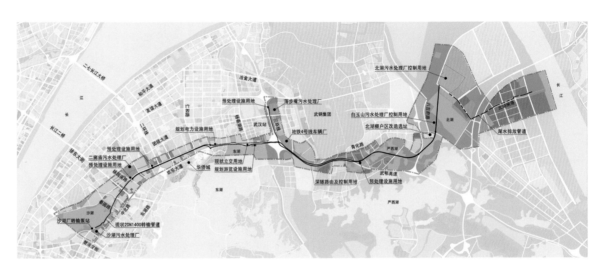

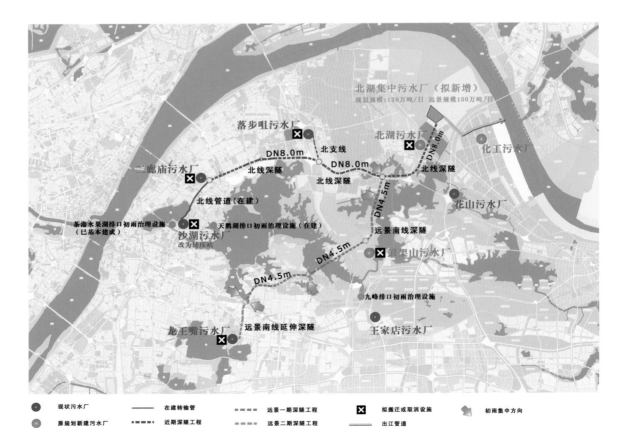

1	3
2	4
	5

1
武汉市主城区水源保护区示意图
Schematic Diagram of Water Conservation Area in Main Urban Area of Wuhan City

2
武汉市主城区污水系统规划图
Planning Map of Sewage System in Main Urban Area of Wuhan City

3
武昌大东湖地区四厂集中方案布局图
Layout of Concentrated Solution of Four Plants in Wuhan East Lake Eco-Tourism Scenic Area in Wuchang

4
武昌大东湖地区深隧路由用地控制图
Map of Routing Land Use Control on Deep Tunnels Wuhan East Lake Eco-Tourism Scenic Area in Wuchang

5
武昌大东湖地区远景大集中方案图
Map of Great Long-term Concentrated Solution for Wuhan East Lake Eco-Tourism Scenic Area in Wuchang

302 CHAPTER FOUR • Planning During Regional Integration

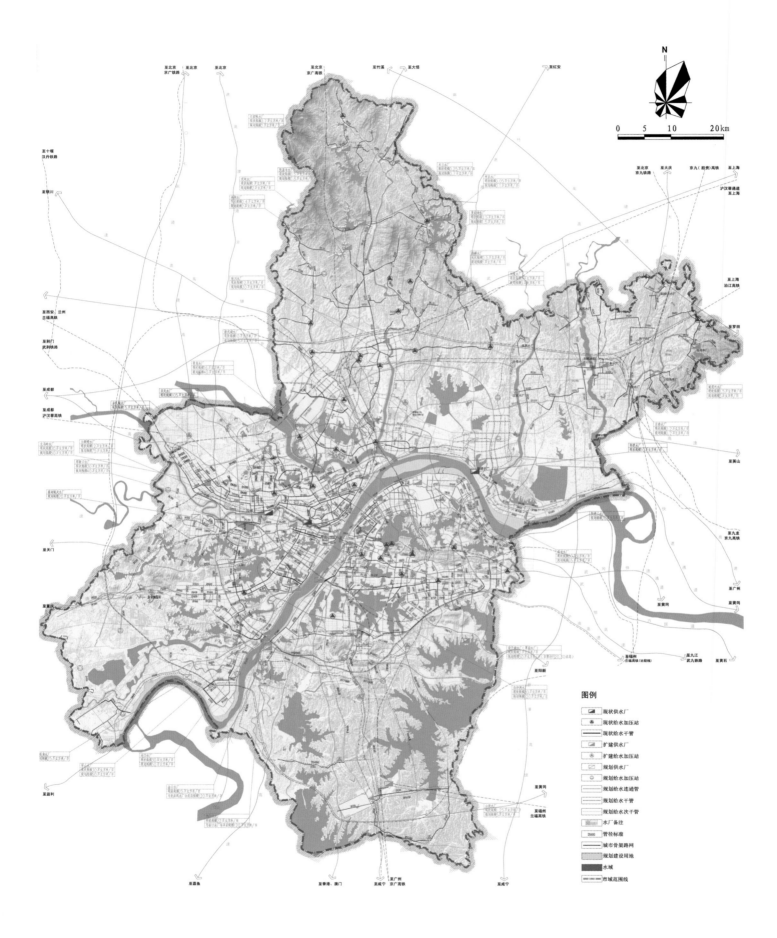

- 保供水：武汉市供水规划

随着"长江大保护""乡村振兴"和"坚决打好污染防治攻坚战"等三大战略相继提出，对供水系统的城乡统筹、布局集约、品质提升都提出了更高要求，在《武汉市城市总体规划（2017~2035）》编制的同期，武汉市水务局组织编制了《武汉市供水规划》。该规划围绕"城乡一体、一网分片、水源地优化、整体提质"对全市供水系统进行合理布局和优化配置，构建了与"三化"大武汉目标相一致、与《武汉市城市总体规划（2017~2035）》《武汉市水资源综合规划（2010~2030年）》相适应的城乡供水系统，以期使武汉市供水水质达到国际先进标准。

- 水环境"共治"规划

2013年12月习总书记提出要大力推进建设自然积存、自然渗透、自然净化的"海绵城市"，要在解决水问题时优先考虑把水留下来、优先考虑利用自然力量排水，这为全国解决城市水问题指明了方向。随后，武汉市作为全国首批"海绵城市建设试点城市"，率先启动了海绵城市示范区的建设，编制出台了《武汉市海绵城市专项规划（2016~2030年）》。规划构建了水生态、水环境、水资源和水安全4个方面的指标体系以及内涝防治、水环境保护和黑臭水体治理等方面的具体策略，将武汉市建成能有效应对50年一遇暴雨的水安全城市，建成水清岸美的滨水宜居城市。为保障海绵城市规划指标的落地，武汉市在国内率先构建了以海绵城市指标取值计算表、下垫面分布图、海绵设施分布图、场地竖向及地面径流路径设计图和建设方案自评表等"三图两表"为核心的海绵城市规划管控技术，开创了武汉市以专项设计制度、自审制度和征信制度为主的海绵城市规划管控制度方法。

为贯彻落实《国务院关于印发水污染防治行动计划的通知》（国发[2015]17号）要求，武汉市于2016年启动了第一批共计18项黑臭水体整治项目。黄孝河、机场河、巡司河均为水环境综合整治的重点工程，是"四水共治"关注的重要河流。随后武汉市陆续出台《黄孝河水环境综合治理规划及系统化方案设计》《机场河水环境综合整治工程修建规划》《巡司河综合整治工程修建规划》。与传统明渠规划不同，该系列规划首先保证河流作为排水廊道的功能需求，同时统筹治污、生态、景观、绿道等系列综合工程，将水污染治理、生态停车场和景观湿地三者结合起来，构建了河道内水资源与周边滨水空间利用及开发的良性互动关系，同时满足水安全、水环境、水空间、水景观、水文化等多重功能，成功地将渠道转型发展作为区域环境整体提升的支点。

1 2

1
供水系统管网规划图
Planning Map of Pipe Network of Water Supply System

2
系统年径流总量控制率管控图
Management and Control Map of Control Rate of Annual Total Runoff of the System

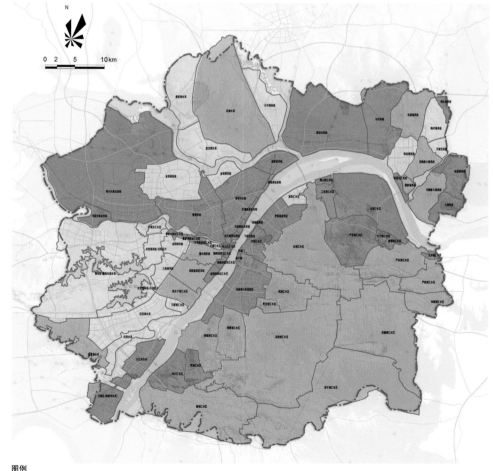

图例
都市发展区范围线 | 年径流总量控制率≥75% | 年径流总量控制率≥65% | 排水系统分区范围线 | 年径流总量控制率≥70% | 年径流总量控制率≥60%

● *Guaranteed Water Supply*

With the three major strategies of "Yangtze River Protection", "Rural Revitalization" and "Resolutely Winning the War on Pollution Prevention and Control" put forward one after another, higher requirements have been proposed for the urban and rural overall plan, intensive layout and quality improvement of the water supply system. During the same period of preparation of Wuhan Urban Master Plan (2017—2035), Wuhan Urban Water Supply Plan was compiled. According to the "Urban and Rural Integration, One Network for All Subareas, Water Source Optimization and Overall Quality Improvement", the Plan provides the urban water supply system with reasonable layout and optimal allocation and offers to set up the urban and rural water supply systems that are consistent with the Wuhan's great goal of "modernization, internationalization and ecologicalization" and fit with Wuhan Urban Master Plan and Plan of Comprehensive Utilization of Water Resource so as to enable the water supply quality in Wuhan to meet the advanced international standard.

● *"Common Governance" Plan for Water Environment*

In December 2013, General Secretary Xi proposed to vigorously promote the construction of a "sponge city" with natural accumulation, natural infiltration and natural purification and upon solving the water problem, priority should be given to retaining water and using natural forces to drain water, which has pointed out the direction for the entire country to solve the urban water problem. Later, Wuhan, as one of the first batch of "pilot cities for sponge city construction" in the country, took the lead in starting the construction of a sponge city demonstration area and formulated and issued the "Special Plan for Sponge City in Wuhan". The Plan establishes an index system for water ecology, water environment, water resources and water safety, as well as specific strategies for waterlogging prevention, water environment protection and black and odorous water treatment. Wuhan will be built into a water safety city that can effectively cope with the 50-year rainstorm, and a waterfront livable city with clean water and beautiful bank. In order to ensure the implementation of sponge city planning index, Wuhan

1

近期建设重点区分布图

Distribution Map of Key Construction Areas in the Near Future

2

黄孝河起端铁桥节点改造方案图

Conceptual Drawing of Tieqiao Node Reconstruction at the Beginning of Huangxiao River

City took the lead in constructing sponge city planning control technology with "Three Maps And Two Tables" comprised of sponge city index value calculation table, underlying surface distribution map, sponge facilities distribution map, site vertical and surface runoff path design map and construction scheme self-evaluation table as the core, and initiated its sponge city planning management and control method with special design system, self-examination system and credit investigation system as the main parts.

In order to implement the requirements of the Notice by the State Council on printing and Distributing the Action Plan for Water Pollution Prevention and Control (G.F. [2015] No.17), Wuhan City started the first batch of 18 black and odorous water remediation projects in total in 2016. Huangxiao River, Jichang River and Xunsi River are all key projects for comprehensive improvement of water environment and are important rivers of common concern for the "Flood Control, Drainage of Stagnant Water, Sewage Treatment and Guaranteed Water Supply". Then, Subsequently, Wuhan City issued the Comprehensive Water Environment Treatment Plan and Systematic Scheme Design for Huangxiao River, Comprehensive Water Environment Treatment and Project Construction Plan for Jichang River and *Comprehensive Treatment and Project Construction Plan for Xunsi River*. Different from the those for traditional open channels, this series of plans firstly ensures the functional requirements of rivers as drainage galleries, and coordinates a series of comprehensive projects such as pollution control, ecology, landscape, greenway, and so on. They combine the water pollution control, ecological parking lot and landscape wetland to construct a benign interactive relationship between the utilization and development of water resources in the river channel and the surrounding waterfront space, and satisfy multiple functions such as water safety, environment, space, landscape, culture and others so as to successfully takes the channel transformation and development as the fulcrum for the overall promotion of the region.

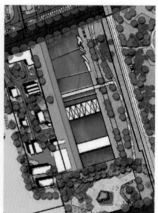

1–2
巡司河综合整治效果图
Effect Picture of Comprehensive Renovation of Xunsi River

3
巡司河综合整治景观结构图
Landscape Structure Map of Comprehensive Renovation of Xunsi River

4–5
机场河明渠景观规划设计效果图
Effect Picture of Landscape Planning and Design of Open Channel in Jichang River

武泰闸历史游园
挖掘历史底蕴，将闸口和地铁作为重大造景元素，提供多样休憩体验。

都市人文景观带
突出巡司河历史风貌，展现当地人文风采，以沿途景观小品，铺装体现巡司河历史文化脉络。

巡司河风情公园
兼顾生态水和市民活动功能。

风景绿廊休闲带
以生态游憩为主要特点，以绿色骑行为主要内容，着力打造沿线风景绿廊。

中环线湿地公园
营建大片湿地，实现生物净化；构筑城市绿肺，同时突出生态教育及科普展示功能。

增强城市防灾减灾能力

● 消防规划

2009年新《中华人民共和国消防法》施行，对防灾减灾、加强应急救援工作、消防设施建设等方面提出了更高要求。伴随着武汉市轨道交通、地下空间和高层建筑的快速发展，消防安全需求也不断提高。为进一步构建适应新形势下的城市消防安全体系，武汉市公安消防局联合规划局开展了《武汉市消防规划（2015～2030年）》的编制工作。

规划区范围划分为两个层次：市域（市行政区范围）和主城（以三环线以内地区为主）。该规划结合主城用地规划布局，对消防站级、规模及用地标准、消防站布局做了周密部署。其中消防站布局包括陆上消防站、水上消防站、航空消防站、消防训练基地、企事业专职消防站等。主城区在现有13个消防站的基础上新增57个，其中陆上消防站54个、水上消防站2个、航空消防站1个。消防规划的实施对加强城市公共消防设施建设、改善城市消防安全布局、建立消防安全体系和确保武汉安全快速发展起了积极作用。

● 地下空间规划

为进一步规范指导武汉市地下空间建设，高质量配置地下空间资源，打造国家中心城市、立体城市的地下工程示范，2018年《武汉市地下空间综合利用规划》编制完成。该规划划定了都市发展区地下空间慎、限、建分区，确定适建区范围1064平方公里，限建区1107平方公里，慎建区1080平方公里，并统筹各类型地下设施布局。通过互连互通、竖向管控等，结合轨道、综合管廊等系统性基础设施建设，弥补解决地上交通、停车、内涝、土地紧缺等城市短板问题，打造多网融合的地下空间网络，以期建成地上、地下空间分工明确，功能互补，地下各系统之间相互协调、集约高效的现代化立体大都市。

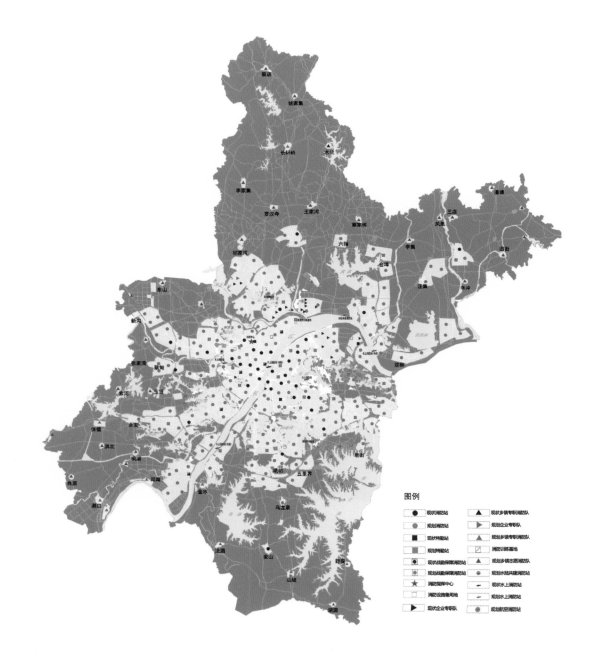

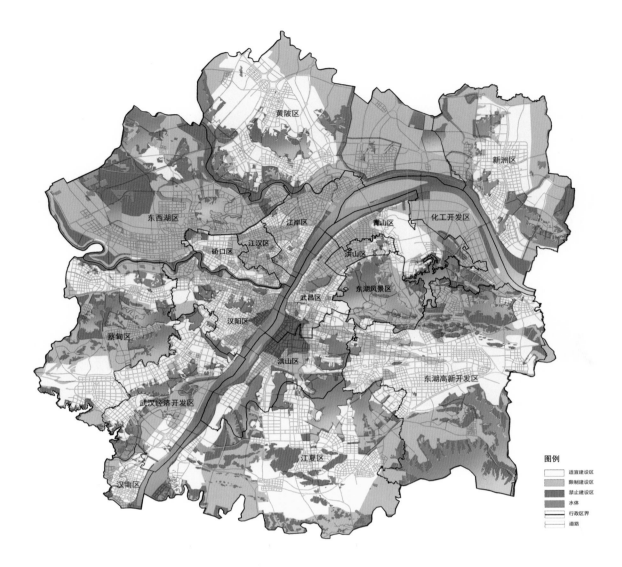

1
市域规划消防站分布图
Fire Station Distribution Planning Map in Urban Area

2
禁限建分区图
Zoning Map of Areas To Be Handled with Caution and Restricted Construction

— Enhancing Disaster Prevention and Mitigation in the City

● *Fire Planning*

In 2009, the implementation of the new Fire Prevention Law of The People's Republic of China put forward higher requirements for disaster prevention and mitigation, strengthening emergency rescue work and building fire protection facilities. With the rapid development of rail transit, underground space and high-rise buildings in Wuhan, the demand for fire safety is also increasing. To further construct the urban fire safety system for the new situation, the government has carried out the compilation of Wuhan Fire Protection Plan (2015—2030).

The government has classified the fire protection into two levels: the fire protection in municipal area (municipal administrative area) and the main urban area (mainly the area inside the Third Ring Road). Combining with the land use planning and the layout of the main urban area, the Plan has carefully stated the level, scale and land use standards, as well as the layout of fire stations. The fire station layout includes land fire station, water fire station, aviation fire station, fire training base, and full-time fire station in enterprises and institutions, etc. With the existing 13 fire stations in the main urban area, the Plan shall build 57 new fire stations, including 54 land fire stations, 2 water fire stations and 1 air fire station. The implementation of fire protection planning has played a positive role in strengthening the construction of public fire protection facilities in the city, improving the layout of urban fire safety, establishing fire safety system and ensuring the rapid development of Wuhan security.

● *Underground Space Planning*

To further standardize and guide the construction of underground space in Wuhan, realizing high-quality allocation of underground space resources and building an underground project demonstration of national central city and three-dimensional city, the government has compiled the Comprehensive Utilization Planning of Underground Space in Wuhan in 2018. The plan classifies the underground space into areas to be handled with caution and restricted construction. The suitable area to be built is 1064 square kilometers. The restricted construction area is 1107 square kilometers. And the area to be built with caution is 1080 square kilometers. The Plan also coordinates the layout of various types of underground facilities. Through interconnection, vertical management and control, combination of the construction of systematic infrastructure such as track and comprehensive pipeline corridor, the Plan can make up for the urban shortages, such as traffic, parking, waterlogging and insufficient land and build a multi-network integrated underground space network. The Plan aims at building a modern three-dimensional metropolis with clear division of role between ground and underground space, complementary functions, coordinated and efficient underground systems.

一 智慧城市建设

武汉市积极应对新经济和新技术发展，开展了前瞻性规划研究与智慧城市建设。借助大数据与 AI 技术开展城市仿真实验室研究，服务国土规划管理决策、推动武汉智慧城市建设。借助微信新媒体传播渠道，在全国率先构建公众版一张图微信发布平台，把"专家们的规划"变成"市民们的规划"。同时，为贯彻落实城乡规划改革"一年一体检、五年一评估"要求，开展规划评估体检报告研制，构建规划体检评估体系，形成可视化评估结果。

● 武汉市城市仿真实验室（2018 年）

城市仿真实验室，旨在运用大数据提升城市治理现代化水平，通过构建空间数学模型，模拟感知城市体征、监测城市活动、预演城市未来，最终构建智慧化的城市治理决策平台。

城市仿真实验室以构建智能化的城市治理决策平台为目标，在纵向上分为"数据汇集、评估预警、仿真模拟、智慧决策"4 个功能层次，采用层层递进的建设路径，分别对应数据采集、城市感知、城市把握和城市引导的应用需求；在横向上分为"人口社会、产业经济、国土规划、公共服务、交通体系、市政设施、自然资源和生态环境、空间形态"8 个专项领域，采用模块化的建设路径，分别对应单要素的系列因子子模块。

目前，城市仿真实验室已完成了"4+8"顶层设计、加速数据汇集、建立检测体系、实施仿真模拟等各项工作。预计用 3～5 年时间建成城市"数据湖"、城市"仪表盘"、规划"预演室"、城市仿真决策"大脑"四大功能板块，支撑智慧化治理。

● "众规武汉"开放平台建设（2015 年）

为贯彻创新驱动城市转型的精神，应对"互联网 +"带来的变革与影响，武汉市国土资源和规划局筹建了"众规武汉"开放平台。围绕"众筹智慧"核心目标，以网站、微网站、微信为载体，建立起一个向社会公众开放的规划平台。该平台自搭建以来，包括东湖绿道、公共停车场等 10 余个规划项目陆续上线，共收到网站留言近 300 条，问卷回收 1000 余份，在线规划参与达到 2000 余人次，微信公众号关注人数达 5800 人。

— **Construction of Smart City**

In response to the development of new economy and technology, Wuhan has carried out forward-looking planning research and smart city construction. With the big data and AI technology, the research of urban simulation laboratory has been set up to serve the decision-making of land planning and management, and to speed up the smart city construction in Wuhan. Through the new media channel of Wechat, Wuhan has taken the lead in the whole country of building a publishing platform in Wechat to seek public opinion, integrating "the planning of experts" with "the planning of citizens". Meanwhile, to implement the requirement of "examination every year and evaluation every five year" in urban and rural planning reform,, Wuhan has carried out the planning assessment medical examination report and established the examination and evaluation system for the planning, making the evaluation results visible.

● *Wuhan City Simulation Laboratory (2018)*

City simulation laboratory aims to improve the modernization level of urban governance by big data. By constructing spatial mathematical model, it can simulate and perceive urban signs, monitoring urban activities. It can also preview the future of the city, building an intelligent decision-making platform for urban governance.

The goal of urban simulation laboratory is to build an intelligent decision-making platform for urban governance. Vertically, the urban simulation laboratory is divided into four functional levels: data gathering, evaluation and early warning, simulation and intelligent decision-making. It is established layer by layer to meet the respective need of data acquisition, urban perception, urban grasp and urban guidance. Horizontally, urban simulation laboratory is divided into eight special fields: population and society, industrial economy, territorial planning, public service, transportation system, municipal facilities, natural resources and ecological environment, and spatial form. It adopts modular construction paths to correspond to a series of factor sub-modules of single factor.

At present, the urban simulation laboratory has completed the design of "4+8" top-level. The data collection has been accelerated. It has developed detection system and implemented simulation and other work. It is anticipated that the four functional modules of urban "data lake", urban "dashboard", planning "rehearsal room" and urban simulation decision-making "brain" will be built in 3-5 years to support intelligent governance.

● *Establishment of the Platform for Public Participation in Planning (2015)*

To put the spirit of innovation driven urban transformation into practice and cope with the changes and impacts brought by "Internet +", Wuhan Natural Resources and Planning Bureau has set up a platform for public participation in planning. Focusing on the core of "crowd wisdom", the platform has opened a channel to the public through the micro-website and Wechat. The platform has received almost 300 messages for more than 10 planning projects from the website, including Donghu Lake Green Road, public parking lot and so on. It has collected more than 1000 questionnaires. More than 2000 people have participated in online planning, and 5800 people have followed Wechat public platform.

1

武汉市城市仿真实验室示意图

Diagrammatic Sketch of Wuhan City Simulation Laboratory

2

"众规武汉" 开放平台

Platform for Public Participation in Wuhan Planning

3

武汉市城市仿真实验室技术架构图

Technical Architecture of Wuhan City Simulation Laboratory

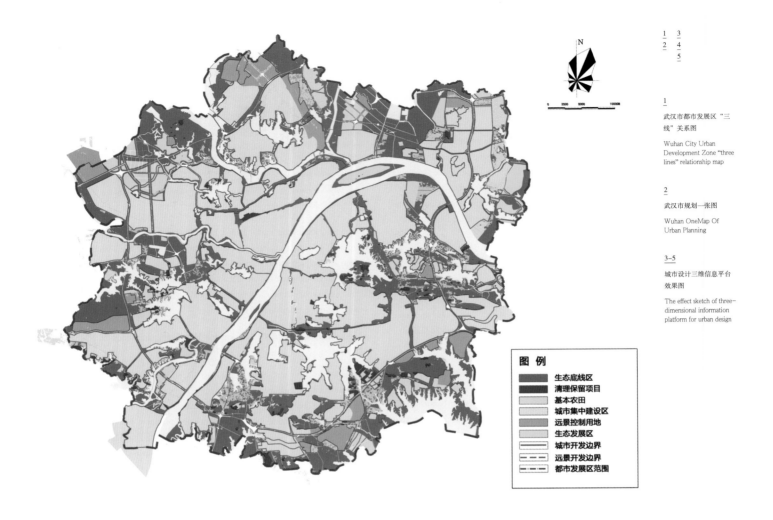

1	3
2	4
	5

1
武汉市都市发展区"三线"关系图
Wuhan City Urban Development Zone "three lines" relationship map

2
武汉市规划一张图
Wuhan OneMap Of Urban Planning

3-5
城市设计三维信息平台效果图
The effect sketch of three-dimensional information platform for urban design

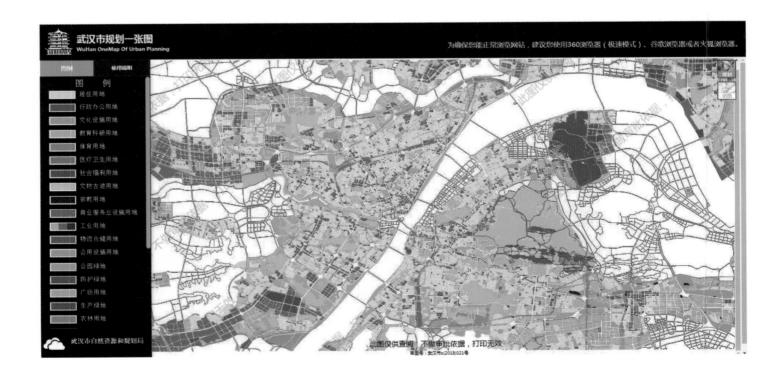

- 公务版、公众版多规融合一张图（2015年）

自2009年"一张图"运行以来，"一张图"在武汉市各项规划管理工作中取得了显著成效，为更好服务于武汉城乡建设工作，面向全市各级部门多渠道主动发布成果，建立联审共管机制，公务版、公众版"一张图"先后于2015年、2018年发布推广，向社会公众实时提供未来道路、医院、学校、公园、轨道交通布局等各类规划的查询。在城镇化发展转型的时代背景下，"多规融合"成为国家和地方空间管控改革的热点，"多规融合一张蓝图"得到进一步推进，以进一步提升规划的综合管控能力。

- 城市设计三维信息平台（2015年）

为大力加强城市设计水平，武汉市新一轮数字三维平台建设启动。按照宏观—中观—微观3个层级建立城市设计模型制作标准，指导模型制作。其中，宏观层面展现城市山水空间格局、整体空间结构、重点功能区分布、重要景观节点的空间布局；中观层面以地块为单元，表现区域整体空间组合关系、开敞空间布局、视线通廊、建筑高度等空间要素；微观层面以重点功能区为基本表现单元，全面展现建筑群体空间关系、建筑高度、建筑外立面材质等要素，同时表达地上、地下建筑空间关系及竖向交通组织和市政管网布局。

● 交通信息系统建设及应用（2015年）

自2011年以来，全市交通信息系统在4年建设期内共建成8个系统，目前系统在提升交通规划编制实力、服务行业（交管、交委、公交、地铁等）、服务社会等方面发挥着重要作用。2015年，在全力推动武汉市交通基础决策支持平台顺利通过专家评估、争取项目尽早落地的基础上，对8个系统进行了更新维护，保证系统正常运行，同时进一步加大交通信息系统的应用拓展工作。

● 武汉市交通市政修规信息系统（2016年至今）

2016年，武汉市规划研究院对交通市政类项目规划、建设、审批信息进行整合，建立了一个集数据查询、分析、共享于一体的综合信息平台。同时还建立了数据的更新维护机制，保障了数据的持续性更新。该系统规范了交通市政类项目规划管理，提高了审批效率，推动了管理审批的信息化和标准化。经过1～3期建设，电脑终端和移动终端平台信息系统已上线使用。

● *Integrating Official Edition, Public Edition and Many Plan in one Chart (2015)*

Since the launch of "One Chart" project in 2009, it has achieved remarkable progress in various planning projects and management in Wuhan. The publication of "integrating official edition, public edition and multi-planning projects in one chart" has been successively launched in 2015 and 2018. It provides real-time inquiries for planning of future road, hospital, school, park, rail transit layout and other planning to the public. The new "One Chart" project aims at better serving the urban and rural construction as well as building a joint examination and co-management mechanism for multi-channel publication of results at all levels of the city. In the era of the development and transformation of urbanization, "multi-planning integration" has become a hotspot of national and local space planning and management reform. The "integrating multi-planning into one blueprint" project has been further developed to improve the comprehensive management and control of planning.

● *Three-Dimensional Information Platform for Urban Design (2015)*

Wuhan has launched a new round of digital three-dimensional platform construction to improve urban design. The platform guides the model making in accordance with the standard of urban design of macroscopic model, mesoscopic model and microscopic model. The macro model shows the spatial pattern of urban landscape, the overall spatial structure, the distribution of key functional areas, and the spatial layout of important landscape nodes. The mesoscopic model takes areas and districts as units to represent the spatial elements such as the overall spatial combination of the districts, the layout of open space, the corridor of sight and the height of the building. The micro model takes the key functional areas as the basic unit to the spatial relationship of building groups, building height, building facade material and other elements in an all-round way. It also shows the spatial relationship between ground and underground buildings, vertical traffic organization and municipal pipeline network layout at the same time.

● *Construction and Application of Traffic Information System (2015)*

Since 2011, the government has built eight traffic information system of Wuhan in four years. At present, these systems play an important role in improving the traffic planning and service industries (traffic management, traffic commission, bus, subway, etc.), serving the society well. In 2015, on the basis of taking every effort to make the third phase of the World Bank project successfully through expert's evaluation and striving for the project to land as soon as possible, Wuhan has decided to upgrade and maintain the eight systems to ensure the normal operation and further expand the application of traffic information systems.

● *Wuhan Municipal Traffic Regulation Revision Information System (2016-present)*

In 2016, Wuhan Planning and Design Institute has integrated planning, construction and approval information of traffic and municipal projects, and established a comprehensive information platform for data query, analysis and sharing. Wuhan Planning and Design Institute has also established a data updating and maintenance mechanism to ensure the continuity of data updating. The system has standardized the planning and management of traffic and municipal projects and improved the efficiency of examination and approval. It has made great progress in promoting the informatization and standardization of management examination and approval. After three phases of construction, the information system of computer terminal and mobile terminal platform has been put into use online.

第肆章 • 区域一体化时期的规划　315

1　2
3
4

1
交通信息系统界面图
Interface of the Traffic Information System

2
交通市政修规信息系统・
综合管廊平面图
Traffic and Municipal Planning Revision Information System —Comprehensive Pipelines and Gallery Plane Graph

3
交通市政修规信息系统・
道路修规平面图
Traffic and Municipal Planning Revision Information System — Road Planning Revision Plane Graph

4
交通市政修规信息系统・
规划路网系统图
Traffic and Municipal Planning Revision Information System — Planning Road Network System Diagram

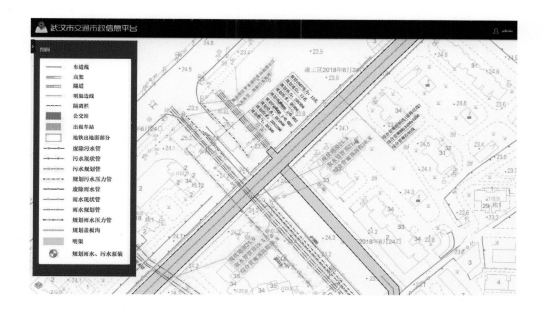

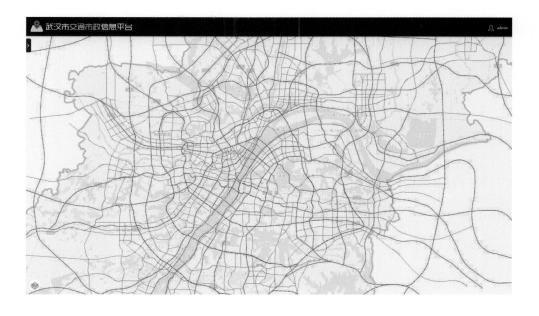

大事记
CHRONICLE OF EVENTS

Timeline

2010 — The "Industrial multiplication" plan was proposed. Four plates of "Greater Optics Valley, Greater Auto City, Greater Airport and Greater Port" were planned.

提出『工业倍增』计划。规划『大光谷、大车都、大临空、大临港』四大板块。

2011 — For the centennial celebration of the 1911 Revolution, Wuhan Avenue, the connection of Donghu Lake and Shahu Lake and many other projects were built.

辛亥革命百年纪念、建设武汉大道、东沙湖连通等一大批工程。

2012 — The century-based Wuhan 2049 Long-Term Development Strategy Study was formulated.

制订立足百年的2049远景发展战略研究。

2013 — Wuhan took the lead in compiling the Wuhan Sponge City Special Plan.

率先启动了《武汉市海绵城市专项规划》编制工作。

2015 — The conceptual planning and implementation planning and design of the expo garden were carried out, turning the city's largest garbage dump into the expo site.

开展了园博园概念规划及实施性规划设计，将城市最大垃圾场变为园博会主场。

2016 — The municipal ecological framework plan was formulated, basic ecological control lines were defined, and regulations on basic ecological control lines were implemented.

编制全城生态框架规划，划定基本生态控制线，施行基本生态控制线管理条例。

2016 — The Wuhan City Comprehensive Traffic Plan was issued.

《武汉市城市综合交通规划》出台。

2016 — Planning of key projects for the comprehensive improvement of water environment was carried out, including the Huangxiao River, the Airport River and the Xunsi River.

开展了黄孝河、机场河、巡司河等水环境综合整治的重点工程规划。

201? — The action plan for "co-governance of four rivers" was proposed.

提出『四水共治』行动计划。

1906	The Beijing-Hankou Railway was fully opened to traffic.	京汉铁路全线通车。
1911	The Revolution of 1911 began in Wuchang and the Hankou Overall Plan was prepared to rebuild the urban area of Hankou.	辛亥革命武昌首义，为重建汉口城区，编制《汉口全镇规划》。
1919	Sun Yat-sen mentioned the future vision of Wuhan in The International Development of China. "Wuhan is indeed one of the largest metropolises in the world, so a scale must be set for the future planning of Wuhan, for example, a scale similar to that of New York and London."	孙中山在《建国方略》提及武汉未来愿景，"确为世界最大都市之一矣，所以为武汉将来计划，必须定一规模，略如纽约、伦敦之大"。
1929	Outline of Public Works Programme for Special City in Wuhan was issued, which became the blueprint of various construction plans of Hankou.	《武汉特别市工务计划大纲》出台，成为其后汉口市各种建设计划的蓝本。
1931	Wuhan was hit by a flood and dike construction has since become an important part of municipal construction.	武汉大水，堤防建设从此成为市政建设的重要内容。
1944	The Draft Plan for the Construction of Greater Wuhan City was issued, proposing "Greater Wuhan" in the planning sense for the first time.	《大武汉市建设计划草案》出台，首次提出了规划意义上的"大武汉"。
1945	The Wuhan Regional Planning Commission was established, which was the first regional planning institution in China.	武汉区域规划委员会成立，是我国第一个区域规划机构。
1947	The Implementation Outline of Wuhan Regional Planning and The Preliminary Study Report on Wuhan Regional Planning were released successively, which started the great development of Wuhan regional planning.	《武汉区域规划实施纲要》《武汉区域规划初步研究报告》相继发布，开启了武汉区域规划的大发展。
1953	To implement the "156" national key projects, the city master plan of the first round for Wuhan after the founding of the People's Republic of China was compiled.	为落实国家"156"项重点工程，编制新中国成立来武汉首轮城市总体规划。

1957	Wuhan Yangtze River Bridge was built, which is the first highway and railway combined bridge crossing the Yangtze River in China.	武汉长江大桥建成，是我国第一条跨越长江的公铁两用大桥。
1959	Under the guidance of "developing large, medium and small industries simultaneously", the Wuhan Urban Construction Plan (Amended Draft) was prepared.	在『大、中、小工业并举』的方针指引下，编制《武汉市城市建设规划（修正草案）》。
1966	The Hankou-Danjiangkou Railway was completed and opened to traffic, and Wuhan initially formed a "T-shaped" railway network.	汉丹铁路建成通车，武汉初步形成了『T』字形铁路网络格局。
1978	The Third Central Urban Work Conference was held, putting forward the policy of "controlling the size of big cities, developing medium-sized cities reasonably and developing small cities actively".	第三次全国城市工作会议召开，提出了『控制大城市规模，合理发展中等城市，积极发展小城市』的方针。
1982	The Wuhan City Master Plan was issued, which paid more attention to the protection of history and culture and the circulation of trade, and corrected the development of single production city in the early stage.	《武汉市城市总体规划》出台，更加重视历史文化保护、商贸流通等，对前期发展单一生产城市进行了纠偏。
1982	The list of China's first batch of famous historical and cultural cities was published, and the reconstruction and repair of historical and cultural buildings in Wuhan, represented by Yellow Crane Tower, Qingchuan Pavilion and Guqin Platform, was launched.	国家第一批历史文化名城名单公布，武汉开启了以黄鹤楼、晴川阁和古琴台等为代表的历史文化建筑重建和修葺工作。
1988	With "traffic" and "circulation" as the breakthrough, the Revised Wuhan City Master Plan of 1988 edition was compiled, aiming to build Wuhan into "an economic center connecting central China and the ocean".	以『交通』和『流通』为突破口，把武汉建成『内联华中、外通海洋的经济中心』，编制了1988版《武汉市城市总体规划修订方案》。
1989	The Huangxiao River comprehensive treatment project and the construction of Jianshe Avenue and Qingnian Road were launched simultaneously.	开启黄孝河综合治理工程和建设大道、青年路的同步建设。
1990	The development strategy of "Rise of Three Districts" was put forward, and the Master Plan for Donghu New Technology District, Master Plan for Wuhan Economic and Technological Development Zone and Master Plan for Wujiashan Taiwan Merchant Investment Zone were successively compiled.	提出『三区崛起』的发展战略，陆续编制《东湖新技术开发区总体规划》《武汉经济技术开发区总体规划》和《吴家山台商投资区总体规划》。

Timeline

1995 — The Second Wuhan Yangtze River Bridge was completed and opened to traffic. After 39 years' efforts, the city broke through the threshold of cross-river development, ending the history of "only one line" connecting the two sides of the Yangtze River.

长江二桥建成通车。时隔39年，武汉突破城市跨江发展门槛，结束了长江两岸『一线牵』的历史。

1995 — Tianhe Airport was officially opened to traffic, and the civil aviation airport facilities were fundamentally improved.

天河机场正式通航启用，民用航空机场设施得到根本性改善。

1998 — Garden city comprehensive planning was carried out.

开展创建山水园林城市综合规划的编制工作。

1999 — The fifth round of city master planning was carried out. The spatial structure of "circle-layer + axis" and the urban town development system of "main city + 7 satellite towns" were constructed.

开展第五轮城市总体规划编制工作，构建了『圈层+轴向』的空间结构和『主城+7个卫星镇』的市域城镇发展体系。

2002 — Improvement along the river bank was carried out. Hankou River Beach, Wuchang River Beach, Hanyang River Beach and Hanjiang River Beach were built successively.

开始对沿江岸线进行治理，先后建成了汉口江滩、武昌江滩、汉阳江滩和汉江江滩。

2003 — The development and construction of new districts in Wuhan was launched.

启动武汉新区的开发建设。

2007 — The construction of Wuhan Railway Station, Hankou Railway Station and Wuchang Railway Station was completed, forming a "triangular" station pattern of passenger transport.

武汉火车站、汉口火车站和武昌火车站建设完成，形成了『三站鼎立』的客运格局。

2010 — The plan for "overcoming difficulties in urban construction" was launched to renovate the first ring road, round the second ring road and improve the third ring road.

启动『城建攻坚』计划，开展一环线整治、二环线划圆、三环线提升等工程。

201? — accelerated. To implement the rise of central China and promote the construction of two-oriented society, Wuhan City Master Plan (2010-2020) was issued.

略，促进两型社会建设，《武汉市城市总体规划（2010—2020年）》出台。

1861	Hankou opened its port for trade.	汉口开埠通商。
1889	In 1889, Zhang Zhidong was appointed Governor of Hu-Guang, and put forward the idea of "cross-shaped" regional traffic network centering on Wuhan.	张之洞调任湖广总督，提出了以武汉为中心的「十字型」区域交通网络构想。
1891	Hanyang Iron Works was founded as the largest iron and steel complex in Asia.	创办汉阳铁厂，时为亚洲最大的钢铁联合企业。
1893	Mathematics School, Mining School and Zi Qiang School were founded.	创办算学学堂、矿务学堂、自强学堂。
1894	the four bureaus of cloth, yarn, silk and hemp were founded, and Wuhan became the second largest textile industry center in China after Shanghai.	创办布、纱、丝、麻四局，成为仅次于上海的中国第二大纺织工业中心。
1899	South Dike and North Dike in Wuchang were constructed.	兴修武昌北堤和南堤。
1905	From 1900 to 1905, commercial ports were built successively in Wuchang and Hankou.	武昌、汉口相继进行商埠建设。
1905	From 1905 to 1907, the wall of Hankou Tower was demolished and rebuilt into Houcheng Road (now Zhongshan Avenue), and the Houhu Dike (also known as Zhanggong Dike) was built.	拆除汉口堡城墙改建为后城马路（今中山大道），修筑后湖大堤（又名张公堤）。

2019	The formulation of the city-wide national land space plan was launched.	开展了全市国土空间规划编制工作。
2018	The construction of the first urban simulation laboratory in China was promoted.	推动国内第一家城市仿真实验室的建设。
2018	Planning for rural revitalization was launched.	开展了"乡村振兴"系列规划。
2018	Comprehensive environmental renovation and improvement were carried out for the World Military Games.	全面开展了迎军运会环境综合整治提升工作。
2018	Planning was carried out for featured parts such as the civilization heart, main axis and new urban area (new district) of the Yangtze River.	开展长江文明之心、长江主轴、长江新城(新区)等亮点区块规划。
2018	To promote the integrated development of Wuhan-Ezhou and Hankou-Xiaogan, coordinated development planning was launched for the metropolitan area.	为促进武鄂、汉孝同城化一体化发展，开展大都市区协同发展规划。
2018	Planning for ecological protection along the Yangtze River was carried out to meet the requirements of "Great Protection of the Yangtze River", and formulation of the 2035 city master plan was launched.	围绕"长江大保护"要求，开展长江生态大保护滨江带规划，开展2035版城市总体规划编制工作。

中文典籍
Chinese References

参考文献
REFERENCES

1. （美）刘易斯·芒福德. 城市发展史起源、演变和前景 [M]. 倪文彦等, 译. 北京：中国建筑工业出版社. 1989.
2. （美）凯文·林奇. 城市形态 [M] 林庆怡, 译. 北京：华夏出版社. 2002.
3. （美）斯皮罗·科斯托夫. 城市的形成：历史进程中的城市模式和城市意义 [M] 单皓, 译. 北京：中国建筑工业出版社. 2005.
4. （美）罗威廉, 鲁西奇. 汉口：一个中国城市的冲突和社区（1796～1895）[M], 罗杜芳译. 北京：中国人民大学出版社. 2008.
5. （美）罗威廉. 汉口：一个中国城市的商业和社会（1796～1889）[M]. 北京：中国人民大学出版社. 2005.
6. （日）水野幸吉. 汉口 [M]. 光绪三十四年（1907）. 武汉地方志馆收藏.
7. 贺业钜. 中国古代城市规划史 [M]. 北京：中国建筑工业出版社. 1996.
8. 钟纪刚. 巴黎城市建设史 [M]. 北京：中国建筑工业出版社. 2002.
9. 杨蒲林, 皮明庥. 武汉城市发展轨迹武汉城市史专论集. 天津：天津社科院出版社. 1990.
10. 皮明庥. 近代武汉城市史 [M]. 北京：中国社会科学出版社. 1993.
11. 皮明庥. 武汉史稿 [M]. 北京：中国文史出版社. 1992.
12. 政协武汉市委文史学习委员会. 武汉文史资料文库 [M]. 武汉：武汉出版社. 1999.
13. 皮明庥. 简明武汉史 [M]. 武汉：武汉出版社, 2005.
14. 邹进文. 武汉通史晚清卷（上、下）[M]. 武汉：武汉出版社. 2006.
15. 邹进文. 武汉通史中华民国卷（上、下）[M]. 武汉：武汉出版社. 2006.
16. 邹进文. 武汉通史中华人民共和国卷（上、下）[M]. 武汉：武汉出版社. 2006.
17. 邹进文. 武汉通史图像卷 [M]. 武汉出版社. 2006.
18. 陈梦雷. 古今图书集成 [M]. 1127 卷：汉阳府部：方舆篇：职方典：备考一.
19. 史火明. 武汉历史地图集 [M]. 北京：中国地图出版社. 1998.
20. 李权时, 皮明庥. 武汉通览 [M]. 武汉：武汉出版社. 1988.
21. 张文彤. 2008 武汉市城乡规划年鉴. 武汉：武汉出版社. 2008.
22. 吴之凌. 拙巧集——武汉市城市规划设计研究院作品选集 [M]. 武汉. 湖北科学技术出版社. 2004.
23. （清）许慎春. 同治汉阳县志校 [M]. 武汉地方志办收藏.
24. （清）徐焕斗修纂, 王夔补葺. 汉口小志：十五卷 [M]. 不详. 民国 4 年（1915）.
25. 武汉地方志编纂委员会. 武汉市志大事记 [M]. 武汉：武汉大学出版社. 1990.
26. 武汉地方志编纂委员会. 武汉市志城市建设志 [M]. 武汉：武汉大学出版社. 1990.
27. 武汉地方志编纂委员会. 武汉市志总类志 [M]. 武汉：武汉大学出版社. 1990.
28. 武汉地方志编纂委员会. 武汉市志政权政协志 [M]. 武汉：武汉大学出版社. 1990.
29. 武汉地方志编纂委员会. 武汉市志工业志 [M]. 武汉：武汉大学出版社. 1990.
30. 武汉地方志编纂委员会. 武汉市志社会志 [M]. 武汉：武汉大学出版社. 1990.
31. 武汉市城市规划管理局. 武汉市城市规划志 [M]. 武汉：武汉出版社. 1999.
32. 袁继成主编. 汉口租界志 [M]. 武汉：武汉出版社. 2003.
33. 武汉市城市规划管理局. 武汉市国土资源管理局. 武汉城市规划志（19802000）[M]. 武汉：武汉出版社. 2008.
34. 武钢志编纂委员会. 武钢志 1952～1981[M]. 武汉：武汉出版社. 1983.
35. 湖北省武汉防汛指挥部办公室. 武汉堤防志 [M]. 武汉：武汉市防汛指挥部办公室. 1986.
36. 吴念椿. 夏口县志 [M], 南京：江苏古籍出版社. 2001.
37. 汉口市政府. 汉口市建设概况 [M]. 汉口：武汉印务馆. 中华民国十九年.
38. 董修甲. 京沪杭汉四大都市之市政 [M]. 上海：上海大东书局. 1933.
39. 刘千俊. 鄂政纪要 [M]. 中华民国 34 年. 中正书局：军政部印刷所. 现武汉地方志办收藏.

40. 潘心藻. 武汉建置沿革 [M]. 武汉：湖北人民出版社. 1956.
41. 苏云峰. 中国现代化的区域研究—湖北省（1860~1916）[M]. 台北：台湾永裕印刷厂. 1976.
42. 石泉. 古代荆楚地理新探 [M]. 武汉：武汉大学出版社. 1988.
43. 李传义. 武汉地区近代建筑史研究论文集 [M]// 汪坦，腾森照信. 中国近代建筑总揽. 北京：中国建筑工业出版社. 1989.
44. 穆和德. 近代武汉经济与社会：海关十年报告——汉口江汉关（1882~1893）[M]. 香港：香港天马图书有限公司. 1993.
45. 武汉市档案馆. 武汉旧影 [M]. 武汉：湖北人民出版社. 1999.
46. 涂文学. 涂文学自选集 [M]. 武汉：华中理工大学出版社. 1999.
47. 匡亚明. 张之洞评传 [M]// 南京大学中国思想家研究中心. 中国思想家评传丛书. 南京：南京大学出版社. 1999.
48. 涂勇. 武汉历史建筑要览 [M]. 武汉：湖北人民出版社. 2002.
49. 李军. 近代武汉城市空间形态的演变（1861~1949）[M]. 武汉：长江出版社. 2005.
50. 韩少林. 城市记忆记者镜头里的武汉 [M]. 武汉：武汉出版社. 2005.
51. 华揽洪. 重建中国 [M]. 北京：生活读书新知三联书店. 2006.
52. 武汉市勘测设计研究院. 武汉地理蓝皮书 [M]. 武汉：武汉市规划局. 2007.
53. 李晓虹等. 武汉大学早期建筑 [M]. 武汉：湖北美术出版社. 2007.
54. 于志光. 武汉城市空间营造研究 [D]. 武汉：武汉大学. 2008.
55. 武汉市东西湖区地方志编纂委员会. 东西湖区简志 [M]. 武汉：武汉出版社. 2012.
56. 武汉城建年鉴编委会. 武汉城建年鉴 [M]. 湖北：湖北人民出版社. 1995.
57. 中国城市规划学会学术工作委员会. 理性规划 [M]. 北京：中国建筑工业出版社，2017.
58. 武汉地方志编纂委员会办公室. 武汉改革开放 40 年鉴 [M]. 武汉：武汉出版社，2018.

期刊报纸文章
Journals and newspaper articles

1. 孙宗汾. 武汉城建往事 [J]. 湖北文史资料，1999（03）.
2. 吴之凌，汪勰. 武汉城市规划思想的百年演变 [J]. 城市规划学刊，2009（4）.
3. 李百浩. 武汉近代城市规划小史 [J]. 规划师，2002，8（5）.
4. 改良市政之难点. 汉口中西报，1921 年 11 月 3 日. 第三版.
5. 方荻生. 理想之大武汉如何实现 [J]. 市政周刊，1927（8）.
6. 皮作优. 私拟汉口市各区市场建造地点表 [J]. 市政周刊，1926，（4-5 合刊）.
7. 市政府行政概况. 汉口民国日报. 1927-07.
8. 武汉特别市工务计划大纲，汉口特别市工务局业务报告，第一卷，第 1 期，1929 年 7 月.
9. 武汉市政月刊. 第一卷，第二号、第五号，1929.
10. 新汉口市政公报. 第一卷，第 12 期，1930.
11. 汉口商业月刊. 第二卷，第 3 期，1935.
12. 中国工程师学会武汉工程分会 [J]. 中国工程师学会武汉分会. 1947.
13. 湖北省汉口市设计委员会集体创作. 重建大武汉方案. 湖北文献，1969. 28-32 期.
14. 万耀煌. 七十年前的武昌 [J]. 湖北文献，1973（28）.
15. 朱介凡. 连雅堂记改造武昌市区的大构想 [J]. 湖北文献. 1974.
16. 李玉堂. 鲍鼎与武汉近现代城市规划 [J]. 华中建筑，2000（2）.
17. 洪亮平，唐静. 武汉市城市空间结构形态及规划演变 [J]. 新建筑，2002（03）.
18. 李军，谢宗孝，任晓华. 武汉市产业结构与城市用地及空间形态的变化 [J]. 武汉大学学报（工学版），2002（5）.
19. 罗名海. 武汉市城市空间形态演变研究 [J]. 经济地理，2004，（24）.

20. 许慧，丁时忠. 武汉城市空间结构演化探析 [J]. 规划师，2004，20（4）.
21. 吴雪飞. 武汉城市空间扩展的轨迹及特征 [J]. 华中建筑，2004，（22）.
22. 何小林. 汉口租界区城市空间演进解读 [J]. 规划师，2004（5）.
23. 吴剑杰. 张之洞与近代中国铁路 [J]. 武汉大学学报（人文社会科学版），1999（3）.
24. 徐渊. 武汉近代铁路发展与城市规划 [J]. 科教文汇，2007（29）.
25. 李明术. 近现代武汉对外交通时代变迁 [J]. 中国水运（下半月），2014，14（4）.
27. 张兵，郝之颖，胡晓华. 武汉城市总体发展战略规划研究 [J]. 城市规划通讯，2006（2）.
28. 黄长义. 张之洞的工业化思想与武汉早期工业化进程 [J]. 江汉论坛，2004（3）.
29. 李治镇. 晚清武汉洋务建筑活动 [J]. 华中建筑，1996（3）.
30. 李百浩，薛春莹，王西波. 图析武汉市近代城市规划（1861~1949）[J]. 城市规划汇刊，2002（6）.
31. 方秋梅. 张之洞督鄂与湖北省府主导汉口市政改革 [J]. 武汉大学学报（人文科学版），2010（1）.
33. 王宜果. 近代（1840—1949）武汉市城市形态演变研究 [D]. 华中师范大学，2012.
34. 涂文学. "湖北新政"与近代武汉的崛起 [J]. 江汉大学学报（社会科学版），2010，27（1）.
35. 童绥宝. 张之洞与武汉教育近代化 [D]. 华中师范大学，2006.
36. 陈宏胜，王兴平，李百浩. "天、地、人"的权衡与平衡——论中国古代"三元"并进的城市规划思想 [J]. 中国名城，2015（3）.
37. 李玉堂，李百浩. 鲍鼎与武汉近现代城市规划 [J]. 华中建筑，2000（2）.
38. 王晓鸣，蔡汉宁. 城市规划发展的佐证与借鉴——评1923年《汉口市政建筑计划书》[J]. 城市规划，1991（5）.
39. 王欣. 董修甲的城市规划思想及其学术贡献研究 [D]. 武汉理工大学，2013.
40. 邱红梅. 董修甲的市政思想及其在汉口的实践 [D]. 华中师范大学，2002.
41. 邱红梅. 董修甲与近代汉口的市政建设述评 [J]. 信阳师范学院学报（哲学社会科学版），2007，27（5）.
42. 田甜，袁园. 汉口原租界片区城市空间形态研究 [J]. 中外建筑，2012（7）.
43. 朱滢. 汉口租界时期城市的规划法规与建设实施 [D]. 清华大学，2014.
44. 李风华. 后城马路的兴筑、发展与近代汉口城市社会发展 [J]. 江汉论坛，2012（5）.
45. 涂文学. 近代汉口市政改革对租界的效法与超越 [J]. 江汉大学学报（社会科学版），2009，26（4）.
46. 吴薇. 近代武昌城市发展与空间形态研究 [D]. 华南理工大学，2012.
47. 王玉霞. 理想与现实——孙中山《建国方略》城市规划思想研究 [D]. 武汉理工大学，2011.
48. 《完善水治理体制研究》课题组. 我国水治理及水治理体制的历史演变及经验 [J]. 水利发展研究，2015，15（8）.
49. 韩春辉，左其亭，宋梦林等. 我国治水思想演变分析 [J]. 水利发展研究，2015（5）.
50. 孙宗汾. 武汉城建往事 [J]. 武汉文史资料，2010（11）.
51. 吴之凌，汪勰. 武汉城市规划思想的百年演变 [J]. 城市规划学刊，2009（4）.
52. 李百浩，徐宇胜. 武汉近代里分住宅研究 [J]. 华中建筑，2000（3）.
53. 李义纯. 武汉市开埠以来城市规划管理历程研究 [D]. 华中科技大学，2015.
54. 闻洪. 辛亥首义之城武汉——《建国方略》蓝图今成现实 [J]. 小康，2016（25）.
55. 彭翔华. 宜昌川汉铁路遗迹的历史价值和保护利用研究 [J]. 三峡论坛，2012（1）.
56. 郭明. 战后武汉区域规划研究 [D]. 武汉理工大学，2010.
57. 陈晖. 张之洞的"商战论"与武汉商业近代化 [J]. 江汉大学学报（社会科学版），2017（4）.
58. 张之洞与近代工业博物馆 [J]. 城市环境设计，2014，80（Z1）.

59. 夏开元. 张之洞与武汉城市建设 [J]. 城建档案，2008（1）.
60. 风文. 张之洞与张公堤 [J]. 中国减灾，2008（3）.
61. 张天洁，李百浩，李泽. 中国近代城市规划的"实验者"——董修甲与武汉的近代城市规划实践 [J]. 新建筑，2012（3）.
62. 郭建. 中国近代城市规划范型的历史研究（1843～1949）[D]. 武汉理工大学，2003.
63. 郭建. 中国近代城市规划文化研究 [D]. 武汉理工大学，2008.
64. 邱瑛. 中国近代分散主义城市规划思潮的历史研究 [D]. 武汉理工大学，2010.
65. 王亚华，胡鞍钢. 中国水利之路：回顾与展望（1949～2050）[J]. 清华大学学报（哲学社会科学版），2011（5）.
66. 李百浩，郭明. 朱皆平与中国近代首次区域规划实践 [J]. 城市规划学刊，2010（3）.
67. 李玉堂，曾真. 租界之于武汉 [J]. 华中建筑，2010，28（7）.
68. 周婕，赵捷，张雨蝉. 规划的力量——武汉市城市空间特色演变 [J]. 中国城市规划年会，2011.
69. 周明长. 新中国建立初期重工业优先发展策略与工业城市发展研究（1949～1957）[D]. 四川大学，2005.
70. 李颖. 城市化进程中工业区的变迁——以武汉市工业区为例 [D]. 华中科技大学，2006.
71. 赵晶晶. 1840年以来武汉工业扩散驱动郊区城镇化的空间过程 [D]. 华中师范大学，2013.
72. 杨晨. 1990年代以来武汉市工业空间演化过程、特征及优化策略研究 [D]. 华中科技大学，2016.
73. 赵煜澄. 通车50年的武汉长江大桥 [J]. 铁道工程学报，2007（10）.
74. 李兆汝. 规划探索"一五"遗产的动态保护 [J]. 中国建设报，2010（3）.
75. 唐文，宋晓菲. 建国初期居民区的苏联建筑模式设计研究——以武汉红钢城九街坊为例 [J]. 艺术与设计（理论），2015（10）.
76. 洪旗，陈静远. 建国初期工业住区的保护与更新（一）——一种尚未充分重视的工业遗产 [J]. 城市规划学刊，2009（7）.
77. 周蜀秦. 中国城市化六十年：过程、特征与展望 [J]. 中国名城，2009（10）.
78. 官卫华. 加入WTO后我国城市化进程所面临的机遇与挑战及对策研究 [J]. 城市规划汇刊，2001（1）.
79. 王曙光. 中国城市化发展模式研究 [D]. 吉林大学，2011.
80. 董菲. 武汉现代城市规划历史研究 [D]. 武汉理工大学，2010.
81. 周麒. 武汉旧城街区改造中的人本探究 [D]. 湖北工业大学，2008.
82. 汪云，刘菁. 特大城市生态空间规划管控模式与实施路径 [J]. 规划师，2016（3）.
83. 本刊综合. 习近平主持召开深入推动长江经济带发展座谈会并发表重要讲话 [J]. 当代兵团，2018，336（09）.

外文资料
Foreign resources

1. Lewis Mumford.The city in history, A powerfully Incisive and Influential Look at the Development of the Urban Form through the Ages, 1961.
2. C. Alexander. A Pattern Language.Oxford Univ.Press,1975.
3. Peter Hall,Colin Ward.Sociable cities: the legacy of Ebenezer Howard,1998.
4. J.Cheng, I. Masser. Urban Growth Pattern Modeling: A Case Study of Wuhan City, P. R. China. Landscape and Urban Planning, 2003.
5. J. V. Henderson.Urban Development: Theory, Fact and Illusion.New York Orford University Press, 1988.
6. Jacobs, J. The Death and Life of Great American Cities, 1972
7. Convention.Ministère des Affaires étrangères de france

网络资源
Network resources

1. http://www.cnki.com.cn/Article/CJFDTotal-SJZJ200105000.htm/ 知网.
2. http://www.mofa.go.jp/mofaj/annai/honsho/shiryo/ 日本外务省外交文史资料馆.
3. http://www.flet.keio.ac.jp/dep/ahist.html/ 日本庆应大学文学部东洋史学专攻.
4. http://www.jacar.go.jp/ 亚洲历史资料中心.
5. http://www.utexas.edu/ 田纳西大学图书馆.
6. http://gdb.u-shimane.ac.jp/neardb/HooverJSP.jsp/ 史丹佛大学胡佛研究所.
7. http://www.whfz.gov.cn/ 武汉方志网.
8. http://www.wpl.gov.cn/pc-3598-253-0.html/ 数字武汉—城乡规划网.
9. http://keropero888.hp.infoseek.co.jp/ 古代世界地图网.
10. http://www.whdaj.gov.cn/ 武汉市档案信息网.
11. Http://bbs.zdic.net/thread-115381-1-1.html/ 汉典论坛.
12. Http://www.laoditu.com/w-laoditu/n1662.html/ 地图收藏网.
13. http://www.lib.whu.edu.cn/ 武汉大学图书馆.
14. http://www.chinamaxicard.com/bbs/index.php/ 中国极限集邮网.
15. http://www.sszx.org.cn/index.asp/ 孙宋资讯网.
16. http://www.gzlib.gov.cn/ 广州图书馆.
17. http://www.findart.com.cn/ 搜艺搜古玩图库.
18. Http://zh.wikipedia.org/ 维基百科.

其他资源
Other resources

中国第一历史档案馆
中国第二历史档案馆
中国地图出版社
北京图书馆
上海图书馆
湖北省图书馆
四川省图书馆
武汉市图书馆
湖北省档案馆
武汉市档案馆
武汉市地方志馆
武汉市博物馆
武汉大学图书馆
武汉大学档案馆
武汉市自然资源和规划局
武汉市规划研究院
武汉市测绘研究院

后记
AFTERWORD

3500年前，盘龙筑城，武汉就此起源，乃至汉水分道，江城自此镶长江、嵌汉水，始有三镇鼎立，继而分合交替，融合发展，终成今日之胜状。历史钩沉、繁冗庞杂，纵观百年以来武汉之发展，城市日新月异，波澜壮阔，江城人民开拓进取的精神展现得淋漓尽致，武汉崛起不可不谓中华民族复兴之缩影。

《武汉百年规划图记》于2009年第一次出版，是从规划角度看待武汉城市百年历程，为武汉规划历史研究奠定了坚实的基础，十年沧桑，本书再版，如登楼远望，极目楚天。在第一版的基础上，本书总结归纳上版内容，通过参阅大量文献、书籍对第一版内容进行了详细考证，充分挖掘历史资料，系统梳理了规划科学与城市建设发展轨迹，同时收集整理了近十年的规划资料，对本书内容进行了详实的补充与完善。

本书编制组成员多次赴湖北省博物馆、武汉市档案馆、张之洞博物馆、宗关水厂博物馆等地学习参观，进一步深入了解武汉规划历史，力求保证本书内容的完整和严谨。

本书在编撰过程中，得到了相关领导和专家的鼓励和支持。武汉市自然资源和规划局局长盛洪涛对本书的编撰工作十分关心，给予了支持和帮助；副局长刘奇志、副局长周强亲自对本书进行了审阅，提出了具体的修改建议；武汉市规划研究院院长陈韦、副院长武洁主导了本书策划、篇章结构制定和内容编写，对全书进行了详细的审稿把关。

最后，我们特别感谢中国建筑工业出版社对本书的出版给予了大力支持,感谢传世比邻翻译公司在较短的时间内为本书进行了精心翻译。

此次再版，时隔十年，陪伴了武汉城市发展的蝶变，见证了城乡规划行业的创新进步。恰逢祖国七十华诞，改革开放四十周年之际，该书出版以期保证其对武汉百年规划历程研究的延续性，更好地促进城乡规划工作者共同思考、共同提升。囿于编者水平有限，本书疏漏之处在所难免，敬请各位读者不吝指正。

武汉市规划研究院
2019年10月

3500 years ago, Panlong City was built, from which Wuhan originated. After the Hanjiang River changed its course, Wuhan was divided by the Yangtze River and the Hanjiang River into the three towns of Wuchang, Hanyang and Hankou. After several periods of separation and integration, Wuhan has finally become prosperous as it is today. History is a jumble of esoteric details. Throughout the development of Wuhan over the past 100 years, the city has changed with each passing day and surged forward with great momentum, vividly displaying the pioneering spirit of Wuhan people. The rise of Wuhan is an epitome of the rejuvenation of the Chinese nation.

First published in 2009, Planning Wuhan: 100 Years examines Wuhan in the past 100 years from the perspective of planning and lays a solid foundation for the study of Wuhan's planning history. A decade later, the second edition of the book was published with great foresight. Based on the first edition, this book summarizes and examines in detail the content of the first edition by referring to a large number of literature and books. Besides, through historical data mining, this book systematically sorts out the development track of planning science and urban construction. Meanwhile, we collected planning data of the last ten years to supplement and improve the content of this book.

The members of the compilation group of this book have visited Hubei Provincial Museum, Wuhan Archives, Zhang Zhidong Museum and Zongguan Waterworks Museum for many times to further understand the planning history, to ensure the integrity and preciseness of this book.

The compilation of this book has been encouraged and supported by relevant leaders and experts. Sheng Hongtao, Director of Wuhan Natural Resources and Planning Bureau, was very concerned about the compilation of this book and gave support and help. Deputy Directors Liu Qizhi and Zhou Qiang personally reviewed the book and put forward specific suggestions for revision. Director Chen Wei and Deputy Director Wu Jie of Wuhan Planning & Design Institute and other leaders actively participated in the planning, chapter structure arrangement and compilation of the book, and reviewed it in detail.

Finally, we would like to express our gratitude to China Architecture & Building Press for its great support to the publication of this book, and to Trans-ocean Translation Company for its careful translation in such a short period of time.

The second edition, ten years after the first edition, has accompanied the rapid urban development of Wuhan and witnessed the innovation and progress of urban and rural planning. Coinciding with the 40th anniversary of China's reform and opening-up, the book was published to ensure the continuity of research on the 100-year planning process of Wuhan, thus encouraging urban and rural planners to think and make progress together. Due to the limited ability of editors, there are inevitably omissions in this book, and your correction will be appreciated.

Wuhan Planning & Design Institute
October 2019